Management for Professio

The Springer series *Management for Professionals* comprises high-level business and management books for executives. The authors are experienced business professionals and renowned professors who combine scientific background, best practice, and entrepreneurial vision to provide powerful insights into how to achieve business excellence.

Samuel O. Idowu • René Schmidpeter

Editors

Case Studies on Sustainability in the Food Industry

Dealing With a Rapidly Growing Population

 Springer

Editors
Samuel O. Idowu
Guildhall School of Business and Law
London Metropolitan University
London, UK

René Schmidpeter
M3TRIX Institute of Sustainable Business
Ingolstadt, Germany

ISSN 2192-8096   ISSN 2192-810X   (electronic)
Management for Professionals
ISBN 978-3-031-07741-8   ISBN 978-3-031-07742-5   (eBook)
https://doi.org/10.1007/978-3-031-07742-5

This Springer imprint is published by the registered company Springer Nature Switzerland AG
The registered company address is: Gewerbestrasse 11, 6330 Cham, Switzerland

Foreword

Progress towards achieving the Sustainable Development Goal 2 of the 2030 Agenda for Sustainable Development to "End hunger, achieve food security and improved nutrition and promote sustainable agriculture by 2030" has stalled in recent years. This stalling predates the COVID pandemic and highlights the massive challenge facing programmes to provide food security across the global population. Failure to meet the goal will have a huge bearing on nutritional intake and physical and psychological outcomes. Increased morbidity and mortality rates are reported with detrimental health outcomes including chronic disease, mental health disorders and obesity in adults and stunting, weak development and decreased academic ability for children. Moreover, when experiencing food insecurity households are less resilient to stress and disasters.

The threats to food security are not universal. There are gross inequalities in the availability of food within and between countries. Poverty and wider inequalities are often the unifying characteristic when exploring why some people are less resilient to food insecurity than others. In other words, there are deeper consequences of low incomes such as the disproportionate impact of food price inflation and living in residential areas which rely on informal food outlets that are more expensive often serving inferior quality and unsafe produce. This level of access tends to follow existing layers of marginalisation with the most vulnerable such as women, children, the unemployed, disabled, geographically remote and migrants likely to be at greatest risk. Alongside income these interweaving characteristics include level of education, female headed households, weak social networks, less social capital, reliability of income source, unemployment and size of family. Moreover, while household income is instrumental in vulnerability to food insecurity, there is a range of regular and seasonal social factors that can either reduce or enhance the likelihood of experiencing nutritional difficulties. These factors include excluded and marginalised populations who are less resistant to stress, hazards and disasters because, without formal welfare support, they lack the capacity and resources for recovery. Urban poorer groups have less access to control of land, financial, information and technological resources and are therefore more vulnerable. The consequences of these interweaving processes are that a large number of people are disproportionately affected by food insecurity while other groups benefit to the point of excess from international supply chains.

These are the deep-rooted issues facing governments, international agencies and communities that need to be addressed if food inequalities are to be reduced and security levels extended. And these are the issues that require much greater engagement by, and solutions from, academics. In this space, Idowu and Schmidpeter's edited collection *Handbook of Cases of Sustainability in the Food Industry* makes an incisive and broad ranging contribution. Chapters pick up on innovations and environmental challenges that surround greater sustainability in agriculture and supply chains. Understandably, the international emphasis in achieving the SDG is directed to the Global South, and regions are well represented in this collection. However, unsustainability of food cycles and waste is most noticeable in the Global North which is often neglected in discussions about food security and it is a neglect that other chapters in this book also address. By bringing together this range of contributors and regions, this book provides important insights into different elements of the food cycle, locational distinctions and underlying issues that have to be considered urgently if there is to be any likelihood of achieving the goal to "End hunger, achieve food security and improved nutrition and promote sustainable agriculture by 2030".

Robert Gordon University   Stephen Vertigans
Aberdeen, UK

Preface

The world population continues to grow at an alarming rate year in, year out. This is putting pressure on all global resources including the food we consume. It has been projected that the world population will grow from its current 7.8 billion to 8.1 billion in 2025. This will continue to add to the existing pressure on global resources. Knowledge as they say is power! We need to understand how to cope and deal with all the challenges that accompany this and other aspects of the projected growth in terms of sustainable development and life in general on planet Earth.

The intention of this handbook is to codify sustainable cases in different areas of sustainability in the food industry which involves food production and consumption, Goal 12 of the United Nations Sustainable Development Goals 2030. The chapters in the book explore cases in different aspects of food provision - farming and fishing, food refining, supply chain - wholesale and retail channels and other relevant aspects that improve our understanding of how sustainability is taking place in the global sector.

It is hoped that our global readers find the information the book provides of value to them up and down planet Earth.

London, UK   Samuel O. Idowu
Bern, Switzerland   René Schmidpeter
Spring 2022

Sustainability in the Food Industry: An Introduction

In the year 2021, the Global Nutrition Report stated that "Diet-related disease and mortality rates are large and increasing in most regions. Deaths attributable to poor diets have grown by 15% since 2010 — and are now responsible for more than 12 million NCD deaths in adults. This is about a quarter (26%) of all adult deaths each year."[1]

In addition to that, the WHO reports in their Global Strategy for Food Safety 2022–2030 that "It is a basic human right to have access to safe, nutritious and healthy food. To guarantee this right, governments must ensure that available food meets safety standards."[2]

Today our food is produced, delivered and consumed in ways that could not have been anticipated two decades ago. In a globalised world such as ours it must be ensured that everybody can access fresh and healthy food to meet their nutritional needs. On the other hand, there have to be procedures to address food waste that happens around the globe.

This book gives an overview of how to meet the existing challenges in food production and food supply.

In chapter one, Ursula Vavrik describes the rationale for sustainable food systems as an urgent necessity due to transgressed planetary boundaries. Climate change, biosphere integrity (biodiversity), land-system changes and biochemical flows have already overshot these boundaries; crucial importance is given to biodiversity and bee preservation. She describes specific projects such as the creation of UNESCO's Man and the Biosphere Programme, organic soy production, urban agriculture, bee protection and organic honey production in and around cities, among many others. Also, new methods of nature-friendly practices to gain higher yields in agriculture are explored in this chapter. Additionally clear recommendations for actions aimed at attaining sustainable global food systems are stated in the conclusion.

---

[1]UN (2021) Global Nutrition Report. https://globalnutritionreport.org/reports/2021-global-nutrition-report/executive-summary/

[2]2 WHO (2022) Global Food Strategy https://cdn.who.int/media/docs/default-source/food-safety/who-global-strategy-food-safety-2022-2030.pdf?sfvrsn=66cdef40_17&download=true

In Chapter two, Adebimpe Lincoln gives insight in a study that was conducted to research the understanding of how sustainability occurs among entrepreneurs, the approach these firms adopt vis-à-vis supply chain sustainability innovation, barriers faced and strategies to overcome these barriers within the Nigerian context. The findings of this study show the missing knowledge about international sustainability frameworks, and many of the practices adopted were often informal and on an ad hoc basis. It also gives insights into the primary barriers of Nigerian entrepreneurs which are affecting supply chain sustainability innovation. The study fills gaps in the literature on supply chain sustainability in the SME sector in Africa and therefore provides valuable work to support Nigerian entrepreneurs.

In chapter three, Boris Christian Herbas-Torrico, Björn Franc, Carlos Alejandro Arandia-Tavera and Pamela Mirtha Zurita-Lara report the big social and economic changes in Bolivia. The Bolivian society has high expectations for a better future such as that their purchased goods are sourced and produced in an environmentally, socially sustainable and safe fashion. They give insights into how supply chains in Bolivia are working and managed and summarising the challenges and opportunities faced by Bolivian firms in the development of sustainable supply chains.

In Chapter four, Mara del Baldo states why circular economy (CE) becomes more and more important in the prevention of further climate disruption. In the agricultural sector, natural resources (land and water) are currently exploited at unprecedented annual levels. In this vein, the empirical analysis is useful to deeply understand why and how companies' business models can incorporate the issue of food safety and food security. Mara del Baldo presents in this chapter a case study relative to an Italian agrifood company (Fileni Group) that is among the leading European chicken farm and meat producers to answer these questions of what firms can do to address the urgent issue of food safety and food security. The chapter addresses the theoretical background, the methods and the case study. Also, the main findings of the study are presented and discussed and conclusions in terms of insights and implications deriving from the study are drawn.

In Chapter Five, Elisa Baraibar-Diez, Maria Odriozola, Ladislao Luna, Ignacio Llorente, Antonio Martin, José Luis Fernández, Ángel Cobo, José Manuel Fernández and Manuel Luna describe the development and planning of the strategy against food waste in the Spanish region of Cantabria. The need to establish actions to reduce food waste is part of the European Union's comprehensive approach to efficient use of resources so that economic development must derive from a linear economy to a circular economy. At a regional level, the Government of Cantabria raised strategic lines and transversal measures within the Social Emergency Plan 2016–2017 to achieve practical proposals aimed at reducing food waste, coordinating efforts that derive in a greater articulation of the agrifood value chain. This chapter exposes the collaborative process of the development and planning of this strategy. It emphasises the need to publicise development processes prior to transversal strategies and can serve as an example for other regions and communities that have concerns regarding food waste.

Chapter Six addresses food security in South Africa and the lessons learned from the COVID-19 pandemic on creating sustainable value chains through corporate social responsibility. Mikovhe Maphiri describes life in South Africa in a pandemic-like environment for more than 25 years where access to adequate healthcare, food security, adequate housing and necessities that make people human is unevenly distributed. Prior to the pandemic, the hunger of millions of South African households has been a constant presence, where families do not have access to food. According to the Global Food Security Index, South Africa is the most food secure country on the African continent. However, millions of South Africans go hungry every day. This chapter seeks to provide a new perspective on re-envisioning the food security system in South Africa and discusses how corporate social responsibility (CSR) could provide avenues for food security and play a role in food equality as a key driver of food security after the pandemic and after the lockdown restrictions have been lifted.

In Chapter seven, Marina Beermann, Patrick Fraund and Felipe Fuentelsaz present a multistakeholder joint project in conventional citrus production in Spain which had the aim to reduce the negative environmental impacts in conventional agriculture without diminishing the amount of harvest. They describe how developing and applying of partly participatory action research and agroecology principles has led to a considerable reduction of negative environmental impacts in citrus cultivation. The reason for this multistakeholder project was an ecological footprint reduction of Germany's retailer EDEKA and its private brand products.

Chapter eight by Sam Sarpong, Ali Saleh Ahmed Alarussi and Faizah Shahudin describes the challenges in Malaysia's sustainability efforts and the role of traceability in the food industry. In recent years, the food industry has made significant strides towards improving its sustainability. Even if the industry continues to play a vital role in improving its own sustainability record, the complex array of issues it is still subjected to yet needs to be addressed. The authors state that especially Malaysia is still grappling with a lot of concerns mainly pertaining to issues like traceability in its meat products, the unsustainable nature of its palm oil production and food waste to which the authorities are urgently seeking solutions. These specific issues are explored in the following chapter and also how food security can be ensured in the country. So the aim of this chapter is to highlight the sustainability issues Malaysia is faced with.

In the next chapter, Silvia Pulu and Nicoleta Mihaela Florea write about food waste in Romania illuminated from the individual and national perspective. Food waste is an important problem in the world, and in order to contribute to the achievement of sustainable development goals (SDG) of the 2030 Agenda for Sustainable Development, mainly the 2nd—Zero Hunger and the 12th—Responsible Consumption and Production, it is urgent to address this topic carefully. The authors of the chapter establish the food waste context in Romania compared to other European countries, identifying the main causes for food waste and the measures that could be implemented to reduce waste and promote a more responsible consumption. For this they had conducted a triple analysis, one focused on individuals, another on non-governmental organisations (NGOs), public authorities and the private sector and the third on the initiatives of the most important retailers in

Romania in terms of turnover and net profit. The results of this analysis are explained in this chapter.

Afterwards, Chapter Ten written by Milan Todorovic gives an insight about the sustainable food production in Serbia. He gives insights of the discursive and the societal structures that impact on the subject matter at hand and makes an exploration of discourse and practice in the early 2020s.

Last but not least, Chapter eleven by Y Alahakoon, Mahendra Peiris and A.D. Nuwan Gunarathane describes the sustainable challenges and the way forward in the tea industry in Sri Lanka. The tea industry in the tea-growing developing countries such as Sri Lanka plays a crucial role in earning foreign exchange, providing employment opportunities for women, and acting as a custodian of the sensitive ecosystems. The tea industry in Sri Lanka (that produces world-famous Ceylon tea) has faced many challenges from various fronts, but nevertheless, these challenges and the responses of the tea industry have not been systematically explored. Due to these circumstances, this chapter explores the sustainability challenges under economic, environmental and social dimensions while discussing the strategies adopted by the Ceylon tea industry. It then presents how the tea plantation companies in Sri Lanka have adopted various measures to address them, including cost-saving initiatives, climate-smart agricultural practices, mechanisation of operations, new product development, community engagement and development and conservation of biodiversity. Even if some taken measurements have shown positive impacts many of the Ceylon tea industry's sustainability challenges, for instance, the effects of climate change, declining land productivity and increase in production cost, still remain unsolved or partially solved. The authors describe in this chapter which measures the Ceylon tea industry needs to focus on to resolve these issues.

All chapters show the diversity and different regional context when it comes to global food production and supply chains. It becomes clear that there is no one-size-fits-all solution when it comes to integrating sustainability to the food industry. Only through stakeholder collaboration, integrative business strategies and international frameworks and guidance we can achieve a more sustainable food supply for 8 billion people in the world. One thing becomes clear we all need to transform our personal, business and consumption lifestyle in order to meet the needs of our planet, its people and our economies!

Acknowledgements

We are grateful to a host of fine scholars spread around the globe for honouring our call to participate in this project. Many of them despite their busy schedules felt obliged to help in putting together this concise edition of *Sustainability in the Food Sector*; to them these two editors are eternally grateful.

We also wish to thank some of our friends who have continued to assist us in many of our professional activities: Nicholas Capaldi, John O Okpara, Richard Ennals, Stephen Vertigans (who wrote a befitting Foreword to the book), David Ogunlaja, Michael Soda, Samson Nejo, Samson Odugbesi, Emmanuel Ayo Osho, Samson Owokoniran, Annett Nike Oke, Bejamin Ade Aina QC, Jonathan Shebioba and Olusheyi Bamgboshe Martin. The lead editor is also grateful to his sister Elizabeth A Lawal. Finally, he is eternally grateful to those who are always on the scene of many of these activities—his wife—Olufunmilola Sarah O Idowu and children—Josiah Opeyemi Idowu and Hannah Ayomide Idowu. God bless you all.

The two editors would like to thank their publishing team at Springer headed by Christian Rauscher, Executive Editor, Barbara Bethke, Associate Editor, and other members of the publishing team who have supported this project and all our other projects.

Finally, we apologise for any errors or omissions that may appear anywhere in the book; no harm was intended to anybody.

Contents

Securing Planetary Health and Sustainable Food Systems with Global Organic Agriculture: Best Practice from Austria 1
Ursula A. Vavrik

Barriers to Supply Chain Sustainability Innovation Amongst Nigerian Entrepreneurs in the Food and Agriculture Industry 49
Adebimpe Adesua Lincoln

Sustainable Supply Chains in Bolivia: Between Informality and Political Instability 81
Boris Christian Herbas-Torrico, Björn Frank,
Carlos Alejandro Arandia-Tavera, and Pamela Mirtha Zurita-Lara

Why Chicken? Fileni (Italy): Between Taste, Circular Economy and Attention to the Territory 101
Mara Del Baldo

Development and Planning of the Strategy against Food Waste in the Spanish Region of Cantabria 119
Elisa Baraibar-Diez, María D. Odriozola, Ladislao Luna, Ignacio Llorente,
Antonio Martín, José Luis Fernández, Ángel Cobo,
José Manuel Fernández, and Manuel Luna

Food Security in South Africa: Lessons from the COVID-19 Pandemic on Creating Sustainable Value Chains Through Corporate Social Responsibility (CSR) 135
Mikovhe Maphiri

Reducing Negative Environmental Impacts in Conventional Agriculture, but Not the Amount of Harvest: A Multi-stakeholder Joint Project in Conventional Citrus Production in Spain 159
Marina Beermann, Patrick Freund, and Felipe Fuentelsaz

**Challenges in Malaysian's Sustainability Efforts:
The Role of Traceability in the Food Industry** 185
Sam Sarpong, Ali Saleh Ahmed Alarussi, and Faizah Shahudin

**Food Waste in Romania from an Individual and a National
Perspective** 203
Silvia Puiu and Nicoleta Mihaela Florea

**Sustainable Food Production in Serbia, an Exploration
of Discourse/Practice in Early 2020s** 229
Milan Todorovic

**Sustainability Challenges and the Way Forward in the
Tea Industry: The Case of Sri Lanka** 271
Y. Alahakoon, Mahendra Peiris, and Nuwan Gunarathne

Index 293

About the Editors

Samuel O. Idowu, PhD, is Senior Lecturer in Accounting and Corporate Social Responsibility at the Guildhall School of Business and Law, London Metropolitan University, where he is currently the course leader for MSc Corporate Social Responsibility and Sustainability. Samuel is Professor of CSR and Sustainability at the Nanjing University of Finance & Economics, China. He is a fellow member of the Chartered Governance Institute, a fellow of the Royal Society of Arts, a Liveryman of the Worshipful Company of Chartered Governance Institute and a named freeman of the City of London. He is the President of the Global Corporate Governance Institute. Samuel has published over fifty articles in both professional and academic journals and contributed chapters in several edited books and is the Editor-in-Chief of three major global reference books by Springer—the *Encyclopedia of Corporate Social Responsibility* (ECSR), the *Dictionary of Corporate Social Responsibility* (DCSR) and the *Encyclopedia of Sustainable Management* (ESM), and he is a series editor for Springer's *CSR, Sustainability, Ethics and Governance* books. Samuel has been in academia for 32 years winning one of the Highly Commended Awards of Emerald Literati Network Awards for Excellence in 2008 and 2014. In 2010, one of his edited books was placed in 18th position out of Top 40 Sustainability Books by Cambridge University Programme for Sustainability Leadership, and in 2016, one of his books won the outstanding book of the year of the American Library Association. Samuel is on the editorial advisory boards of the *International Journal of Business Administration* and *Amfiteatru Economic Journal*. He has been researching in the field of CSR since 1983 and has attended and presented papers at several national and international conferences and workshops on CSR. Samuel has made a number of keynote speeches at international conferences and workshops and written the *foreword* to a number of leading books in the field of CSR and sustainable development. And he has examined a few PhD theses in the UK, Australia, South Africa, New Zealand and the Netherlands.

René Schmidpeter is professor for sustainable management at the IU—International University in Munich and Research Professor at the BFH in Bern. He is a visiting professor at ISM in Vilnius as well as a guest professor at various universities worldwide. He is a series editor for Springer's *CSR, Sustainability,*

Ethics and Governance books, editor of the *Encyclopedia of Sustainable Management* (ESM) and an editor of the *Dictionary of Corporate Social Responsibility* (DCSR). His research and teaching activities focus on the management of corporate social responsibility, international perspectives on CSR, social innovation and sustainable entrepreneurship and the relationship between business and society. He has published widely and is the editor of the German *management series on Corporate Social Responsibility* at Springer Gabler and is member of various think tanks, sustainability initiatives and advisory boards in Europe and abroad. He has conducted research for the Federal Ministry of Education and Research, the European Union, the Bavarian Ministry of Social Affairs, the Anglo-German Foundation, the Bertelsmann Foundation and the Deutsche Forschungsgesellschaft (DFG). He works closely with chambers of commerce and businesses in Europe to develop innovative strategies and entrepreneurial solutions for the sustainable transformation of business and society.

Securing Planetary Health and Sustainable Food Systems with Global Organic Agriculture: Best Practice from Austria

Attaining 40% by 2030 and 100% by 2040: In Combination with Other Measures

Ursula A. Vavrik

Abstract

As outlined by the UN Sustainable Development Goals, the world needs to move toward sustainable production and consumption patterns, also in the agricultural and food sectors. This involves ecologically sound, nonpolluting practices safeguarding ecosystem services, combating climate change, preserving biodiversity, and contributing to a circular economy. Studies suggest that the world population can be fed with food stemming exclusively from organic agriculture, in combination with reduced food wastage and diet changes. However, worldwide only about 1,5% of the agricultural land is organic. With an organic share of 26,1% of the agricultural land in 2019, Austria can be considered a frontrunner. While the European Green Deal of 2019 aims at raising the share of organic agriculture from 8,1 to 25% in the EU by 2030, the EU Commissioner for Agriculture even aspires to make organic farming the norm.

Transgressed planetary boundaries including climate change, biosphere integrity (biodiversity), land system change and biochemical flows plea for an urgent transition to sustainable food systems, agriculture herein appearing to be a key driver. Biodiversity conservation and bee preservation are of crucial importance, as are interconnections with and implications for other policy areas such as health, tourism, development and migration, research and finance.

The chapter was finalized in February 2021.

U. A. Vavrik (✉)
NEW WAYS Center for Sustainable Development, Vienna, Austria
e-mail: director@newways.eu

The chapter provides best practices from Austria, new methods of nature-sound practices, and recommendations for actions aimed at attaining sustainable global food systems.

Keywords

Organic agriculture · Sustainable food systems · Planetary boundaries · Global sustainability · Intensive agriculture · Bees

1 Introduction: The Rationale for Upscaling Sustainable Food Systems

Latest reports from the Food and Agriculture Organization of the United Nations (FAO) point to the fact that food security and supply worldwide are getting more and more under pressure, due to several key drivers, i.e., raising population, advancing climate change and biodiversity loss, unsustainable production and consumption methods, unsustainable land use, depreciation of soils, and differing geographic conditions leading to unequal opportunities (*see* Dury et al., 2019a). Food systems are impacted by those drivers such as, i.e., climate change but also contribute to climate change. Latest research suggests that agriculture can be considered a key policy for the resilience of all planetary boundary issues (see Lade et al., 2019a). There are many interrelations and interconnections between different policies and drivers mentioned above, so that their management requires comprehensive strategies and holistic approaches. A multitude of measures will be necessary to secure a sustainable food supply and food security worldwide (*see* Dury et al., 2019b).

The United Nations Sustainable Development Goals (UN SDGs) outline that the world needs to move toward sustainable production and consumption patterns, also in the agricultural and food sector (*see* FAO, 2021a, b). This implies that agricultural production practices and food consumption will also have to be adapted to reducing climate change, to preserving biodiversity, and should contribute to the UN SDGs to the highest possible extent. So far, organic agriculture is considered to be the most environmentally compatible form of agriculture (BMNT, 2021), and thus the most promising method in this respect, alongside with other agri-ecological farming methods not harming the environment such as permaculture. Assuming that by 2050 the world's population will amount to at least 9,8 billion people (United Nations, 2017), studies suggest that organic agriculture could feed the world more sustainably than intensive agriculture, albeit with some additional conditions (Muller et al., 2017) (*see also* Loury, 2016). These would comprise reducing waste and food wastage drastically, combined with changing diets to less meat-containing and more plant-based ones (Muller et al., 2017), since livestock grazing and livestock feed production require about 77% of the land used for agriculture, agriculture being a

Table 1 Simplified overview of the crossed planetary boundaries (Steffen et al., 2015)

	Planetary boundary (zone of uncertainty)	Current value
Climate change	350 ppm CO_2 (350–450 ppm)	396.5 ppm CO_2
Change in Biospere Integrity	<10 E/MSY (10–100 E/MSY)	−1000 E/MSY Extinction/Mio Species/year
BIT: Biodiversity Intactness Index	BTI: 90%	84% in S. Africa
Ocean acidification	≥80% of the Pre-industrial aragonite saturation	~84%
Biochemical flows (P and N cycles)	Global: 11 Tg P yr–1 (11-100)~ 62 Tg N yr–1 (62-82 Tg N yr–1)	22 Tg P yr–1 150 Tg N yr–1

The table is a simplified summary of the author based on Steffen's research findings

major factor for several planetary boundaries (Ritchie, 2019) (*see also* Lade et al., 2019a).

Currently, worldwide agricultural practices mainly rely on conventional or intensive agriculture with high levels of synthetic fertilizer and pesticide use. On the one hand, the intensification of agriculture has led to increased food availability, but on the other hand it *"has shown considerable adverse environmental impacts such as increase in reactive nitrogen over-supply, eutrophication of land and water bodies"* entailing dead zones in oceans, greenhouse gas (GHG) emissions, and biodiversity losses (Muller et al., 2017). Research on planetary boundaries clearly depicts that biochemical flows, heavily impacted by intensive agriculture, are exceeding sustainability levels, e.g., the nitrogen cycle by a factor of almost 4 (*see* Table 1, Steffen et al., 2015). Whereas the United States, China, and India alone account for 46% of the world's nitrogen emissions (University of Sydney, 2016), even *"the European limit for the nitrogen losses has been exceeded by a factor of 3.3,"* and the one for phosphorus losses by a factor of 2 (EEA, 2020a). In 2015, Steffen and colleagues estimated *"that humanity has already transgressed four planetary boundaries: climate change, the rate of biodiversity loss, land system change, and changes in the global nitrogen cycle"* (Steffen et al., 2015). *"Transgressing one or more planetary boundaries may be deleterious or even catastrophic due to the risk of crossing thresholds that will trigger non-linear, abrupt environmental change within continental- to planetaryscale systems"* (Rockström et al., 2009). Biodiversity—being the foundation for our food systems (*see* Dury et al., 2019c)—is particularly under threat, with a threshold of the rate of biodiversity loss being probably exceeded by a factor of up to ten (*see* Table 1, Steffen et al., 2015). Further, *"the planetary boundaries are densely interconnected"* (Lade et al., 2019b) and *"interactions shrink the safe operating space for future sustainability governance"* (Lade et al., 2019b). The latest research provides evidence that *"biophysically-mediated interactions have more than doubled human impacts on planetary boundaries,"* most interactions *"lead to increased impacts on other planetary boundaries,"* and *"interactions between planetary boundaries lead to trade-offs between planetary boundaries."* On the other hand, countermeasures for one planetary boundary will presumably also be beneficial for another planetary boundary (Lade et al., 2019b).

Table 1 illustrates the four trespassed control variables with their planetary boundary. Planetary boundaries and their current values, as well as the zone of uncertainty.

In this context, it should be noted that *"new regimes of radiation or radioactivity, or bioengineered life-forms may constitute novel entities that have not been assessed yet"* (EEA, 2020b). Genetically Modified Organisms (GMOs) have a negative consequence for at least the boundary biodiversity loss (*see The Insectatlas* h). From this perspective, but also because of their adverse effects for the environment and human health outweighing benefits, GMOs cannot be considered as a viable solution for achieving sustainable global food systems (Neal, 2017) (see also Jacobsen et al., 2013).

This chapter discusses advances in sustainable food systems and sustainable production and consumption patterns worldwide, taking the example of Austria, considered to be a frontrunner in this field. With a percentage of 26,1% of organic agriculture out of total agricultural area in 2019 (Eurostat, 2021; FIBL, 2021), Austria ranks amongst the top three nations worldwide, while this percentage at the international level only amounts to 1,5% (Willer & Lernouds, 2020) and at European Union (EU) level to 8,1% (Eurostat, 2021; FIBL, 2021), respectively. At the same time, research reveals (Muller et al., 2017) that organic farming practices are by far the most sustainable way of food production, considering the negative consequences of intensive agriculture as outlined above. Given the complexity of the issue and the interconnection of main policy fields, research equally reveals that holistic policy approaches will have to be applied, the range of solutions will therefore be multifaceted. The most recent reports from IIASA and the International Science Council conclude that *"deep transformations of food system architecture is needed to ensure the long-term sustainability"* (IIASA, 2020a, b). In Austria, some of these changes have been already taking place for quite some time. Selected projects and initiatives contributing to the achievement of sustainable food systems are described, including the UNESCO Man and the Biosphere parks, the Donausoya/ Europe-soya production launched in Austria, Austrian protein-rich superfoods, bee protection and organic honey production, or urban agriculture and gardening. Further, we present food recycling initiatives, socially sustainable food stores for people in need, as well as new methods of nature-sound practices to gain higher yields in agriculture.

A holistic approach implies that policies connected to agriculture are designed to attain positive synergies. A policy coherence concept for sustainable agriculture is needed. Particular attention is therefore placed on interrelations with public health, tourism, development, and migration policy as well as with sustainable finance. Finally, policy recommendations for the international community steering the transition to sustainable food systems are proposed.

2 The Necessity for a Transition to Organic Agriculture or Other Environmentally Sound Agricultural Practices to Sustain Human Well-being on our Planet

The world is not endless, it has boundaries. Humanity is currently living in a period where human impact on the planetary ecosystems has crossed several boundaries and without a transition back to a safe operating space may have detrimental consequences for the survival of the entire human race. Scientists have named it the Anthropocene era.

In response to these transgressed planetary boundaries, the world agreed on the UN SDGs and other international agreements regarding climate change, biodiversity loss, etc. Most recently, in January 2021, at the One Planet Summit in Paris, countries signed the agreement to preserve 30% of worldwide water and land resources (One Planet Summit, 2021). This is an important step, however, it will not suffice to halt biodiversity loss and to guarantee food supply and food security in the long term. An international soil convention to protect high value agricultural soil or a binding agreement on a rapid transition to sustainable—organic—agriculture worldwide should be considered urgently in this respect. Meanwhile, regional or national policies will have to be directed accordingly.

2.1 Respecting Planetary Boundaries: The Earth System Needs to Be Urgently Navigated Back toward the Safe Operating Space for Human Impacts

2.1.1 Planetary Boundaries Concept: Four Boundaries Transgressed (Steffen et al., 2015)

According to research from 2015, our planet has trespassed already four out of nine planetary boundaries, namely the sustainable levels of climate change, biodiversity loss (biosphere integrity), land system change, and the biochemical flows (nitrogen and phosphorus cycles) (see Fig. 1, Steffen et al., 2015). To curb this development, if at all possible, a swift transition—within the next two decades—from conventional farming methods to more environmental friendly ones appears to be the most promising solution to save the earth from further unsustainable developments in the wider sense, while at the same time guaranteeing sufficient food supply and food security for the projected nine billions of human beings by 2050 and beyond.

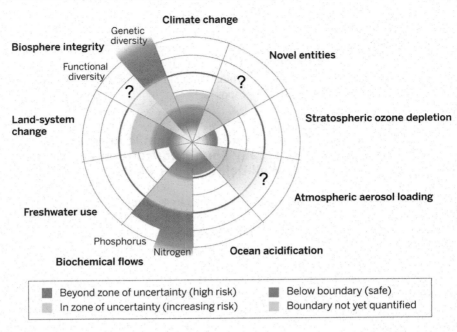

Fig. 1 The planetary boundaries concept and its control variables (Steffen et al., 2015)

Figure 1 pictures the nine control variables for the planetary boundaries concept and highlights a particular emergency situation for biosphere integrity (biodiversity) and biochemical flows (nitrogen cycle), but also an increasing risk for climate change and land system change. Novel entities like chemicals, endocrine disruptors, or GMOs, that may also have detrimental consequences on the Earth's ecosystem have not yet been assessed.

2.1.2 Interactions of Planetary Boundaries—Current State of the Earth System Well Outside Safe Operating Space, Seven Boundaries Transgressed (Lade et al., 2019c)

However, when taking interactions between planetary boundaries into account, the situation is much more alerting: seven planetary boundaries are trespassed. This means that *"the current state of the Earth system as represented by the planetary boundaries is well outside the safe operating space for human impacts"* (Lade et al., 2019b) (Fig. 2).

Lade et al. investigated amongst others *"two climate mitigation measures that involve agricultural activity: large-scale bioenergy production with carbon capture and storage (BECCS), where carbon dioxide from the combustion of rapidly growing crops is geologically sequestered, and a global transition to low-meat diets."* They come to the conclusion that *"reduced agricultural activity triggers interactions that further lower carbon emissions."* Additionally, *"low-meat diets, alongside other transformations of the food systems, are an important strategy for navigating*

Fig. 2 Interactions of planetary boundaries—seven boundaries transgressed (Lade et al., 2019c)

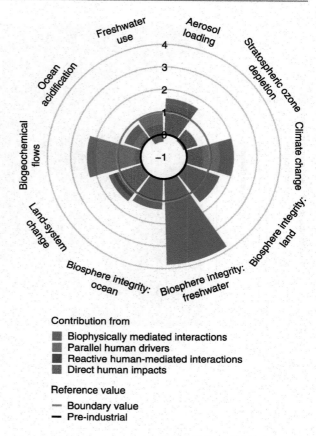

Contribution from

■ Biophysically mediated interactions
■ Parallel human drivers
■ Reactive human-mediated interactions
■ Direct human impacts

Reference value

— Boundary value
— Pre-industrial

back towards the safe operating space for humanity in the Earth system" (Lade et al., 2019b).

Since agricultural activity entails significant negative impacts on several planetary boundaries such as climate change, biodiversity loss, and biochemical flows, its reduction or transition to a more sustainable state can therefore be considered crucial for the earth system and for sustainable food systems. Hence, other agricultural systems that do not have such negative effects on these particular planetary boundaries, particularly organic agriculture, permaculture, or other agri-environmental farming methods, should urgently be prioritized over conventional and intensive agriculture. The international community could think about an international mechanism for agriculture including a cap for agricultural land, as well as an Emission Trading Scheme and a cap for nitrogen emissions, similar to the Emission Trading Scheme and the cap on CO_2 emissions. As shown in Fig. 3, the current state of agricultural activity has already trespassed the safe operating space (*see* Lade et al., 2019a).

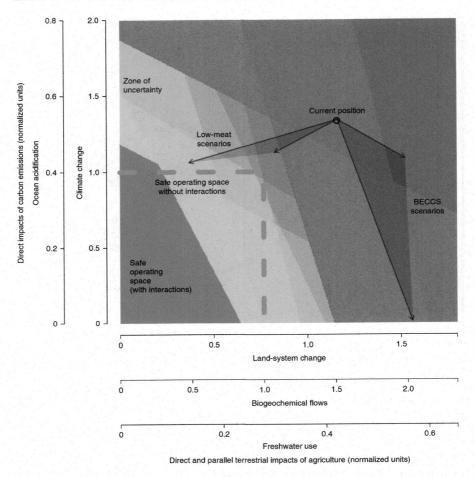

Fig. 3 Effects of interactions between planetary boundaries on the shape of the safe operating space for human impacts on the Earth system (Lade et al., 2019a)

Box 1. Global Policy Actions toward Sustainable Agriculture and Sustainable Food Systems Could Therefore Include:

*** Creating an *International Sustainable Agriculture—Mechanism* to observe and control negative effects from agriculture, keeping them within the safe operating space on the earth system, and agree on a transition to organic agriculture globally to 40% by 2030, and to 100% by 2040.

*** Launch the *Action Plan "Transition to Sustainable Agriculture by 2030"* by introducing a system that leads to and promotes

(continued)

Box 1 (continued)

(a) The reduction of conventionally farmed agricultural land in general, but especially for large-scale bioenergy production with carbon capture and storage (BECCS) purposes.

(b) The conversion from conventional farming to organic farming/other agri-environmental methods.

(c) The creation of organic farms, or other farms applying environmentally sound practices.

(d) Diminishing the synthetic pesticides use by 50% by 2030.

(e) The reduction of agricultural land used for meat production by 20% by 2030.

(f) The creation of protein-rich crops as meat alternative.

(g) Governments/institutions/hotels, etc. offering mainly or to a huge extent low-meat and/or organic dishes (at policy level low-meat and organic policy initiatives should be considered as important as public transport).

(h) The preservation of high-quality agricultural soil.

(i) The enhanced preservation of indigenous crops and species, preferably also an enlarged inventory and storage in gene banks.

*** Setting a *cap for the global agricultural surface managed.*

*** Introducing an *International Emission Trading Scheme and a cap for nitrogen emissions* (similar to the one for CO_2 emissions).

*** Establishing a *"Policy Coherence for Sustainable Agriculture"—Action Plan.*

2.2 The Benefits of Organic Agriculture and Other Agro-Environmental Farming Methods in Search of the Most Promising Strategy for Feeding the World in a more Sustainable Way

2.2.1 Significant Contribution to Attaining the UN SDGs

The global organic market has been steadily growing, representing today already a 106 billion US dollars market (*see* FIBL, 2021), and its contribution to the UN SDGs is remarkable, as illustrated by IFOAM (*see* IFOAM, 2018). Organic agriculture particularly helps achieve:

** *goal #2* to end hunger, achieve food security and improved nutrition, and promote sustainable agriculture: *"The yields of organic farms, particularly those growing multiple crops, compare well to those of chemically intensive agriculture."*

** *goal #3* to ensure healthy lives and promote well-being for all at all stages: *"Organic agriculture does not use harmful chemicals. WHO estimates three million cases of pesticide poisoning each year, and up to 220,000 deaths."*

** *goal #6* to ensure availability and sustainable management of water and sanitation for all: *"Eliminating the use of toxic chemicals in agriculture will hinder pesticide runoff into waterways, which can affect the fish we eat and the water we drink."*

** *goal #12* to ensure sustainable production and consumption patterns: Public procurement policies can be crucial. Denmark has announced a share of 60% of organic food in government institutions.

** *goal #13* to take urgent action to combat climate change and its impacts: *"A key part to organic farming practices is soil management. The degradation of soil due to unsustainable farming has released huge amounts of carbon into the atmosphere.* Organic farming practices such as *"cover crops, compost, crop rotation and reduced fillage"* can *"sequester more carbon than is actually emitted."*

** *goal #15* to protect, restore, and promote sustainable use of terrestrial ecosystems, sustainably manage forests, combat desertification, and halt and reverse land degradation, and halt biodiversity loss: *"organically managed lands have up to 46 and 72% more semi-natural habitats and host 30% more species and 50% more individuals than non-organic farms"* (IFOAM, 2018).

2.2.2 Organic Agriculture Would Significantly Help to Stay within a Safe Operating Space with Respect to Planetary Boundaries

The main added value of applying organic agriculture as a systemic and multifunctional policy approach consists in the scientifically proven fact that it has a positive effect on a multitude of subjects of protection, including (i) soil fertility (humus), (ii) the protection of the climate, (iii) the biodiversity, (iv) groundwater or surface water, and (v) food security. The meta study Thünen Report 65 "Leistungen des ökologischen Landbaus für Umwelt und Gesellschaft" from 2019 (Sanders & Hess, 2019a) investigated 528 studies with 2.816 comparison pairs (organic—conventional) based on 33 indicators. Organic agriculture comes out better with 26 indicators, especially in the protection of water protection, soil fertility, biodiversity, climate adaptation and resource efficiency, and equal in six indicators (translated from *Fertl T., 2020, quoted from* Sanders & Hess, 2019a, b). It is particularly stressed that organic farming reduces nitrogen discharge into water bodies and, together with its contribution to biodiversity protection, thus also provides a model character for an integrated water protection policy. According to FOLDAL, KASPER, ECKER, ZECHMEISTER-BOLTENSTEIN, a transition from 25% conventional agricultural surface to organic farming would entail a reduction of emissions of 5–12 kg N per ha yr.$^{-1}$ (Fertl, 2020, *quoted from* Sanders & Hess, 2019a, b). The Thünen Report 65 measured 39% less nitrogen discharge for organic agriculture (Weisshaidinger & Hörtenhuber, 2020).

As outlined above, organic agriculture scores much better than conventional agriculture with respect to the four trespassed planetary boundaries: climate change, biodiversity loss, land system change, and biochemical flows. A major problem of conventional and intensive agriculture consists in the high use of chemical fertilizers, pesticides, and herbicides which contaminate water, soil, and also the agricultural products we eat. This is why the European Green Deal has set the goal—within its Farm2Fork Strategy—to reduce the use of chemical pesticides by 50% and of chemical fertilizers by 20% by 2030 (European Commission, 2019).

Such an initiative would also be necessary at a global level, since the production and consumption of chemical pesticides and fertilizers has risen dramatically over the last decades. Massive use of chemical pesticides and fertilizers entails losses in soil fertility and biodiversity, and negatively affects biochemical flows. For organic farming, however, high-quality soils are essential, especially for future generations. Focus, therefore, needs to be given to the preservation of high-quality agricultural soil. This can either be achieved by conversion to organic agriculture or by the revitalization of depleted soils, i.e., via permaculture.

Consequently, due to its multifunctional positive aggregated effects on many environmental and planetary boundary constraints, organic agriculture can be considered as—best performing—key technology for sustainable land use and sustainable food systems (Sanders & Hess, 2019b). In this respect, it is interesting to note that Cuba, i.e., experiences *"the largest conversion in world history from conventional agriculture to organic or semi-organic farming"* (Rosset, 2014).

As already conveyed by FAO in 2007 (FAO, 2007), the research considers that organic agriculture alone could feed the world, in combination with reducing food waste and reshaping consumer habits toward low-meat diets and the enhanced cultivation of different protein-rich crops. Muller et al., i.e., depict that *"a food system with a combination of 60% organic production, 50% less food competing feed, and 50% reduced food wastage would need little additional land and have an acceptable N-supply"* (Muller et al., 2017). A study for Austria comes to a similar conclusion: A conversion to 100% organic agriculture would already be feasible now, ideally accompanied by a reduction of meat consumption by 10% or of food waste by 25% (Schlatzer & Lindenthal, 2018). The announcement of the European Commissioner for Agriculture, Mr. Jamusz Wojciechowski, end of 2020, that he will be *"campaigning for organic farming and strive to make organic farming the norm"* (organic-market.info, 2020) certainly ushers in a new epoch for agricultural policy, potentially also beyond Europe.

2.3 The Importance of Preserving Existing Bee Populations for the Ecosystem Earth

The Ecosystem Earth is uniquely diverse, uniquely complex, and fragile. If a crucial species would disappear, disruptions in the Ecosystem Earth may be disastrous and restoration may be difficult if not to say impossible. Attempts by scientists to reconstruct the Ecosystem Earth from scratch failed, as prominently showcased in a vitrine of the Museum of Natural History in Vienna.

Bees represent an indispensable jigsaw piece of our ecosystem—a fact probably still completely underestimated by the wider public. Bees are essential for our ecosystems and biodiversity, for food supply and food security—since they are our main pollinators of food crops—and hence crucial for the survival of mankind. According to the FAO, bees support the production of 87 of our main food crops, and three out of four crops around the globe producing fruits or seeds equally depend, at least partly, on pollinators. *"The vast majority of pollinators are wild, including over 20,000 species of bees."* This is why the FAO plays a leading role in supporting the International Pollinators Initiative 2.0 (FAO, 2018). *"Fewer pollinators mean many plant species could decline or even disappear along with the organisms that directly or indirectly depend on them. In addition, the decline in numbers and diversity of pollinator populations affects food security with potential losses in agricultural yields"* (European Parliament, 2019).

Over the last decades, the world has experienced a tremendous bee population decline. In the USA, beehives declined by 60% between 1947 and 2008, and if counted per hectare even by 90% since 1962 (Greenpeace, 2020). Pollination often does not happen in a natural way anymore; for example, 31 billion honeybees are transported by lorries each February to California to pollinate the almond trees. Then they are brought to New York for the pollination of apples, and to Massachusetts to converge on blueberries. In this process, many American beekeepers lose between 30 and 60% of their bee colonies yearly (Jabr, 2013). Also China's bee diversity is threatened (Teichroew et al., 2017). In many areas pollination needs to be done by hand since bees mostly disappeared in areas with intensive agriculture, and farmers cannot afford to rent mobile beehives (China Dialog, 2012). In Europe, bee populations also suffered a significant decline, amounting to about 30% (European Parliament, 2019). A 2013 report from Greenpeace still gave the figure of about 20% of honeybee colony mortality in Europe, with a wide range of 1,8%–53% between countries (Greenpeace, 2013). In 2019 alone, various countries including the USA, Russia, and Italy reported a massive loss of bee populations of up to 50%: Honey production in Italy fell by 50% (The Local, 2019), in Russia by 20%, in France by more than 50% in 2016 (BBC, 2019), and in Switzerland about 10% of bees are disappearing yearly (NZZ, 2017).

2.3.1 Intensive Agriculture and Pesticides Are Important Factors behind Bee Starving

The European Food Safety Authority estimates that intensive agriculture and pesticide use, poor bee nutrition, viruses, invasive species, and environmental changes such as habitat loss and fragmentation are possible factors for bee starvation (EFSA, 2021). A German long-term study reveals that the population of flying insects in Germany have diminished by 75% (The Insectatlas, 2020k) due to a multitude of factors, the most important driver for the decline in pollinators being intensive agricultural production involving insect killing chemical fertilizers, pesticides, and herbicides such as Neonicotinoid and Glyphosate (The Insectatlas, 2020a, b) (*see also* PNAS, 2018). The European Parliament equally stresses the need to reduce the use of pesticides in order to better be able to protect bees, which are on the verge of extinction. Wild and domesticated pollinators are vital for food security and biodiversity, and contribute in many ways to the economy. *"About 15 billion euros of the*

EU's annual agricultural output is directly attributed to insect pollination" (European Parliament, 2019). Global economic benefit of pollination is estimated at about 265 billion euros, assessed as the value of crops dependent on natural pollination (Greenpeace, 2013).

2.3.2   Pesticides Affect the Entire Ecosystem

"Pesticides are one of the main sources of insect mortality because they affect the entire ecosystem." The high levels of insect mortality can be at least partly explained by the rapidly rising figures of pesticide use worldwide. Conventional farms apply about four million tonnes of chemical pesticides yearly. The global pesticides sales have turned into a 56,5 billion euros market (2018), expected to grow to about 82 billion euros by 2023, whereas four producers share two-thirds of the market, namely BASF and Bayer in Germany, Syngenta in Switzerland (Chinese), and Corteva (USA) (The Insectatlas, 2020b). China alone is using about three times as much pesticides as the global average, and in total about a third of all pesticides worldwide (The Insectatlas, 2020b), while other major users are the USA, Brazil, and Argentina (see The Insectatlas, 2020c).

The Food and Agriculture Organization of the United Nations (FAO) predicts an increase in demand for agricultural products worldwide by 60% until 2050 (The Insectatlas, 2020d). Current policies are, however, insufficient to reduce pesticide use and to control bee starvation. *"Even though the EU has promoted integrated pest management systems and the reduction of the use of pesticides, the amount of pesticides used is not decreasing"* (The Insectatlas, 2020a). Most worryingly, pesticides banned in the European Union are still used in other countries of the world (The Insectatlas, 2020b). This is why farming subsidies of the EU Common Agricultural Policy *"need to be shifted to support high nature value farming, organic agriculture, and agro-ecological systems. Stricter regulations for pesticide use are also needed"* (The Insectatlas, 2020e). Fortunately, the European Commission with its Green Deal from 2019 aspires to halve pesticide use in the EU by 2030 (European Commission, 2019).

In the last 50 years, global crop production has tripled, and the consumption of nitrate fertilizers (mainly for intensive agriculture) has grown tenfold (The Insectatlas, 2020f). About 40% of synthetic fertilizers are washed out and released into the air, whereas organic fertilizers stay in the soil for longer. Some countries introduced pesticide taxes with some success. Denmark's pesticide tax led to halting the use of pesticides between 2013 and 2015, and with the 70 billions of euros collected, farmers were compensated for shortfalls in production (The insectatlas, 2020g).

2.3.3   Excessive Meat Production Is a Major Driver for Unsustainability

"Worldwide demand for meat sparks a chain reaction of deforestation, monocultures, and chemical sprays. Nature is being destroyed fastest in those areas rich in insects." "No other type of farming has more influence on ecosystems than intensive livestock farming" (The Insectatlas, 2020a, p. 24). "The soya boom goes hand in hand with the use of pesticides. Both Brazil and Argentina grow mainly genetically modified soya . . . .and use pesticides that are banned in the European Union because of their negative effects to the environment" (The Insectatlas, 2020h).

2.3.4 Protecting 40% of Land and Sea Territories for Biodiversity by 2030

Protected areas now cover 15% of the world's surface, but they still cannot make up for the large-scale negative effects of intensive agriculture entailing insect- and biodiversity loss (The Insectatlas, 2020i). And *"organic farming focuses on maintaining soil fertility and biodiversity, but for an insect-friendly agriculture the whole farm-landscape has to change"* (Insectatlas, 2020j). Even the most recent international agreement on biodiversity signed by 50 countries in Paris in January 2021, aimed at protecting 30% of land and water bodies by 2030 (One Planet Summit, 2021) (*see also* France 24, 2021) would probably not suffice to significantly protect biodiversity and bees. We are therefore advocating for a higher percentage of 40% worldwide.

2.4   The Threats of GMOs

The main threats of GMOs in combination with toxic pesticides and/or herbicides such as Round-Up consist in their negative effects on both biodiversity and health (*see also* The Insectatlas, 2020h). Results thereof are the often irreversible loss of indigenous plants and insects, and contamination of agricultural soil and crops, especially in the neighborhood of GM crops, as well as the negative health effects on persons applying GMOs or related pesticides or herbicides, or living in the vicinity of GM crops. In 2020, the German pharmaceutical giant Bayer had to pay $10 billion for about 95,000 cancer-related lawsuits linked to the herbicide Round-Up from the US firm Monsanto, the worldwide biggest producer of GM seeds, acquired by Bayer in 2018 (see The New York Times, 2020). In 2006, Monsanto dominated the GM-seed market with 70–100% market share and a global revenue of $US 7.344 billion (Guinnessworldrecords, 2006). Furthermore, GM seeds often create dependency of farmers on those GM seeds, in several cases even driving them into financial ruin or even suicide (Gingesh et al., 2017). Finally, the scientific uncertainty of long-term effects of GMOs on the gene pool of the earth as a whole still prevails (*see* UNCTAD, 2002) (*see* FAO, 2003).

In the planetary boundaries concept, GMOs would be treated under ‚novel entities' which have not been assessed so far. As outlined above, biodiversity loss or species extinction happens at alarming speed all over the world, especially regarding birds, butterflies, bees, and other insects, a major factor being intensive agriculture and pesticides or herbicides which are often combined with GMO use. The United States of America (USA) together with Brazil and Argentina are leading the list of countries for both the highest pesticides' use and the biggest area of GMO crops (Statista, 2019); for the latter, China ranks place 7. The USA and China representing the biggest users of both pesticides and GMOs, also share the same destiny of worrisome decline in bee populations (*see also* The Insectatlas, 2020c).

In 2016, GM crops (such as maize, soybean, oilseed-rape, and cotton) amounted to about 3,4% of globally used agricultural area (IFOAM-EU, 2017a, b). In its assessment "World Agriculture Towards 2015/2030—An FAO Perspective" the FAO reminds the scientific community about the responsibility to address ethical concerns such as to ensure that biotechnology does not adversely affect food safety or risk damaging the environment (FAO, 2002). About 150 million hectares of world crop acreage are planted with GM crops. Already in 2004, the Council of Europe referred to the controversial scientific debate and advocated for "*stricter regulation of labelling, liability, good farming practice and GM-free zones*." In 2017, IFOAM-EU states that coexistence between GMO and non-GMO breeding and seed production is considered to be unfeasible (IFOAM-EU, 2017a). Despite huge amounts of money being spent in research on GMOs, experts conclude that natural breeding methods are still more sustainable. The European Commission and the EU Member States are therefore requested to provide a regulatory framework that allows for competitive GMO-free farming (IFOAM-EU, 2017b), since there is a strong civil society movement for a GMO-free Europe (GMO-free-regions, 2021).

Thus, applying the precautionary principle, a basic principle in sustainable development policy, implies that—for the time being—there seems to be no more room for GMOs in representing a solution for future sustainable food systems, and other, more sustainable solutions to guarantee sustainable food supply and food security will need to be found and applied.

2.5 Linking Sustainable Agricultural Practices to Improving Public Health

2.5.1 Promote Organic Farming to Counter Hunger and Obesity

Hunger and obesity are two rising main health issues worldwide closely linked to our food system. While on the one hand about 820 million people suffered from hunger in 2018, out of which about nine million died, on the other hand almost 700 million people were obese, leading to the death of four million of them (*see* Health Policy Watch, 2019, FAO, 2019, WHO, 2019, NPR, 2020). Food insecurity and malnutrition are two underlying reasons for these developments.

Organic farming is considered to provide a solution to both problems at the same time. The UNCTAD Report from 2008 describes *organic agriculture as "a holistic production system based on agro-ecosystem management rather than on external inputs" utilizing both traditional and scientific knowledge.* On the same line as a FAO Report, UNCTAD concludes that "*organic agriculture can be more conducive to food security in Africa than most conventional production systems, and that is more likely to be sustainable in the long term*" (UNCTAD, 2008), since conventional systems failed in the past half-century to solve the hunger problem (UNCTAD, 2008).

Certainly, malnutrition and obesity in developed countries have different underlying causes. However, there is scientific evidence that people who favor organic food over other types of food tend to also have healthier lifestyles and have healthier dietary patterns overall. Moreover, several human health benefits are documented for organic food in comparison to conventional food such as a lower content of cadmium and antibiotics, a higher content of omega 3-fatty acids in milk and probably meat, as well as higher phenolic compounds in fruit and vegetables. Further, *"epidemiological studies have reported adverse effects of certain pesticides on children's cognitive development at current exposure"* (EU Parliament, 2016) (Mie et al., 2017).

2.5.2 There Is Growing Evidence that Cancer Is Linked to Pesticides Exposure

Similarly, with respect to cancer, a 2018 French cohort study with almost 70,000 adults detected that a higher frequency of consumption of organic food is associated with a lower risk of cancer. There is further growing evidence about cancer being linked to pesticide exposure (Baudry et al., 2018) (*see also* IIARC, 2015). In this respect it appears to be interesting to look at the profile of the French organic food consumer: Besides some socio-economic characteristics (higher education, higher physical activity levels, less smoking, and higher incomes), he shows healthier dietary patterns, less chemical exposure, a lower obesity level, and lower disease rates (Organic without boundaries, 2019). Should these results be confirmed by other studies, the beneficial links between organic food and health should be incorporated into education for children and adults as well as into other policy areas. IFOAM starts a relevant initiative in schools in February 2021 (IFOAM, 2021).

2.5.3 COVID-19 Boosts Organic Food Consumption

With the COVID-19 impact, consumer perceptions toward good health have been directed toward healthy eating. The organic food market is therefore expected to grow at a CAGR of 12,2%, reaching a value of $272 billion by 2027 (Meticulous Research, 2020). In an interim report on the Impact of COVID-19 on Food Systems, the IFAO-HLE Panel recommends governments amongst others to support an increase in local food production, including in home and community gardens (urban gardening), enhance food resilience, and minimize food waste (FAO, 2020).

2.6 Fostering Organic and Sustainable Agricultural Practices As Well As Upscaling Subsistence to Semi-Subsistence Farming in Low-Income Countries, as Approaches to Prevent International Migration

More than 75% of the world poor live in rural areas and depend on agriculture. *"Rural poverty, food insecurity, lack of employment opportunities, access to social protection, natural resource depletion and deterioration in rural livelihoods,"* together with climate change, *"are important drivers of migration." "The challenge in the next decades will be to generate enough employment to absorb a booming labour force"* (FAO, 2017).

Migration happens when people dream of better chances for a prosperous life elsewhere. Contributing to a decent life and decent work conditions across the globe appears therefore as a prerequisite for successful development and migration policy. Supporting subsistence of local and organic agriculture as well as other types of small-scale and medium-sized sustainable farming practices, together with upscaling initiatives toward semi-subsistence farming could prove to be promising approaches to prevent international migration.

Whereas in 2006 Kilcher attributes organic farming to playing a major role in sustainable development in poorer countries (Kilcher, 2006), one year later, the FAO appraises organic farming for having the potential to feed the world (FAO, 2007). At the UN level, experts point out that "*agriculture must be directed to environmentally sound, socially just production methods to address the food and energy crises, hunger, poverty and climate change*" (de Schutter, 2012). Another study conducted by the FAO and INRA "*undertook a survey of innovative approaches that enable markets to act as incentives in the transition towards sustainable agriculture in developing countries.*" Among others, e.g., new technologies for sustainable agriculture (effective microorganisms and biopesticides) were the result (FAO and INRA, 2016). The latest IIASA report on Resilient Food Systems also favors sustainable agricultural practices, particularly for underdeveloped countries (IIASA, 2020a, b).

While in Europe about two-thirds out of 14 million farms are under 5 ha in size, two-thirds of those are practicing subsistence farming. The remaining third relies on semi-subsistence farming. The latter means that production does not only suffice to feed the farm family, but also leads to a surplus and a regular income (Dower, 2009). In poorer countries, the share of subsistence farming is often much higher. Ways need to be found to upscale subsistence farming to semi-subsistence farming and policies directed accordingly.

2.7 Sustainable Finance as a Multiplier for Sustainable Food Systems

With its new sustainable finance strategy, the EU aims at steering public and private capital toward sustainable investment. A taxonomy was agreed upon to define which activities would qualify as sustainable investments. Some activities related to the preservation of biodiversity and ecosystems would also qualify for the agricultural sector (European Commission, 2020a, b, c, d, e). Sustainable finance policy could and should be used more intensely to foster sustainable food systems.

Within the EU Multi-Annual Financial Framework 2021–2027, the EU has set the possibility to financially support organic agriculture and other agri-ecological schemes with new instruments called eco-schemes (European Commission, 2021), mainly in the wider framework of the European Green Deal. The latter constitutes a transition to a more sustainable food system with ambitious goals such as reducing 50% of chemical pesticides and achieving at least 25% of agricultural land under organic farming by 2030. The so-called environmental pillar of the CAP has been strengthened, and the right steps toward more sustainable food systems have been taken. Similar measures need to be proposed at the international level, the European Green Deal could serve as an example.

Given the urgent planetary boundary constraints as outlined above, especially regarding transgressed biochemical flows, it remains to be evaluated and studied continuously, whether the measures taken will suffice to drive human activities back to a safe operating space, both in Europe and worldwide. More in-depth studies in this area would certainly be desirable. Stricter CAP measures would probably be necessary and sustainable finance could be used even more effectively to ensure a safe transition to sustainable food systems within a safe operating space for the Ecosystem Earth.

3   Austria: A Frontrunner in Organic Agriculture and Sustainable Food Systems

As outlined above, the main factors for achieving a transition to sustainable food systems lie in a boost of organic agriculture accompanied by measures to reduce chemical pesticides, foster biodiversity and bee preservation, diminish food wastage, and support a change to more low-meat, but equally protein-rich diets. In many ways, Austria has been at the forefront of such developments.

While the European Commission in its Green Deal of 2019 has set the goal to reach 25% of organic agriculture at the EU level by 2030 (European Commission, 2019), Austria overfulfills this target already recording a share of 26,1% of organic agriculture in 2019 (Eurostat, 2021) (FIBL, 2021). Biodynamic and organic farming labelled as Demeter as well as permaculture add to the sustainable production methods, whereas a percentage of roughly 20% of production over the last years was distributed directly including within social markets. Austria, therefore, has a model character regarding both organic production and distribution methods. Several other interesting initiatives such as UNESCO Man and the Biosphere reserves, bee protection and organic honey production, the development of GMO-free and partly organic Donausoya and Europasoya, food recycling initiatives as well as the cultivation of local superfoods such as pumpkinseeds-oil and linseeds are part of a comprehensive policy set directed toward more sustainable food production and consumption patterns.

3.1   Organic Farming in Austria: 100% Organic Cultivation Would Already be Feasible Now

In 2021, Austria has one of the highest shares in organic agriculture worldwide. With 26,1% of total agricultural area, Austria belongs to the world leaders together with Liechtenstein (41%, 2019) and Sao Tome and Principe (24,9%, 2019) (FIBL, 2021). In the province of Salzburg, e.g., the organic share amounts to more than 50%. The geographical condition of the small alpine country with its mountainous landscapes and a vast majority of small-scale farmers was probably a major driver of this development. Looking back to a history of organic farming of more than a hundred years, Austria enjoys today a high reputation as an exporting country, also with respect to the high quality of organic agricultural products, worth more than \$2 billion (2018) (USDA, 2020).

A study from 2018 investigated the possibilities and consequences of a conversion to 100% organic agriculture with current nutrition habits. The results confirm the feasibility with either a 10% reduction in meat consumption or a 25% less food waste. Interestingly, recommendations of the Austrian Association for Nutrition regarding healthy diets would require diminishing meat consumption by 64% (Schlatzer & Lindenthal, 2018).

3.1.1 History of Environmentally Friendly and Organic Agriculture

As early as 1927, the first organic farm worldwide was registered in Austria. Based on the findings of the famous Austrian researcher and anthroposophist, Rudolf Steiner, who's educational philosophy and holistic understanding of the world is still today being taught at more than 600 schools (Rudolf Steiner schools or Waldorf schools) in 75 countries, Austria's organic agricultural policy developed into a success story. Austria, having been at the forefront of promoting "environmentally friendly" methods of farming, was also the first country in the world to establish national regulations for organic farming in 1991 (2xxbioforschung.at), ten years earlier than the European Union. In 2020, about 26,1% of the farmland is managed under the high environmental standards of organic farming by more than 24,000 organic farmers (Bio-Austria, 2020a) (see also *Organicexport-info*, 2019) (FIBL, 2021) (Fig. 4).

Figure 4 depicts the steady rise of both the number of organic farmers and the surface of organic agricultural areas. In 2019, more than 24,000 organic farmers were registered in Austria, and 26,1% of the total agricultural area was organic (FIBL, 2021), revealing a rise of more than 20% both regarding the number of farmers and the surface area between 2014 and 2019.

About two-thirds of more than 165,000 farms are located in alpine mountain regions which are much less productive than conventional agricultural farms in flat areas. Due to these specific topographical and climatic conditions, a specific subsidies scheme for agriculture in mountainous regions was developed taking account of the multiple functions of agriculture for the environment. The latter particularly relate to water and landscape management, avoidance of erosion, and the preservation of high soil quality and the cultural landscape, as well as to culture and traditions in general. Austrian farmers, although accounting for only 1,3% of GDP (2019), also make a substantial contribution of about 35% to renewable energy production in Austria. About two-thirds of supported farms under organic regime are in mountainous areas, whereas the government support per farm can amount to up to 86% of total farm income in the highest support category (see Hovorka, 2006) (*BMIRT*, 2021). In 2019, payments for agriculture accounted for 1281.3 million Euros, out of which 53.9% were direct payments (*see* European Commission, 2020b).

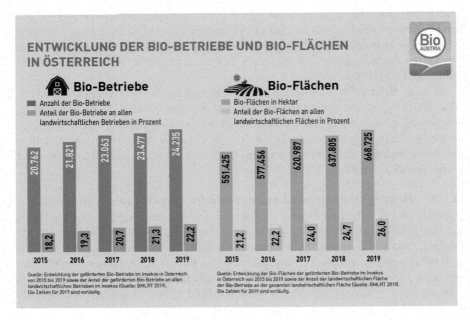

Fig. 4 Development of organic agriculture in Austria: farms and surfaces (Bio-Austria, 2020a)
Source: Bio-Austria

Figure 5 shows the development of the organic market in Austria which rose from a value of 1.338 million Euros in 2014 to 1.833 million Euros in 2018, documenting an increase of more than 30% in that period (*see* Bio-Austria, 2020a).

Between 2014 and 2018, the share of organic products in terms of quantity rose by 35% and in terms of value even by 47% (*see* Bio-Austria, 2020a).

Market shares of organic products (Fig. 6) are ranging from 25,5% to 23,7%, respectively, for milk and natural yoghurt to less than five percent for meat/poultry products and ham/sausages. Between 2014 and 2019, household expenditures for organic products grew by about 40% (Bio-info, 2021a, b).

In addition to certified organic production, many farmers apply eco-friendly methods without being certified, since the demand for high-quality agricultural products from Austria is very high. The Austrian consumer has traditionally high expectations of environmentally sound high-quality food, free from GMOs or disputed pesticides like Glyphosate. Non-certification is a consequence of often cumbersome and costly certification processes.

The average size of an Austrian organic farm comprises about 22 ha (BMIRT, 2020), whereas about 90% of agricultural production is provided by SMEs (European Commission, 2020b). About 10% of the products sold in supermarkets are organic (Tasteofaustria, 2021).

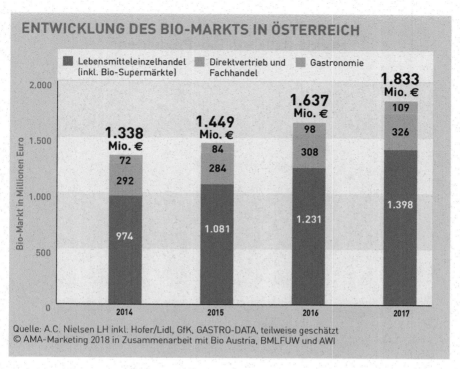

Fig. 5 Development of the organic market in Austria. Source: Bio-Austria

3.1.2 Legal and Institutional Framework

The success of the organic movement developed from a good mix of stakeholder initiatives, the creation of organic farmers associations, critical consumers close to nature, and government support going hand in hand with a subsidiary legal and institutional framework including, e.g., also certification and consumer protection bodies and institutions. The Ministry for Sustainability and Tourism being responsible for agriculture is the competent authority for organic farming. The government supports and fosters organic farming through organic action programs. The fifth "Bio-Aktionsprogramm 2015-2022" aims at maintaining Austria's position as organic country #1, continuously raising the surface of organic farmland and raising the share of organic products in the retail sector (BMIRT, 2020). *"Some of its responsibilities like the authorization and supervision of control bodies is delegated to the province governors (Bundesländer)"* (see also Organicexportinfo, 2019). Among several organic farmers' associations, BIO-Austria, established in 2005, is the largest interfacing about 15,000 farmers and more than 420 partners (Bio-Austria, 2020b).

Austria adopted the first legislation on organic agriculture in 1991 and was, since its entrance into the European Union in 1995, very active in shaping an

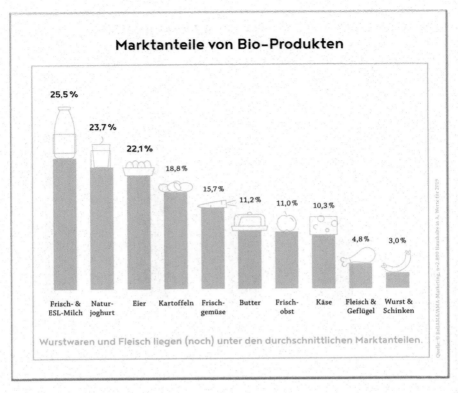

Fig. 6 Market shares of organic products (Bio-info, 2021a, b). Source: Bio-Info

environmentally friendly Common Agricultural Policy. Having fully implemented the EU legislation on organic farming, additional national rules apply. *"The EU Quality Regulation Implementation Act (EU QuaDG) defines requirements concerning the implementation of EU organic legislation, e.g. regarding the control system. Some national decrees related to organic production as well as the annotated versions of Council Regulations (EC) No 834/2007 and Council regulation (EC) No. 889/2008 aim at the harmonization of the application of EU organic legislation at national level. Additional requirements not covered by EU organic legislation concerning animal husbandry, organic cosmetics and organic catering are laid down in an Austrian guideline on organic production. The national rules are only relevant for operators in Austria, not for exporting companies from other countries"* (Organicexportinfo, 2019). Austria's legislation on liability has model character at the EU level and partly served as a guideline for EU legislation.

3.1.3 Organic Labels in Austria

Labelling of organic products is of utmost importance for customers enabling them to make informed decisions. Organic products have to be labelled as such and

Fig. 7 Organic Labels used in Austria—The Demeter Label (*Demeter*, 2021)

JA! NATÜRLICH.   **Zurück zum Ursprung**   **SPAR Natur*pur**

Fig. 8 Organic Labels used in Austria—Supermarkets (Ja-Natürlich, 2021) (Zurück zum Ursprung, 2021) (Spar, 2021)

undergo clear legal regulations and controls. The nomination Bio is protected by the EU regulations 834/2007 and 889/2008. These regulations are binding for all EU member states and govern crop production, animal husbandry, processing, trade and labelling of organic food. Producers and fabricators of organic food need to comply with them and only products from controlled organic agriculture can be labelled as such. *"Pre-packed organic products sold in Austria have to be labelled with the EU organic farming logo. For products imported from third countries the use of the EU organic farming logo is optional. A mandatory national organic label does not exist"* (Organicexportinfo, 2019). Besides the EU Organic label, the AMA Gütesiegel as national label and other labels are used (Fig. 7).

Demeter

The first Demeter symbol was introduced as a nomination for common biodynamic quality standards and controls in 1928. Meanwhile, Demeter is an international label, used for more than 3500 food products, cosmetics, and fashion articles in over 60 countries (Demeter, 2021a, b, c) (Fig. 8).

JA! NATÜRLICH. Zurück Zum Ursprung SPAR Natur*Pur

Austria's probably biggest organic label was launched already in 1994 by Werner Lampert for the retail business group Billa. While Billa started with 30 products, today the organic product line comprises about 1000 products ranging from dairy products, fruits and vegetables, bread and bakeries, over staple food, and frozen products until baby food. In 2010, the first Austrian urban gardening line was initiated, offering customers to grow their own organic food in gardens, on balconies or terraces. In 2011, the pioneers introduced sustainable packing in Austria and in 2016, the first national rice was marketed. Since 2018, the products are free from palm oil and the first sustainable advertisement spot was launched. The brand

Fig. 9 Organic Labels used
in Austria—The Bio-Austria
Label (Bio-Austria, 2020b)

cooperates with about 7000 organic farmers and about 80 partner firms. Moreover,
Ja-Natürlich offers fair-trade products like coffee and bananas (Ja - Natürlich, 2021).

Other certified organic supermarket labels followed such as "Zurück zum
Ursprung" issued by the supermarket chain Hofer. The brand, e.g., saves about
20,000 tons of carbon dioxide per year which led to the Austrian Prize for Climate
Protection 2009, about 10,6 billion liters of water per year. The production of
hay-milk in some regions entails a 26% better protection of biodiversity than
conventional farms (Zurück zum Ursprung, 2021).

SPAR's organic line "SPAR Natur*pur" started in 1995 with 10 products and
500 organic farmers and today comprises more than 900 mainly national products,
being provided by over 7000 organic farmers. SPAR's sustainability goals perfectly
outline how dealing with organic products can be embedded in a holistic
sustainability concept. They include (Spar, 2021):

- Production of vegetable crops on humus soils (1600 hectares) which allowed
  saving 60,000 tons of carbon dioxide by 2020.
- Production 100% free of palm oil.
- By 2016 SPAR banned all hormone-reactive chemicals in its own cosmetic
  brands.
- Reduction of sugar in own brands of 2000 tons within five years by 2022,
  fostering of sugar alternatives.
- Management of satisfaction and participation of employees to be kept at high
  levels (80%).
- Energy consumption to be reduced per m^2 sales area by 2050 by 50% (basis year
  2009 607 kWh/m^2).
- CO_2 reduction of 90% per m^2 sales area to be achieved by 2050 (basis year 2009
  130 kg of CO_2/m^2) (Fig. 9).

BIO-AUSTRIA-Label
Issued by the organization BIO-AUSTRIA—Association for the promotion of
organic agriculture, established in 2005 as umbrella associations (Bio Austria,
2020b) (Fig. 10).

Fig. 10 Organic Labels used in Austria—The national AMA Label (Ama, 2021)

Fig. 11 Organic Labels used in Austria—The EU Organic Label (European Commission, 2020c)

AMA: National BIO-Label

This national label of the Austrian Ministry for Agriculture is operated by AMA (AgrarMarktAustria) was established in 2014 by law as a followerof the AMA-Biozeichen (AMA, 2021) (Fig. 11).

The European Organic Label

Introduced by law in July 2010, the new organic label of the EuropeanUnion stands for food products with at least 95% organic standards whereby genetically modified organisms, chemical-synthetic pesticides and fertilizers are strictly prohibited.

3.1.4 Distribution Systems

About 20% of organic products are distributed directly, either to the customer or via local markets. The major part of organic food is sold in supermarkets, and a small portion through restaurants. Direct distribution systems have many benefits: they entail higher gains for farmers, fresher products for clients and less food waste. A study issued by Adamah characterizes distribution through local markets as most sustainable, followed by direct customer delivery, supermarkets representing the third best option.

More than 10 Shops in Vienna Exclusively Organic

The city of Vienna accommodates several weekly organic markets, and all big supermarkets and drugstores have dedicated areas for the meanwhile huge variety of organic products. Three supermarkets developed their own organic product line, DENNs-biomarkt sells exclusively organic products. Products from other important brands like Alnatura, Sonnentor (herbs, tees, etc.), or EZA can be found in almost all supermarkets. More than ten shops, as well as restaurants in Vienna, are exclusively organic (Wien, 2021a) including some that only sell their own brands like

Sonnentor. More than 80 independent Fair-Trade shops all over Austria offer the greatest variety of EZA products including organic edibles, organic cosmetics, and fair fashion made out of organic. Established in 1975, the EZA Fairer Handel GmbH developed mainly out of a catholic movement with the aim to support marginalized producers in Africa, Asia, Latin America, and the Middle East. About 140 associations of craftspeople and farmers represent their partners in more than 40 countries (EZA, 2021).

32 Exclusively Organic Supermarkets in Austria
Denn's-biomarkt opened in Austria in 2005 and nowadays hosts about 6000 organic products which are sold in 31 stores (2019), operated by denree Naturkost GmbH, a subsidiary company of the German denree Group. In 2018, turnover was more than 55 million Euros, the company employed more than 400 staff. The German parent enterprise was founded in 1974, and supplies 1400 independent bio-markets and bio-supermarkets with an assortment of 13,000 products (Denns, 2021).

3.2   Biodynamic Organic Agriculture: Demeter

The origin of organic agriculture lies in Rudolf Steiner's anthroposophical research. Also deployed for the biodynamic cultivation of land, it was first applied in Austria in 1927 at the Wurzerhof in Carinthia (Demeter, 2021a). The first Demeter association was established in Austria in 1969. Later, in 1981, the first regulations on biodynamic and organic agriculture were adopted. The movement spread out at international level right from the beginning, with the launch of the first biodynamic and organic coffee farm in 1928 in Mexico.

Today, about one percent of organic agriculture in Austria is managed by biodynamic standards on about 6900 ha, and sold under the international brand name Demeter. *"Biodynamic farming is a holistic approach to agriculture in which vitality has the highest priority."* *"Biodynamic farmers return more to the soil than they remove in the process of cultivating crops and animals, the farm is considered an organism, in which plants, animals and humans are integrated together. The significant difference is that the biodynamic method attempts to work with the dynamic energies in nature and not solely with its material needs. One aspect of this is the use of cosmic rhythms, for instance cultivation, sowing and harvesting are scheduled, if possible on favorable days"* (Demeter, 2021b).

Worldwide there were 6429 Demeter certified farms in 2020, out of which 214 in Austria, in total cultivating more than 220,000 ha of land (Please find a detailed list of countries and areas) (Fig. 12).

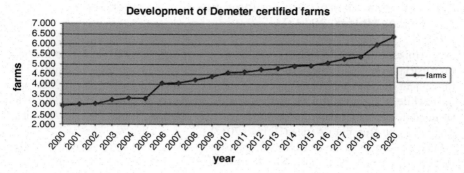

Fig. 12 Development of Demeter certified farms (Demeter, 2021c)

3.3 Eco-Social Agricultural Policy: The Premise of Austrian Agricultural Policy

In Austria, farmers have always thought about the generations to come. Sustainability and eco-social agriculture have been the premise of agricultural policy (Landwirtschasftskammer, 2016a). Environmentally friendly production being a specific goal of the national agricultural policy assures residue-free production meeting baby-food standards (Landwirtschasftskammer, 2016b). In 2016, the goal was to double the retail share of organic products from 8% at that time to at least 15% by 2025 (Landwirtschasftskammer, 2016c). Currently, the share lies at about 9,3%, which ranks Austria among the four European countries with the highest organic retail share together with Denmark (12,1%) and Switzerland (10,4%) (FIBL, 2021).

Due to relatively high certification costs representing a major hurdle for small farmers, a lot of farmers produce organic or almost organic quality without being certified. The quality aspect of agricultural products always played a predominant role at national level and this is probably why Austrian agricultural products are highly appreciated abroad.

Regarding pesticide use, Austria could presumably do better, since sales went up slightly between 2011 and 2018 and were about double as high as in Switzerland e.g. (Eurostat, 2018). With respect to Glyphosate, however, the legislation foresees its use in agriculture only to kill pest plants. The pesticide, therefore, does not come into contact with cultivated plants. Austria was even the first country worldwide who decided to ban Glyphosate in 2019 (Foote, 2020). This decision was challenged by the European Commission though, and the use of the pesticide is now allowed until the end of 2022, subject to further approval thereafter (European Commission, 2020d).

3.4   Permaculture/Agro-Ecology: Austrian Pioneer Sepp Holzer (*See* Holzer, 2013a)

Sepp Holzer developed another ecologically-sound agricultural production method in Austria from 1962 onwards, the Holzer Permaculture or agro-ecology,. The movement spread and gained international recognition already in the 1980s and meanwhile enjoys worldwide the reputation to be an important contribution to sustainable agriculture, from Europe and the United States, South America, Thailand, and Russia to Africa (Holzer, 2013b).

The concept was first developed for fruit crops at high altitudes (1000–1500 meters above sea level) and relies on working with the circular developments and interactions of nature. Agro-ecology is further based on traditional, local, and indigenous knowledge and cultures combined with knowledge from modern science, involves all stakeholders and draws its complex solutions from thereon. The strength of this holistic concept lies in linking ecology, biology, and agricultural sciences as well as medicine, anthropology, social- and communication sciences. *"It is also about assuming responsibility for our environment and for all beings, and to advocate with civic courage for the preservation of a healthy planet earth worthwhile to live at."*

Holzer Permaculture/agro-ecology comprises:

- Landscape architecture (the construction of terraces, the laying of hills and raised beds, water gardens, ponds, detention reservoirs for humus, arid and wet biotopes, small climate zones, etc.)
- Agroforestry (including trees and bushes for agricultural use).
- Fisheries.
- Growing of water plants.
- Animal husbandry.
- Fruit plantations.
- Cultivation of medicinal plants.

Permaculture promotes the development of self-sustaining systems, a "symbiotic" cultivation of land allows each individual to design their personal strategy for life. Holzer's concept is applied by many farms and firms in Austria and abroad, by individuals, institutions, and in the framework of large-scale projects. It offers alternative solutions for land utilization in extreme zones (ranging from alpine zones in Austria to very dry areas in Spain or Portugal), suggests remedies for the recultivation and restoration of land damaged by intensive agriculture, deals with the prevention of catastrophes (flood, erosion, damage due to storms, etc.), the planning of holistic projects, as well as with the creation of nature-adventure sites. Special attention is given to the water balance of the territory.

Permaculture landscape design essentially involves restoring a partially destroyed natural landscape. It offers us a viable alternative to the monoculture system that destroys our soil and pollutes our groundwater. A permaculture landscape is designed so that all of the plants and animals living there will work in harmony with each other. This is the only way to manage land in a stable and sustainable way (Holzer, 2014).

Restoration projects include the project Varnita/Moldavia where land degraded from intensive agriculture could be changed into a flourishing permaculture landscape, or a similar project called Project P near Moscow. The project AsGrad in Omsk/Russia turned degraded land from intensive agriculture into a huge ecological site for about 2000 people.

3.5 Specific Projects and Initiatives Aimed at Achieving more Sustainable Food Systems and more Sustainable Production and Consumption Patterns

3.5.1 UNESCO man & Biosphere Parks (Biosphaerenparks, 2021a)

"Biosphere reserves are ecosystems, recognized on the basis of uniform, internationally agreed criteria, where models of sustainable use of the biosphere are being developed, tested and implemented. As so-called 'Learning sites for sustainable development,' they serve not only to protect and maintain certain ecosystems, but also carry out ecological research, environmentally aware land use and education for sustainable development." The three main functions of BRs are to preserve biological diversity, protect natural resources (protection function), and promote regional development toward a model region for sustainable action (development function) such as, e.g., ecological agriculture. At the same time, they also fulfill a research and education function, foster the implementation of sustainability principles, and are meant to be inclusive as well as strengthen public awareness and increase responsibility for the impact of human activities.

In Austria, there are four UNESCO Man and the Biosphere Parks, the Biosphere reserve Großes Walsertal (19.200 ha, founded in 2002), the Biosphere reserve Wienerwald in Vienna and Lower Austria (105,004 ha, founded in 2005), the Biosphere reserve Salzburger Lungau and Kärtner Nockberge in Salzburg and Carinthia (148.914 ha, founded in 2012) and the Biosphere reserve Unteres Murtal in Styria (13.177 ha, founded in 2019) (Biosphaerenparks, 2021b).

The Biosphere reserve Wienerwald, e.g., comprises several ecological agricultural and tourism partner-businesses such as the Demeter Bio-Imkerei Hinterhölzl, the Biogärtnerei Austropalm, the Biohof Edibichl, the Umwelthotel Gallitzinberg, the Bioimkerei und Bienenschule 4 Kids, the Bioimkerei der Bienenfreund, as well as the Bioschafhof Sonnleitner (Biosphärenpark, 2021c).

3.5.2 Bee Protection and Organic Production of Honey in and around Vienna

Vienna is not only the greenest city in the world with about 280 imperial parks and gardens and over 2000 green spaces open to the public (*see also* Viljoen, 2018), but it is also the city of bees (Wien, 2021b). Aware of the critical importance to preserve bee populations, there are over 700 beekeepers, managing over 6000 bee hives in

Vienna, some of them being placed on the roofs of most renowned buildings such as the Kunsthistorisches Museum Vienna, the University of Vienna, the Kunsthaus Wien, or the Vienna State Opera, the Secession Hall, the City Hall, and the Austrian Mint, but also on the roofs of many hotels. In the Botanical Garden of the University of Vienna, almost 130 species of wild bees have been documented. In the world-famous Schönbrunn Zoo, about 100,000 bees can be watched in their honeycombs in a unique apiary. While the Wiener Bezirksimkerei sells honey in each district, there are several organic bee sites such as (a) the one the Honigstadt Organic Beekeeping Center (Bio-Imkerei Honigstadt) whose bees collect their nectar at the Schwarzenbergpark or (b) the one of the organic beekeeping company Villa Erbse (Wien, 2021b). New research on wild bees and green roofs in Vienna reveals that there is a need to increase the wildflower diversity on green roofs (Kratschmer et al., 2018).

3.5.3   Donausoja and Europe-Soya: Proteins without GMOs

In the framework of a sustainable food system for the world, the supply of sufficient protein-rich food will play a prominent role. Whereas previously, the big soy producers could be mainly found overseas, Europe has started to develop its own production thanks to several initiatives. The Donausoja Association (Donausoja Association, 2021a), e.g., was founded in the year 2012 by Matthias Krön in Vienna, Austria, with the aim to promote GMO-free soya production in the Danube region and later in the whole of Europe. The initiative thus created momentum for a new European Protein Strategy and offers European alternatives for a sustainable protein supply, making European farmers less dependent on foreign soya production that often contains GMOs. The two brands Donausoja and Europe-Soya can also be certified as organic brands complying with European standards.

Today, the association counts 280 members and is supported by prominent institutions and organizations as well as numerous governments. The Donausoja declaration was signed by fourteen EU governments in 2017, and the Europe-Soya Declaration by five more governments including Switzerland in 2019. The latter is the only country worldwide importing soya 100% sustainably certified GMO-free, and has raised imports from Europe to 40% of its total imports (Donausoja Association, 2019).

In 2018, 10 Mio tonnes of soybeans were cultivated in Europe. As the acreage of soya has been doubled over seven years between 2011 and 2018, reaching 4,three million ha in 2018, the potential for further growth appears to be high, especially in Central and Eastern Europe. By 2025, a rise of 40% is forecasted. The highest average yields are reported from Europe, with the additional benefit for consumers of a GMO-free production. Main soya producers in Europe are Russia and Ukraine, covering two-thirds of the market. Austria ranks in place eight (Donausoja Association, 2021b).

3.5.4 Electrostatic Fields Promise Higher Yields in Agricultural Production

About 10 years ago, an Austrian farmer experimented successfully with this innovative agricultural technology. He simulated a pathbreaking discovery made at the Pharma Group Cuba Geigy in the 1980s where cereal crops and fish eggs were exposed to electrostatic fields. These were established between the two plaques of a capacitor in which the crops were kept for some time and then replanted into their natural environment (Bürgin, 2007a). These experiments were realized with spores, seeds, germs, ferns, watercress, and the results were always similar: Both germination and growth could be enhanced. In wheat crops, the maturation period could be significantly shortened without any additional use of fertilizers or pesticides, since pests did not have the time to develop. With some maize crops, yields could be tripled or even become up to six times higher (Bürgin, 2007a). It seemed as if through the electrostatic fields the archetype ("Ur-code") of the plant could be found again and then be reactivated (Bürgin, 2007b). Bürgin, therefore, calls this method a real ecological alternative to genetic engineering.

A huge number of research papers similarly suggest that exposure to electromagnetic fields could enhance germ and plant growth, and also increase productivity (*see* Barman & Bhattachanja, 2016; Ahmad et al., 2016).

3.5.5 Reducing Food Waste: Austrian Initiatives

As outlined above, food waste still occurs at unsustainable levels. Initiatives to reduce food waste are therefore critical to achieving more sustainable food systems. Alarmingly, the United Nations Report on Reducing Food Loss and Waste determines about 35% of food to be lost, about 14% between harvest and retail, and about 20% in retail and at the consumption level. The use of surface and groundwater resources (blue water) attributable to food lost or wasted represents about six percent of total water withdrawals. Food waste amounts to 1.3 billion tonnes per year. Lost food waste has also a huge negative effect on our CO_2 balance. In 2011, FAO asserted that food waste was responsible for about 8% of total anthropogenic greenhouse gas emissions annually, which is almost equivalent (87%) to global road transport emissions (FAO, 2011).

"*Austrians throw away one fifth of all food they buy of which 14,5% are avoidable or at least partly avoidable and amount to one billion Euro per year*" (Pavlik, 2019). Out of the many Austrian initiatives that focus on contributing to the reduction of food loss and waste, the following are presented:

+++ Food Recycling—Unverschwendet—An Award Winning Business Idea

The company Unverschwendet, established in 2015, aims to transform surplus fruits, vegetables, and herbs into gourmet products like jam, sirup, chutneys, sauces, and much more. At the beginning of the undertaking in 2015, about 500 glasses of chutney, etc. were produced, in 2018 already 100,000 glasses. Whereas about 100,000 kg of edibles were handled in 2018, nowadays Unverschwendet has already been offered more than five million kg of high-quality food. The demand for such activities is growing and highlights the urgency for such initiatives allowing to cope with the food waste problem in Austria. Worldwide, healthy, circular, and resource-efficient food systems need to be established (*see* Unverschwendet, 2021).

+++ Direct Distribution of Organic Food—Adamah
Whereas direct distribution of organic products slightly fell between 2014 and 2018 from over 20% to about 18% (see figure xxx Bio-Austria), the Adamah study determines local markets and home/office delivery to be the most sustainable ways of distribution systems. Existing for more than 20 years, considered to be one of the most experienced organic food suppliers, Adamah has become famous thanks to its direct distribution system of "organic-boxes" ("Biokistl"). These boxes meanwhile are available in various forms such as the "regional box," "fruit box," "vegetable-box," "fruit & vegetable box," "regional fruit & vegetables box," "mother & child box," etc. They are prepared with fresh products at the farm and delivered during the day to homes and offices (Adamah, 2021).

+++ Social Markets—Supermarkets for Socially Disadvantaged People Also Offer a Platform to Rescue Food
Founded in 2008, the nonprofit association Sozialmarkt Wien provides food for socially disadvantaged people in its social supermarkets, at strongly reduced prices. Meanwhile, more than 25 social markets are managed in Vienna by different associations (Wien, 2021c). These social markets often get food donations from "normal" supermarkets or other providers that would not be able to sell within their premises. SPAR Österreich alone donated edibles worth 18,6 millions of Euros in 2019 to social institutions. In August 2019, the Ministry for Sustainability and Tourism in Austria launched the action program "Food is precious." The yearly generation of about 12.250 tons of donated comestibles for social organizations is unique in Europe and qualifies Austria as frontrunner in the field (Spar, 2021).

3.5.6 The City of Vienna: The Prototype of an Environmental City—Urban Organic Agriculture and Gardening, Also in Raised Beds and on City Roofs

Urban organic agriculture and gardening has been listed by the FAO and others as one potential measure to enhance both biodiversity in cities and to contribute to more sustainable lifestyles. The City of Vienna has been enjoying an unbowed trend of urban agriculture and gardening for almost the last 20 years. Keeping a high portion of organic farming is part of the city's policy. *"Urban agriculture covers 17% (5900 ha) of the city area, 25% being cultivated organically." "Organic farming is also fostered by purchasing an obligatory 30% (monetary) of organic food for municipal kitchens. 100,000 meals daily are provided for hospital patients, elderly people, kinder-garden children and pupils, with yearly costs for organic quota totaling 15 mio Euros"* (IFOAM, 2020). In November 2015, the first European Urban Green Infrastructure Conference took place in Vienna (Vienna, 2015), awarding a city garden project (NL) and The Green Living Room project with 140 m^2 of vertical green space on top of the town hall's car park (DE). In 2016, the City of Vienna created the competence center "Urban gardening in Vienna" within Bio-Forschung Austria (Austria's first research laboratory for organic farm-ing, founded in 2006) (Bioforschung, 2021), central information and service center, as part of its program "The Prototype of an Environmental City" (Wien, 2021d).

Raised Beds for Individual Urban Organic Gardening
Raised beds have been existing in Austria for more than 40 years. The Land Niederösterreich has been fostering urban gardening with its program "nature in the garden" for more than 20 years. In 2010, the Austrian Embassy at the Vatican was awarded for its eco-garden in the raised bed. Magenta presented its first "smart-beet," a digitalized raised bed, in Vienna in 2017 (Magenta, 2021). For the school year 2020/2021 the acquisition of raised beds for planting and nurturing fruits and vegetables in connection with the conveyance of knowledge on organic agriculture is even promoted and financially supported by the European Commission (Ama, 2021b).

Green Walls and Green Roofs: *Organic Plantations on City-Roofs*
Whereas already in the 1980s, the Austrian artist Friedensreich Hundertwasser built the meanwhile famous Hundertwasserhaus with a green rooftop garden, the city of Vienna has been fostering and subsidizing the installation of green roofs since 2003. The study shows that green roofs provide a microclimate, a "natural air-condition effect" for the rooms below, and can even prolong the lifespan of roofs from 20 to 35 years (*European* Commission, 2013). Other cities like Salzburg, Linz and Graz followed. Since 2010, developing standards for green roofs do exist (ÖNORM L 1131:2010 06 01). The city's Urban Heat Island Strategy Plan of 2018 makes a prominent contribution to creating microclimates while at the same time enhancing greeneries, public space, and open water bodies in the city, among others with green walls and facades as well as green roof tops. Part of this strategy is the preservation and the management of agricultural land in Vienna in accordance with green space planning and the continued development of environmentally friendly (sustainable) agricultural production (Vienna, 2018).

3.5.7 Sustainable Procurement Programs "ÖkoKauf Wien. Think Green: Buy Green," EcoBusiness Vienna & "Natürlich Gut Teller": The City of Vienna as International Role Model (Wien, 2021e)

Within the wider context of sustainability policy, public procurement can play a major role in promoting sustainable production and consumption patterns constituting thus a decisive element for more sustainable food systems.

More than a decade ago, the City of Vienna launched the sustainable public procurement program ÖkoKauf Wien (Ecobuy), nominated as one of the three most outstanding European public procurement programs recently. In buying goods and services according to ecological criteria, the annual carbon dioxide emissions could be cut by approximately 15,000 tonnes and thus saved about 1,5 millions of Euros per year. *"More than half of the food purchased in school catering originates from certified organic agriculture"* (European Commission, 2015).

Another program, EcoBusiness Vienna, has involved more than 740 locals businesses since 1998. The significant reduction in energy, raw materials, and water use translated into savings in operating costs of about 55,seven million of Euros in total (Wien, 2021e).

The "Natürlich gut Teller" (NGT) was developed in 2010 with the goal *"to identify dishes that stand out through their environmental friendliness and (organic) quality and thus make a contribution to the sustainable development of community catering in Vienna."* The NGT needs to contain organic and seasonal ingredients and reduced amounts of (organic) meat and sustainable raised fish. The impact assessment report highlights that around 20% of the guests choose the "Natürlich gut Teller" which equals around 800,000 NGT portions a year, out of which 56% were vegetarian, 24% contained fish and 20% meat, respectively. In the period 2011–2016, around 1,420,000 kg of seasonal organic fruits and vegetables and 79,000 kg of organic meat were processed entailing a curtailment of 78,000 tonnes of greenhouse gases due to avoidance of transport. The criterion *"reduced meat portions"* saved up to 53,000 kg of meat and hence about 1,390,000 m^3 of virtual water. Other positive consequences of the NGT were the awareness-raising effect and the effect on procurement behavior. The latter even multiplied by a factor from 2–4 with regard to purchase of organic food and meat reduction (Wien, 2021e).

3.5.8   Changing to more Sustainable Consumption Patterns and Diets: More Organic in General, Less Meat, More Vegan and Vegetarian Food, Protein-Rich Meat Alternatives, and Mushrooms

In order to achieve more sustainable food systems, several changes will be necessary, such as switching from conventional food to more organic food, or the search for healthier and more sustainable diets containing less meat and fish. Austria is on the right track.

Transition to more Organic Food Consumption

In 2019, about 26% of total agricultural area in Austria was organic. Organic food consumption and per capita expenditure of organic sales in Austria are among the highest in the world (see figures.....). Since 2013, sales of organic food products have increased by over 50%, which corresponds to a five times higher growth than conventional food products. In 2018, about 52,7 of goats, 29,7% of sheep, 22% of cattle, and 19% of poultry operations were organic (USDA, 2020).

Less Meat, More Vegan, and Vegetarian Food

The production of 327 million tonnes of meat per year take up almost 80% of the global agricultural area (FAO, 2019b)(*see* Ritchie, 2019), whereas all but 1,5% are managed in a conventional unsustainable way, entailing high land use, high water and energy use and provoking biodiversity loss, soil depletion and the emission of CO_2 emission. This is why diets with less meat and fish intake should be favored in the future.

As a general trend, a slowdown of meat consumption could be observed in Western economies during the last 10 years. Consumers are more and more changing from meat- to plant-based diets. As a result, food-related companies need to respond to these demand changes and invent new product lines. The booming mushroom business can be explained from this perspective. Research suggests that the percentage of vegetarians in selected EU countries such as Sweden, Germany, and Austria reaches about 10% (Ploll et al., 2020) (Vou, 2019). Recently, the famous red Michelin guide offered the first Michelin Star to a vegan restaurant.

In Austria, meat consumption per capita peaked in 2010 with over 100 kg/person, whereas in 2019 with 93,8 kg per capita, 2002 levels were reached (Statistics Austria, 2020). For a healthy diet, the consumption level however is still too high, and should be reduced by 64% according to the national Association for Nutrition.

Protein-Rich Meat Alternatives: Innovative Products from Austria
Proveg names ten vegan meat alternatives: tofu, tempeh, seitan/wheat, soya, lupin, green spelt, oat flakes, black beans, chickpeas, and pea protein (Proveg, 2021).

Within the EU-funded Phytamin project, several plant protein products emerged from the research including *"a dry, 100% sunflower protein that can be used to manufacture vegetarian and vegan products as purely vegetable minced-meat substitute and alternative to soy." "The product was the winner of the World Innovation Food Award 2018."* Total investment for the project reached EUR 524430 (European Commission, 2018).

Mushrooms: The Future of Agriculture
Rich in proteins and vitamin D, mushrooms are becoming a trendy meat alternative, and are also excelling as model for a circular farm economy. Farming mushrooms is organically efficient and allows for an economical and considerate exploitation of land. The cultivation of organic mushrooms may be considered as the "agriculture of the future," providing vitality for soils, plants, animals, and humans.

Mushroom plantations equally represent an important pillar for the creation of closed nutrient cycles, being essential for the optimized utilization of agricultural waste and byproducts and for avoiding the loss of valuable biological resources. Waste dump from plant and animal breeding serves as a substrate for the cultivation of edible and medicinal mushrooms. After the production cycle, the "leached out" mushroom substrate acts as a fertilizer to build the humus layers of the fields. About 10 kg of leached-out mushroom substrate does have the same ingredients as 1 kg of synthetic fertilizer. The advantage of the mushroom substrate in comparison to the synthetic fertilizer consists in forming a symbiosis with the plants, in combination with natural soil mushrooms, the mycorrhiza mushrooms, which releases nutrients slowly and regularly. This allows for just the necessary nutrient intake for the soil the plant really needs at the moment, thus avoiding a toxic intake of a nutrients overdose into groundwater and land (*translated from Oe24, 2017*) (Tiroler Glückspilze,

2021). Mushrooms also play an important role in the EU project "From value chain to value loop with biogas and mushrooms" (European Commission, 2020e).

National mushroom production in Austria represents an innovative niche—also for organic farmers—and amounted to almost 20,000 tonnes in 2019 (Statista, 2020), equalling almost 10% of the total market (90% were imported) (xxxx Bio-Austria). Meanwhile, a wide range of organic farms are active in cultivating mushrooms like Adamah (Adamah, 2021), existing since 1997, or Tiroler Glückspilze (Tiroler Glückspilze, 2021), specialized since 2014 in the production of organic gourmet mushrooms. Most important varieties sold include oyster and king oyster mushrooms, white mushrooms, Portobello, and Shiitake mushrooms, but also Igel-Stachelbartpilz and Reishi mushrooms.

3.5.9   Austrian Superfood: Pumpkinseeds-Oil from Styria and Linseeds
Local crops are often neglected or underestimated. Nevertheless, they are key to sustainable food and nutrition systems.

Pumpkinseeds-Oil from Styria on the Rise
Since 1735, Styria has been producing pumpkinseeds-oil which developed into a renowned EU regionally certified Austrian brand (GGA). Only from about 1970 onward, the oil crops were also farmed in other parts of Austria (Wikipedia). The relatively thick, green oil with a nutty, spicy taste is rich in Vitamin E, as well as in antioxidants and has a high share of polyunsaturated fatty acids, especially linolenic acid. Pumpkinseeds contain minerals including zinc, magnesium, potassium, calcium, copper, selenium, manganese, and phosphor as well as the vitamins E, A, C, D, B1, B2, and B6 which qualify both seeds and oil as "superfood" (Gesundheit, 2021).

In 2016, on about 100,000 ha of land almost 30,000 tons of pumpkinseeds-oil were harvested, out of which about 40% were exported. In 2020, the cultivation area was enlarged by 30% in Styria alone. Within Austria, the pumpkinseeds-oil occupies a share of about 6,5% of all oilseeds (APA, 2020).

Organic Linseeds/Flaxseed
While linseeds count among the oldest cultivated plants existing already 7500 before Christ, they were still important during the Austrian Danube-Monarchy in 1875 with a crop area of about 94,000 ha, the third biggest worldwide. Today, only about 500 ha are farmed in Austria (BMIRT, 2021b). Rich in omega-3 fatty acids, proteins, vitamins and minerals, linseeds and linseed oil are predestined for healthy diets, disposing of antioxidant, anticoncerogen, and anti-inflammatory properties. Linseed products can be found nowadays in many organic shops in Austria.

3.5.10   Ecotourism and Organic Agriculture: Pioneer BIO HOTELS Offer Exclusively Organic Food
While organic agriculture, especially in Austria often involves operations at small scales, the average size of an organic farm being about 22 ha, a combination with

ecotourism enhances income opportunities. At the same time, the combination offers a unique experience of recreation in a close-to-nature setting for guests. Ecotourism is also booming as stand-alone economic sector, not only in Austria.

BIO HOTELS, founded in Austria in 2001 and probably the most advanced and biggest association on the market of this kind, comprises more than 80 certified members in Austria, Germany, Italy, Greece, Slovenia, and Switzerland. These hotels feature sustainable holidays embedded into a holistic approach with the same philosophy and standards such as 100% organic food, natural cosmetic, eco-electricity and climate neutrality as well as green weddings (BIOHOTELS, 2021). Other initiatives of that kind include "GREEN PEARLS unique places" (Greenpearls, 2021) or ECOHOTELS (Ecohotels, 2021). Within their sustainability concept, Steigenberger hotels, e.g., offer a green meetings program as well as locally and regionally sourced food products (in Austria at least in some hotels mainly organic, steigenberger.com/Krems), have rooftop organic gardens, use LED lamps, water limiters and biodegradable detergents, etc. (Steigenberger, 2021). The Boutiquehotel Stadthalle in Vienna was the first climate-neutral and zero-energy balanced hotel worldwide and of course spoils its guests with an organic breakfast (Hotel Stadthalle, 2021). The EU funded project "Green Ideas for Tourism for Europe" designed as an educational tool, provides comprehensive information about best practices for sustainable tourism, but unfortunately does not refer to organic food—yet (Green ideas, 2021).

4 Recommendations and Solutions for International, Regional, and National Policy Approaches Aimed at Achieving Sustainable Food Systems Worldwide

On the basis of the above analysis, we therefore propose the following recommendations:

Recommendation 1: Establish an International Soil Convention for the Preservation of all High Valuable Agricultural Soil by 2025.
This is ever more important since high-value agricultural soil is being depleted by both conventional agriculture and sealing thus diminishing possibilities for future generations to practice organic farming, unless at very high remediation costs.

Recommendation 2: Agree on a worldwide net of naturally protected areas/national parks at land and at sea/coastline of 40% of the total area by 2030.
At the One Planet Summit early 2021, 50 countries launched the initiative to preserve 30% of the total land and sea territory for biodiversity which is a remarkable step. However, it will not suffice to save biodiversity and threatened species such as bees and other insects.

Recommendation 3: Create and launch an international mechanism for agriculture, setting a cap for land used for agriculture as well as setting a cap for nitrogen emission within an N-Emission-Trading-Scheme (similar to the ETS for CO$_2$ emissions).

See also elaboration of the mechanism above.

Recommendation 4: Establish an international Agreement for the transition to more sustainable farming methods with a special goal to reach 40% of organic agriculture worldwide by 2030 and 100% by 2040.

This could facilitate shifting investments and ODA into this particular area and would raise political and public awareness of the necessity to do so which currently does not seem to be the case.

Recommendation 5: Work toward an international agreement to protect and preserve indigenous plants and crops, while at the same time making an inventory of them with a special focus on species threatened by extinction or in danger of extinction, which would significantly contribute to halting biodiversity loss.

Very often, archetypes of species of plants and crops are rare, have disappeared or have been replaced by less sustainable and less healthy species as it is the case in Austria for certain wheat species.

Recommendation 6: Make the reduction of the environmental footprint of agriculture and food systems a policy priority embedded in a corresponding legal framework.

France, e.g., has recently inaugurated a law requiring information on the environmental footprint on food products in order to allow for informed decisions at consumer level.

Recommendation 7: Foster full-cost pricing in transport and foster regional trade as well as regional and local markets as political priority, especially for agricultural products aimed at diminishing environmental adverse effects from—exaggerated—transport.

Full-cost pricing for transport would inspire regulation through markets and price policy. It should be avoided that prawns are caught in Northern Europe, transported to Africa for processing and transported back to Northern Europe for sale for example.

Recommendation 8: Promote policies that align agricultural production processes to a circular economy approach thus avoiding waste at all levels.

Recommendation 9: Realign distribution channels into an environmentally friendly and zero-waste policy fostering direct distribution systems, regional and local organic markets as well as social stores and cooperation with food recycling organizations and companies.

Recommendation 10: Enhance education and training on ecologically sound food systems and healthy diets at all educational levels (from kinder garden to University) as well as within the public and private sector as lifelong learning theme.

Recommendation 11: Prohibit advertisement for meat and meat products similar to the advertisement ban for cigarettes.

Unhealthy diets do not only affect individuals negatively, but even cause disturbances of planetary scale.

Recommendation 12: Promote self-sustained living and housing concepts with enough green space for food production within own gardens aimed at providing sufficient food supply, also, e.g., in enhancing urban agriculture.

Austria is famous for its "Schrebergärten"—small-scale gardens for city dwellers on the periphery of cities—which offer a the same time urban gardening, a pleasant lifestyle and leisure opportunities.

Recommendation 13: Encourage enhanced statistics and research on sustainable food systems and food security, in particular, on organic agriculture, and regarding benefits from organic food and permaculture, veganism, and vegetarianism as well as regarding the link between organic food and public health.

The European Commission has just started to collect more statistics on organic agriculture, but the latter should be a legal obligation for member states and also requested at international level.

Recommendation 14: Increase research on planetary boundaries and policy coherence for sustainable development.

As shown by Lade, climate-related policies could have adverse effects from a more holistic sustainability point of view.

References

Adamah. (2021). Accessed Dec 14, 2021, from https://www.adamah.at

Ahmad, H., et al (2016, May). Ahmad, H., Ahmad, M. H., Awang, N. A., Zakaria, I. H., *Quantitative analysis of plant growth exposed to electric fields*, p. 2. Accessed Jan 12, 2021, from https://www.researchgate.net/publication/304714181_Quantitative_Analysis_of_Plant_Growth_Exposed_to_Electric_Fields

Ama. (2021). *Checklists Audit Ama Biosiegel*. Accessed Feb 17, 2021, from https://docplayer.org/82557520-Checkliste-audit-ama-biosiegel.html

Ama. (2021b). *Hochbeete*. Accessed Dec 22, 2021, from https://www.ama.at/Fachliche-Informationen/Schulprogramm/Kommunikationsmassnahmen/Hochbeete

APA. (2020, October 14). *Steirisches Kürbiskernöl Championat 202072021*. Accessed Jan 17, 2021, from https://www.ots.at/presseaussendung/OTS_20201014_OTS0095/steirisches-kuerbiskernoel-championat-202021-jungbaeuerin-legte-debuetsieg-hin

Barman, P., & Bhattachanja, R., (2016, February). *Impact of electric and magnetic field exposure on young plants–a review*, p.186, Accessed Jan 12, 2021, from https://www.researchgate.net/publication/295542783_Impact_of_Electric_and_Magnetic_Field_Exposure_on_Young_Plants-A_Review

Baudry, J., Assmann, K. E., Touvier, M., et al. (2018). Association of frequency of organic food consumption with cancer risk. *JAMA Network, JAMA Intern Med, 178*(12), 1597–1606. Accessed Feb 13, 2021, from https://jamanetwork.com/journals/jamainternalmedicine/fullarticle/2707948

BBC. (2019, July 19). *BBC News, Russia alarmed by large fall in bee populations*. Accessed Feb 06, 2021, from https://www.bbc.com/news/world-europe-49047402

Bio-Austria. (2020a). *Statistik, 2020*. Accessed Dec 26, 2021, from https://www.bio-austria.at/bio-bauern/statistik/

Bio-Austria. (2020b). *Über uns*. Accessed Dec 26, 2021, from https://www.bio-austria.at/bio-partner/marketing-gmbh/ueber-uns/

Bioforschung. (2021). Accessed Dec 17, 2021, from https://www.bioforschung.at

BIOHOTELS. (2021). *Über uns*. Accessed Feb 06, 2021, from https://www.biohotels.info/de/urlaubsregionen/griechenland/

Bio-info. (2021a). *Bio in Zahlen*. Accessed Dec 02, 2021, from https://bioinfo.at/bio-in-zahlen

Bio-info. (2021b). *Bio-Siegel*. Accessed Feb 16, 2021, from https://bioinfo.at

Biosphaerenparks. (2021a). *The concept*. Accessed Dec 08, 2021, from biosphaerenparks.at, http://www.biosphaerenparks.at/index.php/en/the-concept

Biosphaerenparks. (2021b). *Biosphaerenparks/model-region*. Accessed Jan 08, 2021, from http://www.biosphaerenparks.at/index.php/en/model-region

Biosphaerenparks. (2021c). *Biosphärenparks/Partnerbetriebe*. Accessed Jan 08, 2021, from https://www.bpww.at/de/partner/partnerbetriebe

BMIRT. (2020). *Organic farming, Broschüre, Publikation des Bundes*, 2010, reprint 2020, Accessed Feb 14, 2021, from www.bmrit.gv.at

BMIRT. (2021a). *Sonderrichtlinie ÖPUL 2015*, Accessed Feb 16, 2021, from https://www.bmlrt.gv.at/land/laendl_entwicklung/foerderinfo/sonderrichtlinien_auswahlkriterien/srl_oepul.html

BMIRT. (2021b). *Leinöl*. Accessed Jan 28, 2021, from https://www.bmlrt.gv.at/land/lebensmittel/trad-lebensmittel/oel/leinoel.html

BMNT. (2021). *Ministry for sustainability and tourism, Austria, organic farming*. Accessed Dec 25, 2021, from https://www.bmlrt.gv.at/english/agriculture/organic-farming.html

Bürgin, L. (2007a). *Der Urzeit-Code, Die ökologische Alternative zur Gentechnik* (p. 49). Herbig Verlag.

Bürgin, L. (2007b). *Der Urzeit-Code, Die ökologische Alternative zur Gentechnik* (p. 40). Herbig Verlag.

China Dialog. (2012). *Decline of bees forces China's apple-farmers to pollinate by hand*. Accessed Feb 06, 2021, from https://chinadialogue.net/en/food/5193-decline-of-bees-forces-china-s-apple-farmers-to-pollinate-by-hand/

Demeter. (2021a). *Demeter, history*. Accessed Dec 04, 2021, from https://www.demeter.net/what-is-demeter/history

Demeter. (2021b). *Demeter, Über uns*. Accessed Jan 03, 2021, from https://www.demeter.at/ueber-demeter/

Demeter. (2021c). *Demeter, Statistics*. Accessed Jan 04, 2021, from https://www.demeter.net/statistics

Denns. (2021). *Denns bio-Markt, presse*. Accessed Feb 17, 2021, from https://www.denns-biomarkt.at/ueber-uns/presse/

Donausoja Association. (2021a). *About us*. Accessed Dec 08, 2021, from https://www.donausoja.org/en/about-us/the-association/

Donausoja Association. (2021b). *Soya cultivation in Europa*. Accessed Feb 17, 2021, from https://www.donausoja.org/fileadmin/user_upload/Downloads/Soya_cultivation_in_Europe_plus_DS_Benefits.pdf

Donausoja Association. (2019, January 28). *Pressrelease*. Accessed Dec 08, 2021, from https://www.donausoja.org/fileadmin/user_upload/Press/Press_Release/PA_Schweiz_unterzeichnet_Europa_Soja_Erkla__rung_012019.pdf

Dower M., (2009). *Forum synergies, subsistence and semi-subsistence farming in Europe: A challenge to EU and national policies*. Accessed Feb 13, 2021, from https://www.forum-synergies.eu/bdf_fiche-proposition-16_en.html

Dury, S. et al. (2019a), , *Food Systems at risk: New trends and challenges*. Rome, Montpellier, Brussles, FAO, CIRAD, and European Commission. p.19ff, https://doi.org/10.19182/agritrop/00080. Accessed Jan 31, 2021, from http://www.fao.org/3/ca5724en/CA5724EN.pdf

Dury, S. et al. (2019b), *Food Systems at risk: New trends and challenges*. Rome, Montpellier, Brussles, FAO, CIRAD, and European Commission. p.35ff. https://doi.org/10.19182/agritrop/00080. Accessed Jan 31, 2021, from http://www.fao.org/3/ca5724en/CA5724EN.pdf

Dury, S. et al. (2019c). *Food Systems at risk: New trends and challenges*. Rome, Montpellier, Brussles, FAO, CIRAD, and European Commission. p.59ff. https://doi.org/10.19182/agritrop/00080. Accessed Jan 31, 2021, from http://www.fao.org/3/ca5724en/CA5724EN.pdf

Ecohotels. (2021). Accessed Jan 27, 2021, from https://ecohotels.com

European Commission. (2013). *National water retention measures, case study: green roofs of vienna, Austria*. Accessed Feb 17, 2021, from http://nwrm.eu/sites/default/files/case_studies_ressources/cs-at-2-final_version.pdf

European Commission. (2015, May). *GPP in practice*, Issue no. 51, "Ökokauf Wien"–Vienna's Sustainable Procurement Programme. Accessed Feb 17, 2021, from https://ec.europa.eu/environment/gpp/pdf/news_alert/Issue51_Case_Study106_Vienna_Okokauf.pdf

European Commission. (2018). *Companies in Austria, Germany develop plant-protein based foods.* Accessed Feb 26, 2021, from https://ec.europa.eu/regional_policy/en/projects/Germany/companies-in-austria-germany-develop-plant-protein-based-foods, accessed on 26022021.

European Commission. (2019). *European green deal.* Accessed Dec 31, 2020, from https://www.dw.com/en/eu-presents-green-deal-farming-plan-to-half-pesticide-use/a-53511139

European Commission. (2020a, June 22). *EU Taxonomy for sustainable activities.* Accessed Feb 14, 2021, from https://ec.europa.eu/info/business-economy-euro/banking-and-finance/sustainable-finance/eu-taxonomy-sustainable-activities_en

European Commission. (2020b, June). *Statistical factsheet Austria.* Accessed Feb 16, 2021, from https://ec.europa.eu/info/sites/info/files/food-farming-fisheries/farming/documents/agri-statistical-factsheet-at_en.pdf

European Commission. 2020c, *EU Bio-Logo.* Accessed Feb 17, 2021, from https://ec.europa.eu/info/food-farming-fisheries/farming/organic-farming/organic-logo_de

European Commission. (2020d). *Glyphosate.* Accessed Jan 17, 2021, from https://ec.europa.eu/food/plant/pesticides/glyphosate_en

European Commission. (2020e, August 06). *From value chain to value loops with biogas and mushrooms.* Accessed Jan 26, 2021, from https://ec.europa.eu/research/infocentre/article_en.cfm?&artid=52945&caller=other

European Commission. (2021, January). *European Green Deal, List of potential agricultural practices that eco-schemes could support.* Accessed Feb 13, 2021, from https://ec.europa.eu/info/sites/info/files/food-farming-fisheries/key_policies/documents/factsheet-agri-practices-under-ecoscheme_en.pdf

EEA. (2020a). *European Environmental agency, is Europe living within the limits of our planet?–An assessment of the Europe's environmental footprints in relation to planetary boundaries, EEA Report No 01/2020, Joint EEA/FOEN Report.* Accessed Feb 02, 2021, from https://www.eea.europa.eu/publications/is-europe-living-within-the-planets-limits, p.37.

EEA. (2020b). *European environmental agency, is Europe living within the limits of our planet?–An assessment of the Europe's environmental footprints in relation to planetary boundaries, EEA Report No 01/2020, Joint EEA/FOEN Report.* Accessed Feb 02, 2021, from https://www.eea.europa.eu/publications/is-europe-living-within-the-planets-limits, p.19.

EFSA. (2021). *European food safety authority, bee health.* Accessed Feb 06, 2021, from https://www.efsa.europa.eu/en/topics/topic/bee-health

European Parliament. (2016). *European parliament research service, scientific foresight unit (STOA), Human health implications of organic food and organic agriculture.* Accessed Feb 13, 2021, from https://www.arc2020.eu/wp-content/uploads/2017/02/EPRS_STU2016581922_EN.pdf

European Parliament. (2019, December 18). *What's behind the decline in bees and other pollinators infographic.* Accessed Jan 10, 2021, from https://www.europarl.europa.eu/news/en/headlines/society/20191129STO67758/what-s-behind-the-decline-in-bees-and-other-pollinators-infographic

Eurostat. (2018). *Eurostat, pesticides sales.* Accessed Jan 03, 2021, from https://ec.europa.eu/eurostat/statistics-explained/images/f/f4/Pesticides_sales-01.jpg

Eurostat. (2021, January). *Statistics explained, organic farming statistics.* Accessed Feb 02, 2021, from https://ec.europa.eu/eurostat/statistics-explained/index.php/Organic_farming_statistics

EZA. (2021). *EZA Fairer Handel GmbH.* Accessed Feb 17, 2021, from https://www.eza.cc

FAO. (2002). *World agriculture: Towards 2015/30–Summary report.* p. 5. Accessed Feb 14, 2021, from http://www.fao.org/3/a-y3557e.pdf

FAO. (2003). *Food and agriculture organisation of the United Nations, weighing the GMO arguments: Against.* Accessed Feb 10, 2021, from http://www.fao.org/3/Y4955E/y4955e0a.htm

FAO. (2007). *Organic agriculture.* Accessed Feb 13, 2021, from http://www.fao.org/organicag/oa-faq/oa-faq7/en/

FAO. (2011). Global food loses and food waste. Accessed Feb 17, 2021, from http://www.fao.org/3/mb060e/mb060e00.htm

FAO. (2017, December 18). *Migrants and agriculture. What do you need to know?* Accessed Feb 13, 2021, from http://www.fao.org/fao-stories/article/en/c/1072891/

FAO. (2018). *Why do bees matter.* Accessed Feb 06, 2021, from http://www.fao.org/documents/card/en/c/i9527en/

FAO. (2019). *The state of food security and nutrition in the world. Safeguarding against economic slowdowns and downturns.* Accessed Feb 12, 2021, from http://www.fao.org/3/ca5162en/ca5162en.pdf

FAO. (2019b). *FAO/OECD, Chapter meat.* Accessed Jan 24, 2021, from http://www.fao.org/3/CA4076EN/CA4076EN_Chapter6_Meat.pdf

FAO. (2020, March 24). *High level panel of experts, on food security and nutrition, interim issue paper on the impact of Covid-19 on food security and nutrition,* v.1. p. 7. http://www.fao.org/fileadmin/templates/cfs/Docs1920/HLPE_2020/New_HLPE_paper_COVID_EN.pdf

FAO. (2021a). *Sustainable development goals, target 2.4.* Accessed Feb 14, 2021, from http://www.fao.org/sustainable-development-goals/indicators/241/en/

FAO. (2021b). *Sustainable development goals, ensure sustainable production and consumption patterns.* Accessed Feb 14, 2021, from http://www.fao.org/sustainable-development-goals/goals/goal-12/en/

FAO and INRA. (2016). *Innovative markets for sustainable agriculture, how innovations in market institutions encourage sustainable agriculture in developing countries.* Accessed Feb 13, 2021, from https://www.alimenterre.org/system/files/ressources/pdf/1002-fao-innovative-markets-sustainable-agriculture.pdf

Fertl T.. (2020, March 20). *Stellungnahme zum Entwurf der SWOT-Analyse, Brief von Bio-Austria an das BMLRT,* Accessed Dec 26, 2020, from bmirt.gv.at, bio-Austria–BMLRT.

FIBL. (2021, February 17). *Research institute of organic agriculture, organic world, global organic statistics and news, world of organic agriculture 2021: Key data* (2019), Accessed Feb 17, 2021, from https://www.organic-world.net/yearbook/yearbook-2021/key-data.html

Foote, N. (2020, January 23). *Euractiv, Austriaon course to become the first EU country to ban Glyphosate.* Accessed Jan 01, 2021, from https://www.euractiv.com/section/agriculture-food/news/austria-on-course-to-become-first-eu-country-to-ban-glyphosate/

France 24. (2021, January 11). *France hosts one planet summit aimed at protecting global biodiversity.* Accessed Feb 10, 2021, from https://www.france24.com/en/france/20210111-france-hosts-one-planet-summit-aimed-at-protecting-global-biodiversity

Gesundheit. (2021). *Superfoods–9 gesunde Lebensmittel, Kürbiskerne,* Accessed Jan 27, 2021, from https://www.gesundheit.de/ernaehrung/gesund-essen/superfoods/kuerbiskerne

Greenpeace. (2020). *Save the bees.* Accessed Feb 06, 2021, from https://www.greenpeace.org/usa/sustainable-agriculture/save-the-bees/

Greenpeace. (2013, January). *Greenpeace research laboratories technical report* (Review). Reyes Tirado/Gergely Simon/Paul Johnston, Bees in Decline, A review of factors that put pollinators and agriculture in Europe at risk, Greenpeace International, Accessed Feb 10, 2021, from https://www.researchgate.net/publication/267639396_Bees_in_Decline_A_Review_of_Factors_that_Put_Pollinators_and_Agriculture_in_Europe_at_Risk, p.4.

Greenpearls. (2021). Accessed Jan 27, 2021, from https://www.greenpearls.com

Gingesh, T., et al. (2017). Farmer-suicide in India: Debating the role of biotechnology. *Journal Life Sciences, Society and Policy,* US National Library of Medicine, National Institute of Health, Accessed Feb 10, 2021, from https://www.ncbi.nlm.nih.gov/pmc/articles/PMC5427059/

GMO-free-regions. (2021). *European GMO-free regions conference.* Accessed Jan 09, 2021, from https://www.gmo-free-regions.org

Green ideas. (2021). *Green ideas for tourism for Europe.* Accessed Jan 28, 2021, from http://www.greentourism.eu

Guinnessworldrecords. (2006). *Largest producer of genetically modified seeds.* Accessed Feb 10, 2021, from https://www.guinnessworldrecords.com/world-records/largest-producer-of-genetically-modified-seeds

Health Policy Watch. (2019, July 16). Hunger & obesity are both raising worldwide says un report. Accessed Feb 12, 2021, from https://healthpolicy-watch.news/hunger-obesity-both-rising-worldwide-says-un-report/

Holzer, S. (2013a). *Holzer Permakultur–Agro-Ökologie.* Accessed Jan 06, 2021, from https://www.seppholzer.at/agro-oekologie

Holzer, S. (2013b). *Projekte in aller welt,* Accessed Jan 08, 2021, from https://www.seppholzer.at/projekte

Holzer, S. (2014). Sepp Holzers's Permaculture, *A practical guide to small-scale integrative farming and gardening.* Chelsea Green Publishing, White river Junction, Vermont, 2004 by Leoopold Stocker Verlag, p. 5, Accessed Jan 07, 2021, from http://www.leotiger.com/omg/sepp.pdf

Hotel Stadthalle. (2021). *Boutique hotel Stadthalle Vienna.* Accessed Jan 28, 2021, from https://www.hotelstadthalle.at/index-en.html

Hovorka, G. (2006, January). *The influence of agricultural policy on the structure of mountain farms in Austria.* p. 15. Accessed Feb 16, 2021, from www.bmnt.gv.at

IARC. (2015, May 20). *International agency for research on cancer. specialised agency of the world health organisation.* IARC Monographs Volume 112: Evaluation of five organophosphate insecticides and herbicides. Accessed Feb 13, 2021, from https://www.iarc.who.int/wp-content/uploads/2018/07/MonographVolume112-1.pdf

IFOAM-EU. (2017a). *IFOAM EU und FIBL, Socioeconomic impacts of GMOs on European agriculture,* p. 6. Accessed Feb 10, 2021, from https://orgprints.org/33084/1/SOCIO-ECONOMIC%20IMPACTS%20OF%20GMOs.pdf

IFOAM-EU. (2017b). *IFOAM EU und FIBL, Socioeconomic impacts of GMOs on European agriculture,* p. 7. Accessed Feb 10, 2021, from https://orgprints.org/33084/1/SOCIO-ECONOMIC%20IMPACTS%20OF%20GMOs.pdf

IFOAM. (2018, April 01). *IFOAM-organics international, how organic agriculture helps achieve the sustainable development goals, Fact sheet.* Accessed Feb 04, 2021, from https://www.ifoam.bio/sites/default/files/2020-04/oasdgs_web.pdf, https://www.ifoam.bio/how-organic-agriculture-helps-achieve-sustainable

IFOAM. (2020). *IFOAM organics europe, leading by example.* Accessed Jan 16, 2021, from https://euorganic2030.bio/initiatives/improve-inspire-deliver/feeding-vienna-organically/

IFOAM. (2021). *How can food-education in school foster a new generation of citizens aware of food sustainability and health?,* 10022021, Accessed Feb 10, 2021, from https://www.organicseurope.bio/news/how-can-food-education-in-school-foster-a-new-generation-of-citizens-aware-of-food-sustainability-and-health-lets-figure-it-out-25-february-2021-online/

IIASA. (2020a, December). *International institute for applied systems analysis, and international science council.* Transformations within reach: Pathways to a sustainable and resilient world. Resilient food systems. Accessed Feb 13, 2021, from http://pure.iiasa.ac.at/id/eprint/16822/1/Food%20%281%29.pdf

IIASA. (2020b, December). *International institut of applied systems analysis and international science council.* Transformations within reach: Pathways to a sustainable and resilient world. Resilient food systems. Accessed Feb 05, 2021, from https://council.science/wp-content/uploads/2020/06/IIASA-ISC-Reports-Resilient-Food-Systems.pdf, p.6.

Jabr, F. (2013, September 1). *The Mind-Boggling Math of Migratory Beekeeping–31 billion honeybees plus 810 000 acres of almond trees equals 700 billion almonds-and one looming agricultural crisis, Scientific American.* Accessed Feb 14, 2021, from https://www.scientificamerican.com/article/migratory-beekeeping-mind-boggling-math/

Jacobsen, S. E., Soerensen, M., Pedersen, S. M., et al. (2013). Feeding the world: Genetically modified crops versus agricultural biodiversity. *Agronomy for Sustainable Development, 33,*

651–662, Springer link, Accessed Feb 02, 2021, from https://link.springer.com/article/10.100
7%2Fs13593-013-0138-9

Ja-Natürlich. (2021). *Meilensteine*. Accessed Jan 03, 2021, from https://www.janatuerlich.at/wir-
sind-bio/wofuer-wir-stehen/unsere-meilensteine/

Kilcher, L. (2006). University of Kassel. *How organic agriculture contributes to sustainable
development*. Accessed Feb 13, 2021, from https://orgprints.org/10680/1/Kilcher_2007_
JARTS_SP_89.pdf

Kratschmer, S., et al. (2018, February 20). *Buzzing on top: Linking wild bee diversity, and
abundance and traits with green roof qualities*, p.1. Springer Verlag, Urban Ecosystems.
Accessed Jan 18, 2021, from https://link.springer.com/article/10.1007/s11252-017-0726-6

Lade S. J., et al. (2019a, December). *Earth system interactions amplify human impacts on planetary
boundaries*. fig. 5, Accessed Feb 02, 2021, from https://www.researchgate.net/publication/33
7953629_Human_impacts_on_planetary_boundaries_amplified_by_Earth_system_
interactions

Lade S. J., et al. (2019b, December). *Earth system interactions amplify human impacts on planetary
boundaries*. Accessed Feb 02, 2021, from https://discovery.ucl.ac.uk/id/eprint/10089893/1/
Article_NATSUST%20v2.pdf,https://www.researchgate.net/publication/337953629_Human_
impacts_on_planetary_boundaries_amplified_by_Earth_system_interactions

Lade S. J., et al. (2019c, December). *Earth system interactions amplify human impacts on planetary
boundaries*. fig. 3, https://discovery.ucl.ac.uk/id/eprint/10089893/1/Article_NATSUST%20v2.
pdf, Accessed Feb 02, 2021, from https://www.researchgate.net/publication/337953629_
Human_impacts_on_planetary_boundaries_amplified_by_Earth_system_interactions

Landwirtschasftskammer. (2016a, August). *Landwirtschaftskammer Österreich, Agrarischer
Ausblick 2025*. p. 5, Accessed Jan 01, 2021, from https://www.lko.at/media.php?
filename=download%3D%2F2016.11.30%2F148049533266992.pdf&rn=Agrarischer%20
Ausblick%20%D6sterreich%202025.pdf

Landwirtschasftskammer. (2016b, August). *Landwirtschaftskammer Österreich, Agrarischer
Ausblick 2025*. p. 43, Accessed Jan 01, 2021, from https://www.lko.at/media.php?
filename=download%3D%2F2016.11.30%2F148049533266992.pdf&rn=Agrarischer%20
Ausblick%20%D6sterreich%202025.pdf

Landwirtschasftskammer. (2016c, August). *Landwirtschaftskammer Österreich, Agrarischer
Ausblick 2025*, p. 54. Accessed Jan 01, 2021, from https://www.lko.at/media.php?
filename=download%3D%2F2016.11.30%2F148049533266992.pdf&rn=Agrarischer%20
Ausblick%20%D6sterreich%202025.pdf

Loury, R. (2016, February 14). Organic farmers could feed the world, Euroactiv. *Journal de
l'Environnement, translated by Samuel White*. Accessed Jan 31, 2021, from https://www.
euractiv.com/section/agriculture-food/news/organic-farmers-could-feed-the-world/; http://
www.fao.org/family-farming/detail/en/c/427118/

Magenta. (2021). *Hochbeet*. Accessed Jan 20, 2021, from https://newsroom.magenta.at/2017/05/0
8/t-mobile-und-wiener-startup-zeigen-ersten-smarten-hochbeet-garten-ueber-den-daechern-
wiens/

Meticulous Research. (2020, December 8). *Organic food market Worth $272,18 billion by 2027,
growing at a CAGR of 12,2% from 2020 with Covid-19 impact*. Meticulous Research ®
Analysis. Accessed Feb 13, 2021, from https://www.globenewswire.com/news-
release/2020/12/08/2141147/0/en/Organic-Food-Market-Worth-272-18-billion-by-2027-Grow
ing-at-a-CAGR-of-12-2-From-2020-With-COVID-19-Impact-Meticulous-Research-Analysis.
html

Mie, A., et al. (2017, October 27). *Human health implications from organic food and organic
agriculture: A comprehensive review*. National Library of Medicine, National Center of Biotechnology Information. Accessed Feb 13, 2021, from https://pubmed.ncbi.nlm.nih.gov/29073
935/

Muller, A., Schader, C., El-Hage Scialabba, N., et al. (2017, January 14). Strategies for feeding the
world more sustainably with organic agriculture. *Nat Commun 8, 1290*(2017). https://doi.org/

10.1038/s41467-017-01410-w. Accessed Jan 24, 2021, from https://www.nature.com/articles/s41467-017-01410-w, p. 5

Neal D. (2017). *The threats of GMOs, Kent State University, 2017 Conference Contributions.* Accessed Feb 02, 2021, from https://oaks.kent.edu/starkstudentconference/2017/Presentations/34

NPR. (2020, May 5). *U.N. warns number of people starving to death could double amid pandemic.* Accessed Feb 12, 2021, from https://www.npr.org/sections/coronavirus-live-updates/2020/05/05/850470436/u-n-warns-number-of-people-starving-to-death-could-double-amid-pandemic

NZZ. (2017, April 05). *Jörg Krumenacher, Diesen Winter sind Tausende Bienenvölker gestorben.* Accessed Jan 09, 2021, from https://www.nzz.ch/schweiz/wintersaison-dezimiert-bienenvoelker-honigbienen-im-milbenstress-ld.155464?reduced=true

Oe24. (2017, October 05). Oe24.at, *Pilzanbau als Landwirtschaft der Zukunft–Tiroler Glückspilze.* Accessed Jan 26, 2021, from https://www.oe24.at/tyroler-glueckspilze/pilzanbau-als-landwirtschaft-der-zukunft/302609774

One Planet Summit. (2021, January 11). *One planet summit for biodiversity.* Accessed Feb 10, 2021, from https://www.un.org/sustainabledevelopment/blog/2021/01/one-planet-summit-2/

Organicexportinfo. (2019). Research institute of organic agriculture. Accessed Dec 26, 2020, from https://www.organicexport.info/austria.html

Organic-market.info. (2020, January 30). *EU Commissioner campaigns for more organic farming.* Accessed Dec 31, 2020, from https://organic-market.info/news-in-brief-and-reports-article/eu-commissioner-campaigns-for-more-organic-farming.html

Organic without boundaries. (2019, February 13). *10 year study shows lower cancer risk for individuals with higher organic food consumption rates, nutrinet-santé study.* Accessed Feb 13, 2021, from https://www.organicwithoutboundaries.bio/2019/02/13/lower-cancer-risk/

Pavlik, A. (2019, July 1). *Impacthub vienna, lets talk food waste in Austria.* Accessed Jan 13, 2021, from https://vienna.impacthub.net/2019/07/01/lets-talk-food-waste-in-austria/

PNAS. (2018, October 9). *Proceedings of the national academy of sciences of the United States of America, Glyphosate perturbes the gut microbiota of honeybees.* Accessed Jan 09, 2021, from https://www.pnas.org/content/115/41/10305.short

Ploll, U. et al. (2020, July 02). A social innovation perspective on dietary transitions: Diffusion of vegetarianism and veganism in Austria. Accessed Jan 25, 2021, from https://www.x-mol.com/paper/1282721796180328448

Proveg. (2021). proveg.com. Plant-based food and lifestyles. Accessed Jan 25, 2021, from https://proveg.com/plant-based-food-and-lifestyle/vegan-alternatives/

Ritchie, H. (2019, November 11). Half of the world's land is used for agriculture. Accessed Jan 14, 2021, from https://ourworldindata.org/global-land-for-agriculture

Rockström, J., et al. (2009). Planetary boundaries: Exploring the safe operating space for humanity. *Ecology and Science, 04*(2), 32. p.1. Accessed Feb 02, 2021, from http://www.ecologyandsociety.org/vol14/iss2/art32/

Rosset, P. (2014). *Organic farming in Cuba,* 15 December 2021, Accessed Feb 04, 2021, from https://www.researchgate.net/publication/265361529_Organic_Farming_in_Cuba; https://www.researchgate.net/publication/232916920_Cuba%27s_Nationwide_Conversion_to_Organic_Agriculture

Sanders, J., & Hess, J. (2019a). Johann Heinrich von Thünen Institut, Bundesforschungsinstitut für Ländliche Räume, Wald und Fischerei, Deutschland, Die Leistungen des Ökolandbaus für Umwelt und Gesellschaft, Thünen Rep. 65, https://doi.org/10.3220/REP1576488624000, Accessed Feb 04, 2021, from https://www.thuenen.de/de/thema/oekologischer-landbau/die-leistungen-des-oekolandbaus-fuer-umweltund-gesellschaft/

Sanders, J., & Hess, J. (2019b). Johann Heinrich von Thünen Institut, Bundesforschungsinstitut für Ländliche Räume, Wald und Fischerei, Deutschland, Die Leistungen des Ökolandbaus für Umwelt und Gesellschaft, Thünen Rep. 65, https://doi.org/10.3220/REP1576488624000;

Accessed Apr 04, 2021, from https://www.thuenen.de/media/ti-themenfelder/Oekologischer_Landbau/Die_Leistungen_des_OEkolandbaus/UGOE_Auf-den-Punkt-gebracht.pdf

Schlatzer, M., & Lindenthal, T. (2018, May 22). *100% Biolandbau in Österreich–Machbarkeit und Auswirkungen.* p. 8, Accessed Feb 14, 2021, from https://archiv.muttererde.at/motherearth/uploads/2018/05/FiBL_gWN_-Bericht_-100P-Bio_Finalversion_21Mai18.pdf

de Schutter O. (2012). Eco-farming addresses hunger, poverty and climate change. *Pacific Ecologist.* Accessed Mar 13, 2021, from https://pacificecologist.org/archive/21/pe21-eco-farming.pdf

Spar. (2021). *Nachhaltigkeit bei Spar.* Accessed Jan 03, 2021, from https://www.spar.at/nachhaltigkeit

Statista. (2019). *Area of genetically modified (GM) crops in 2019, by country.* Accessed Feb 10, 2021, from https://www.statista.com/statistics/271897/leading-countries-by-acreage-of-genetically-modified-crops/

Statista. (2020, April). *Annual consumption volume of cultivated mushrooms in Austria.* Accessed Jan 27, 2021, from https://www.statista.com/statistics/482255/consumption-volume-mushrooms-austria/

Statistics Austria. (2020, August 28). *Meat consumption declined in 2019, manufacturing of animal production sunk by 1%, Press Release.* Accessed Jan 26, 2021, from http://www.statistik.at/web_en/press/124188.html

Steffen W. et al. (2015, February 13). *Planetary boundaries: Guiding human development on a changing planet.* Science. Accessed Feb 04, 2021, from https://science.sciencemag.org/content/347/6223/1259855.full

Steigenberger. (2021). *Green at a glance.* Accessed Jan 28, 2021, from https://www.steigenberger.com

Tasteofaustria. (2021). *Embassy of Austria in the United States, organic country #1, food quality.* Accessed Dec 19, 2021, from https://www.tasteofaustria.org/organic-country-no-1

Teichroew J. L., et al (2017, June). *Is China's unparalleled and unstudied bee diversity at risk?* Ecological conservation, Vol. 210, Part B., p. 19–28, Elsevier. Accessed Feb 06, 2021, from https://www.sciencedirect.com/science/article/abs/pii/S0006320716302038

The Insectatlas. (2020a). *Facts and figures about friends and foes in farming, Heinrich Böll Stiftung/Bund für Umwelt und Naturschutz Deutschland/Le Monde Diplomatique.* Accessed Jan 10, 2021, from https://www.boell.de/sites/default/files/2020-05/insectatlas2020_web_200525.pdf?dimension1=ds_insect_atlas, 2. Auflage, p.16, accessed on 10012021.

The Insectatlas. (2020b). *Facts and figures about friends and foes in farming, Heinrich Böll Stiftung/ Bund für Umwelt und Naturschutz Deutschland/Le Monde Diplomatique.* Accessed Jan 10, 2021, from https://www.boell.de/sites/default/files/2020-05/insectatlas2020_web_200525.pdf?dimension1=ds_insect_atlas, 2. Auflage, p.20.

The Insectatlas. (2020c). *Facts and figures about friends and foes in farming, Heinrich Böll Stiftung/Bund für Umwelt und Naturschutz Deutschland/Le Monde Diplomatique,* Accessed Jan 10, 2021, from https://www.boell.de/sites/default/files/2020-05/insectatlas2020_web_200525.pdf?dimension1=ds_insect_atlas, 2. Auflage, p.21.

The Insectatlas. (2020d). *Facts and figures about friends and foes in farming, Heinrich Böll Stiftung/Bund für Umwelt und Naturschutz Deutschland/Le Monde Diplomatique.* Accessed Jan 10, 2021, from https://www.boell.de/sites/default/files/2020-05/insectatlas2020_web_200525.pdf?dimension1=ds_insect_atlas, 2. Auflage, p.15.

The Insectatlas. (2020e). *Facts and figures about friends and foes in farming, Heinrich Böll Stiftung/Bund für Umwelt und Naturschutz Deutschland/Le Monde Diplomatique.* Accessed Jan 10, 2021, from https://www.boell.de/sites/default/files/2020-05/insectatlas2020_web_200525.pdf?dimension1=ds_insect_atlas, 2. Auflage, p.17.

The Insectatlas (2020f). *Facts and figures about friends and foes in farming, Heinrich Böll Stiftung/Bund für Umwelt und Naturschutz Deutschland/Le Monde Diplomatique.* Accessed Jan 10, 2021, from https://www.boell.de/sites/default/files/2020-05/insectatlas2020_web_200525.pdf?dimension1=ds_insect_atlas, 2. Auflage, p.30.

The Insectatlas. (2020g). *Facts and figures about friends and foes in farming, Heinrich Böll Stiftung/Bund für Umwelt und Naturschutz Deutschland/Le Monde Diplomatique*. Accessed Jan 10, 2021, from https://www.boell.de/sites/default/files/2020-05/insectatlas2020_web_200 525.pdf?dimension1=ds_insect_atlas, 2. Auflage, p.44.

The Insectatlas. (2020h). *Facts and figures about friends and foes in farming, Heinrich Böll Stiftung/Bund für Umwelt und Naturschutz Deutschland/Le Monde Diplomatique*. Accessed Jan 10, 2021, from https://www.boell.de/sites/default/files/2020-05/insectatlas2020_web_200 525.pdf?dimension1=ds_insect_atlas, 2. Auflage, p.5.

The Insectatlas. (2020i). *Facts and figures about friends and foes in farming, Heinrich Böll Stiftung/ Bund für Umwelt und Naturschutz Deutschland/Le Monde Diplomatique*. Accessed Jan 10, 2021, from https://www.boell.de/sites/default/files/2020-05/insectatlas2020_web_200525. pdf?dimension1=ds_insect_atlas, 2. Auflage, p.45.

The Insectatlas. (2020j). *Facts and figures about friends and foes in farming, Heinrich Böll Stiftung/ Bund für Umwelt und Naturschutz Deutschland/Le Monde Diplomatique*. Accessed Jan 10, 2021, from https://www.boell.de/sites/default/files/2020-05/insectatlas2020_web_200525. pdf?dimension1=ds_insect_atlas, 2. Auflage, p.46.

The Insectatlas. (2020k). *Facts and figures about friends and foes in farming, Heinrich Böll Stiftung/Bund für Umwelt und Naturschutz Deutschland/Le Monde Diplomatique*. Accessed Jan 10, 2021, from https://www.boell.de/sites/default/files/2020-05/insectatlas2020_web_200 525.pdf?dimension1=ds_insect_atlas, 2. Auflage, p.18.

The Local. (2019, October 4). *Climate crisis: Italian beekeepers suffer worst honey harvest ever*. Accessed Feb 06, 2021, from https://www.thelocal.it/20191004/climate-crisis-italian-beekeepers-suffer-worst-honey-harvest-ever

The New York Times. (2020, June 24). *Patricia Cohen, roundup maker to pay $10 billion to pay cancer suits*. Accessed Feb 10, 2021, from https://www.nytimes.com/2020/06/24/business/roundup-settlement-lawsuits.html

Tiroler Glückspilze. (2021). *Über uns*. Accessed Feb 17, 2021, from https://gluckspilze.com/UEBER-UNS

UNCTAD. (2002). *United nations conference on trade and development, key issues in biotechnology*, p.3. Accessed Feb 10, 2021, from https://unctad.org/fr/system/files/official-document/poitetebd10.en.pdf

UNCTAD. (2008). *United nations conference on trade and development, organic agriculture and food security in Africa*, p. iii, Accessed Feb 12, 2021, from https://unctad.org/system/files/official-document/ditcted200715_en.pdf

United Nations. (2017). *United Nations, world populations prospects: The 2017 revision*. Accessed Feb 01, 2021, from https://www.un.org/development/desa/en/news/population/world-population-prospects-2017.html

University of Sydney. (2016, January 25). *Global nitrogen footprint mapped for first time, Science news*. Accessed Feb 02, 2021, from https://www.sciencedaily.com/releases/2016/01/16012 5114124.htm

Unverschwendet. (2021). Accessed Jan 14, 2021, from https://www.unverschwendet.at/english/

USDA. (2020, January). *Roswitha Krautgartner, United States department of agriculture, foreign agricultural service, GAIN, Austrian organic production and consumption continues upward trend, report*. Accessed Dec 26, 2020, from https://apps.fas.usda.gov/newgainapi/api/Report/DownloadReportByFileName?fileName=Austrian%20Organic%20Production%20and%20Consumption%20Continues%20Upward%20Trend_Vienna_Austria_01-29-2020

Vienna. (2015). *Green Urban Infrastructure conference*. Accessed Jan 20, 2021, from https://urbangreeninfrastructure.org

Vienna. (2018). *City of Vienna, Urban Heat Island Strategy*, p.19, Accessed Jan 18, 2021, from https://www.wien.gv.at/umweltschutz/raum/pdf/uhi-strategieplan-englisch.pdf

Viljoen, S. (2018). *Urban Farming in Vienna, Diplomarbeit*, p.3., Accessed Jan 09, 2021, from https://repositum.tuwien.at/bitstream/20.500.12708/1976/1/UrbanFarmingVienna%20Urban%

20farming%20as%20a%20component%20of%20Urban%20development%20strategies%20 with%20reference%20to%20Vienna%20Austria.pdf

Vou, A. (2019). *European data journalism network, Europe is going veg.* Accessed Feb 24, 2021, from https://www.europeandatajournalism.eu/eng/News/Data-news/Europe-is-going-veg

Weisshaidinger, R., & Hörtenhuber, S. (2020, August 07). *Bio-Austria, Was macht der Bio-Landbau für den Wasserschutz.* Accessed Feb 04, 2021, from https://www.bio-austria.at/ was-macht-der-bio-landbau-fuer-den-wasserschutz/

WHO. (2019, July 15). *World hunger is still not going down after three years and obesity is still growing, press release.* Accessed Feb 12, 2021, from https://www.who.int/news/item/15-07-2019-world-hunger-is-still-not-going-down-after-three-years-and-obesity-is-still-growing-un-report

Wien. (2021a). *Wien Bio-Lokale.* Accessed Feb 17, 2021, from https://www.wien.info/de/ einkaufen-essen-trinken/restaurants/bio-lokale

Wien. (2021b). *Vienna, The city of bees.* Accessed Jan 09, 2021, from https://www.wien.info/en/ sightseeing/green-vienna/urban-beekeepers

Wien. (2021c). *Sozialmärkte.* Accessed Jan 15, 2021, from https://www.wien.gv.at/sozialinfo/ content/de/10/SearchResults.do?keyword=Sozialmärkte

Wien. (2021d). *Umwelt, Umweltmusterstadt Wien.* Accessed Jan 21, 2021, from https://www.wien. gv.at/umwelt/natuerlich/

Wien. (2021e). *City of Vienna, Ökokauf Wien–programme for sustainable public procurement.* Accessed Jan 20, 2021, from https://www.wien.gv.at/english/environment/protection/oekokauf/

Willer, H., & Lernouds, J. (2020). *FIBL & IFOAM–organics international, the world of organic agriculture–statistics and emerging trends 2020,* p.19, Accessed Dec 28, 2021, from https:// www.fibl.org/fileadmin/documents/shop/5011-organic-world-2020.pdf

Zurück zum Ursprung. (2021). Accessed Jan 03, 2021, from https://www.zurueckzumursprung.at/ nachhaltigkeit/

Dr. Ursula A. Vavrik holds a Master's degree and a PhD (1990) in Economic and Social Sciences, her doctoral thesis was awarded with the Rudolf Sallinger Prize of the Austrian Chamber of Commerce for outstanding research. She pursued an international career having worked for multinational organizations (UN, OECD), EU institutions, academia, the private sector and NGOs. Areas of specific expertise encompass international and EU policy, particularly in the fields of environment, development and sustainable development. She served during all three Austrian EU Council Presidencies, 2006 as senior advisor and 2018 also at the Austrian Ministry of Foreign Affairs. As international consultant, she facilitated the Water Sector Review in Uganda, advised the European Environment and Sustainable Development Advisory Councils on EU Budget Reform and the CAP, and acted as EU and OSCE election observer. She held numerous positions as Guest professor and lecturer (BOKU/Vienna, Sciences-Po/Paris, ITAM/Mexico, Diplomatic Academy Vienna, etc.), launched the first seminar cycle on Sustainable Development at the Vienna University of Economics and Business Administration, and published in Europe, USA and Mexico. Since 2009 she leads the NEW WAYS Center for Sustainable Development. In 2018, she initiated the European Award "Excellence in the Implementation of the UN SDGs" and awarded 50 companies. More recently, the focus of her activities lies in research on planetary boundaries and sustainable and organic agriculture.

Barriers to Supply Chain Sustainability Innovation Amongst Nigerian Entrepreneurs in the Food and Agriculture Industry

Adebimpe Adesua Lincoln

Abstract

The rationale for this research stems from the fact that there is a need to improve our understanding of how sustainability occurs among entrepreneurs, the approach these firms adopt vis-a-vis supply chain sustainability innovation, barriers faced and strategies to overcome these barriers within the Nigerian context. The study adopts both the quantitative and qualitative approach to data collection involving the use of questionnaire surveys and semi-structured interviews with entrepreneurs in the food and agriculture industry. Questionnaire surveys were carried out with 396 entrepreneurs in the food industry and 63 entrepreneurs in the agriculture sector. In addition, 20 semi-structured interviews were carried out with some of the entrepreneurs who took part in the questionnaire survey with a view to elicit further information on some of the main issues relevant to the overarching research aim. The findings show that many of the entrepreneurs were unaware of international sustainability frameworks and many of the practices adopted were often informal and on an ad hoc basis. The finding also shows that the primary barrier affecting supply chain sustainability innovation among Nigerian entrepreneurs are economic and financial barriers, technological, legal regulatory and institutional barriers. The entrepreneurs advocate for sustainable technology development, regulatory and environmental strategies as well as enhanced networking strategies as ways of overcoming some of the barriers identified.

There is now a growing recognition of the importance of promoting sustainable innovative practices as part of a firm's core business model. The study, therefore, makes empirical contributions to the burgeoning literature in this area thus filling gaps in the literature on supply chain sustainability in the SME sector

A. Adesua Lincoln (✉)
University of Liverpool, Liverpool, UK

© The Author(s), under exclusive license to Springer Nature Switzerland AG 2022
S. O. Idowu, R. Schmidpeter (eds.), *Case Studies on Sustainability in the Food Industry*, Management for Professionals,
https://doi.org/10.1007/978-3-031-07742-5_2

in Africa. From a Nigerian standpoint, it provides a theoretical perspective on which future research and policy initiatives can be developed. The practical implication for Nigeria specifically and Africa more generally is that the study shows countries on the African continent lag behind their Western counterparts vis a vis advancement in sustainability measures. There is a need for better education amongst businesses on the continent in order to bring them in line with current thinking in the field. This will enable entrepreneurs to incorporate innovative strategies in their business practices thus enhancing competitiveness and growth and also allowing them to make substantial contributions to global sustainable drive. There is also a need to ensure these firms are better supported by various governmental and non-governmental institutions such as NGOs, to enable them to flourish.

Keywords

Entrepreneurship · Small and Medium Enterprise · SMEs · Nigeria · Food industry · Agriculture sector · Innovation · Supply chain · Sustainability · Barriers · Strategies

1   Introduction

In recent times, there has been a clear shift in the extant literature from a compromise often between expenses and economic growth to a more nuanced perspective, which views sustainability as being integral to business growth and competitive advantage (Pantouvakis et al., 2017; Lincoln et al., 2020). A firm's competitive advantage is often said to be derived from its core competence, which in turn could result in efficiency and profitability. Thus, competitive advantage can be achieved if a firm is able to either control cost drivers thereby gaining a cost advantage over their competitors or they are able to rebuild the value chain to attain cost advantage by finding alternative ways to design, produce, distribute or sell their products. So, for example timeous product delivery and excellent customer service can result in public image convalescence and an augmented competitive market positioning, thus enhancing customer satisfaction and loyalty.

On a global scale, there is increased recognition of the need to manage risk and achieve competitive advantage. Companies in many countries are actively striving to create shared value with relevant stakeholder groups to achieve product and service differentiation, whilst also taking advantage of innovation in technologies, supplier diversity and partnership synergies within their supply chains. The approach to supply chain sustainability is still at its infancy in Nigeria, due in part to the nature of the sector involved, size of the firm, organizational culture and leadership commitment, lack of resources, expertise and infrastructure as well as supply chain complexity. The Nigerian food industry as with many other industries in the country is faced with an escalating conundrum; achieve economic success, whilst also ensuring they focus on improving energy savings and ensuring social and environmental performance. In order to succeed there is a need to optimise costs and adopt effective strategies which support socially responsible and sustainable business practises. Entrepreneurs in the food industry should also endeavour to evolve and respond to changing consumer trends, effectively identify and manage risks and opportunities and leverage available resources and technology to innovate and improve their business operations and models.

This research seeks to explore supply chain sustainable innovation in the Nigerian food and agriculture industry. The research focuses in particular on barriers to effective engagement by entrepreneurs and strategies which can be adopted to overcome the barriers. The rationale for the study stems from the fact that entrepreneurship is a dynamic field, particularly as it relates to the process of change creation and vision. The entrepreneurs' vision to recognise opportunities and willingness to take calculated risks are critical factors needed for success (Kuratko & Hodgetts, 2007; Lincoln et al., 2016; Lincoln, 2017). Furthermore, the growth of sustainability entrepreneurs in recent times has helped make significant strides towards a greener way of life in many countries around the world. Consequently, the important role entrepreneurs play in sustainability advancement and the appreciation of their role cannot be overemphasised. In addition to the research aim mentioned above, the study sought to answer the following research questions:

- Are Nigerian entrepreneurs in the food and agriculture industry aware of national and international goals on sustainability?
- What are the sustainability measures currently adopted by entrepreneurs in the Nigerian food and agriculture industry?
- What are the perceptions of the entrepreneurs on the barriers that hinder the adoption of sustainable supply chains in the Nigerian food and agriculture industry?
- What are the major barriers and what strategies can be adopted in alleviating the barriers identified?

2   Food Industry and Entrepreneurship

Food is a vital aspect of human life. The role of the food industry worldwide includes ensuring customer satisfaction and the provision of safe goods which are nutritionally adequate. In addition, the food industry is tasked with ensuring product information is clearly provided and maintaining commercial viability of the goods produced. The food industry comprises a complex network of activities relating to the supply, consumption and catering of food produce and service. According to Economy Watch (2010), the food industry is highly diverse and comprises several important components, each adding a distinct value to holistic food chain by improving sustainability and ensuring production of better produce. The various activities which make up the food industry include:

1. Agricultural activities for growing crops, raising livestock and sea food.
2. Manufacturing and processing of fresh products into canned, packed goods and frozen foods.
3. Marketing, packaging, advertising and distribution of food including wholesale and retail (Economy Watch, 2010).
4. Grocery stores, convenience stores, food and drinks services and restaurants, catering, bakeries, confectionaries, fast food sector, drug stores, market stalls and supermarkets (Statista Research department, 2020).
5. Research and development on food technology.
6. Manufacturing fertiliser, farm machinery and hybrid seeds to facilitate agricultural production.
7. Financial service including insurance and credit to facilitate food production and distribution.
8. Regulation on food production and distribution to ensure quality and safety (Economy Watch, 2010).

The importance of the food industry in recent times cannot be overstated. The food industry is continuously growing. Factors that have amplified the demand in the global food industry include wealth distribution, varied lifestyles, health awareness, the growing demand for organic food and the ever-increasing population levels. The food supply drivers include the quality of the supply chain, level of competition in the industry and the composition of the target consumers (Economy Watch, 2010). The food industry employs a massive number of skilled and unskilled work force and many of the businesses operating in the global food industry are Small and Medium Scale Entrepreneurs. For example, in the UK, the UK government in its 26 October 2020 publication notes that there are approximately 7130 Micro, Small and Medium Enterprises in the food and drink sector with a turnover of around £21 Billion and 135,000 employees as of 2019. In the food sector (excluding beverages), SMEs accounted for 79% of businesses, 27% of employment and 17% of turnover. Around a third of the SMEs are manufacturers of bakery products. Furthermore, the 'other food products' category which includes prepared meals, confectionary, condiments and seasonings was the largest manufacturing category accounting for £6.3 Billion in 2018 and contributing 22% to the total food and drink manufacturing Gross Value Added. While the meat and meat products had a Gross Value-Added contribution of £4.1 Billion, contributing 14% to the total food and drink manufacturing Gross Value Added (UK National Statistic, 2020). Furthermore, the UK government reports that the total consumer expenditure on food, drink and catering rose by 2.5% in 2019 to £234 Billion. Expenditure on food including non-alcoholic drinks increase by 2.1%, alcoholic drinks by 2.5% and catering by 3.1% (UK National Statistics, 2020). The UK is not alone in its identification of the unique role played by the SME sector and its importance in the food industry. EU Data and Trend on the Food and Drink Industry (2019) provides similar statistics. According to the EU Data and Trend the food supply chain employs 23 million people, total turnover amounts to €3.8 Trillion and €732 Billion Value Added. The EU's extensive food supply chain, from agriculture and the input industry to food and drink service employs 33 million professionals (EU Data and Trends, 2019). The food and drink industry is the largest manufacturing industry in the EU, employing 4.72 million people and generating a turnover of €1.2 Trillion and €236 Billion in Value Added. Small and Medium-Sized Companies account for around 47.5% of food and drink turnover and 60.8% of food and drink employment. The Food and drinks industry accounts for more than 290,000 SMEs.

Credible statistics on the various industries/sectors in Nigeria is often not readily available, which makes effectual comparison with other countries a challenging feat. Available data from the World Bank (2020) shows that agriculture which includes forestry, hunting, and fishing, as well as cultivation of crops and livestock production remains the most important source of employment for the population and the poor, and accounts for 90% of rural livelihoods (World Bank, 2020). Most of the country's agricultural production is carried out by SMEs in the Southern part of Nigeria. Even though agriculture still remains the largest sector of the Nigerian economy and employs two-thirds of the entire labour force, the production hurdles have significantly stifled the performance of the sector. Over the past 20 years, value-added per capita in agriculture has risen by less than 1% annually (UN, no date). Nigeria also boosts a variety of SMEs involved in other food processes, including food wholesale and retail, fresh food processing, grocery stores, convenience stores, food and drinks services and restaurants as well as the fast-food sector. According to the Association of Fast Food and Confectioners of Nigerians (AFFCON), the Nigerian food industry is estimated to be worth over One Trillion Naira (NIPC, 2016). As of 2014, there were over 800 Quick Service Restaurant outlets in Nigeria, the vast majority of them branded by about 100 small and medium-sized local companies, according to the Association of Fast Food and Confectioners of Nigeria (AFFCON), they generated about N200 Billion ($1.22 Billion) in revenue and employed more than 500,000 workers (NIPC, 2016).

SMEs in Africa represent a large proportion of businesses contributing immensely to the mobilisation of resources, employment and poverty alleviation. Consequently, these firms are seen as dominant forces for economic development and industrialisation in the African region. Due to their small and flexible character, SMEs are able to withstand adverse economic environments and survive where many large enterprises fail (Lincoln, 2018). In Nigeria, entrepreneurship is often the only option for those seeking an income to overcome poverty and unemployment. The National Bureau of Statistics states that the unemployment rate in Nigeria is a staggering 27.1% (National Bureau of Statistics, 2020). The importance of the SME sector is matched by the appreciation of its unique role as one of the most vital contributors to people's income and to economic development and innovation. SMEs are perceived to be the seedbed for indigenous entrepreneurship and are known to generate many small investments, thus encouraging entrepreneurship is a key policy instrument for promoting economic growth and employment creation including social and environmental development in Nigeria (Woldie et al., 2008; Lincoln et al., 2016; Lincoln, 2017). In spite of such promising reports about SMEs, the Nigerian small and medium enterprise sector is, however, characterised by low productivity, inability to compete with imports, lack of diversification, unfavourable business environment, infrastructural deficiencies, corruption, low access and high cost of finance and weak institutions, lack of skills and expertise (OECD, 2015).

3 Sustainability and Sustainable Development

Cruz and Marques (2014) define sustainability as a social objective on achieving the triple bottom line performance of profit, planet, and people. Corporate sustainability is now considered of utmost importance in terms of long-term value delivery from a financial, social, environmental and ethical perspective (UN Global Compact, 2015). According to the UN Global Compact (2015), it is imperative for companies to ensure that they are able to operate in a responsible manner and endeavour to align their activities with universal principles and take actions, which support local communities and the wider society. In so doing, companies must commit at the highest level, engage in annual reporting on their efforts and engage locally where they have a presence.

The rise of the term 'sustainability' is often attributed to the 1987 report by Brundtland. According to the Brundtland Report, in order to achieve sustainable development, the level of 'development' must meet the needs of the present without compromising the ability of future generations to meet their own needs (World Commission on Environment Development, 1987). The sustainability concept was defined to include three main pillars namely, economic, environmental and social. Since then, sustainability has been defined in varied ways encompassing the triple bottom line notion of people, planet and profits as posited by John Elkington in 1994. The triple bottom line is a sustainability framework that examines a company's social, environment and economic impact.

3.1 Environmental Pillar

The focus is usually on reducing carbon footprints, packaging waste, water usage and overall effect on the environment. From a cost-effective perspective, lessening the amount of material used in packaging usually reduces the overall spending on those materials. Linking the environmental pillar to Elkington (1994) planet measurement, the focus is on how environmentally responsible a firm is and how it measures environmental variables, for example air and water quality, energy consumption, natural resources, solid and toxic waste and land use.

3.2 Social Pillar

The importance here is on securing the support and approval of key stakeholder groups, for example employees, suppliers, customers and the community. Aspects of significance include fair treatment of employees, introduction of community support initiatives such as investment in local projects, fundraising, sponsorship, scholarships, awareness of how the supply chain operates relative to your product/ services, e.g. forced labour, child labour and ensuring people are being paid fairly. Linking this pillar to Elkington (1994) people measurement, the focus is on how socially responsible an organization has been throughout its history. The social dimensions measures factors such as education, equity and access to social resources, health and well-being, quality of life and social capital.

3.3   Economic Pillar

This involves aspects of profitability and wealth maximisation as well as overall corporate governance, e.g. compliance, effective governance and risk management, alignment of boards of directors and shareholders' interests as well as that of the company's value chains and customers. Linking this pillar to Elkington (1994) profit measurement, the profit measures flow of money, corporate profit accounts and the financial health of the firm as well as economic aspects. Economic measurements include income and expenditures, business diversity and climate factors, taxes and employment.

More recently, John Elkington in his 2018 review questions the ability to achieve the sustainability goals. He asserts that success or failure on sustainability goals cannot be measured only in terms of profit and loss. It must also be measured in terms of the well-being of billions of people and the health of our planet. Accordingly, Elkington (2018) postulates that there cannot be mixed messages on sustainability goals and that there are terrible ramifications to ignoring sustainable innovative practices ranging from destruction of the rainforest and the ozone layer to encouraging exploitation of labour and use of child labour. Fundamentally, to be effective a new wave of sustainable innovation and deployment is needed which sees sustainability initiatives having a clear and effectual correlation to climate change, and improvement in water resources, oceans, forests, soils and biodiversity (Elkington, 2018).

3.4   The United Nations Sustainable Development Goals (SDGs)

The SDGs focus on sustainable agenda and action plan for people, planet and prosperity (UN, 2015). It proposes 17 Sustainable Development Goals which align with academic thinking on sustainability vis-a-vis economic, social and environmental dimensions. The 17 Sustainable Development Goals include:

Goal 1. End poverty in all its forms everywhere.
Goal 2. End hunger, achieve food security and improved nutrition and promote sustainable agriculture.
Goal 3. Ensure healthy lives and promote well-being for all at all ages.
Goal 4. Ensure inclusive and equitable quality education and promote lifelong learning opportunities for all.

Goal 5. Achieve gender equality and empower all women and girls.

Goal 6. Ensure availability and sustainable management of water and sanitation for all.

Goal 7 Ensure access to affordable, reliable, sustainable and modern energy for all.

Goal 8. Promote sustained, inclusive and sustainable economic growth, full and productive employment and decent work for all.

Goal 9. Build resilient infrastructure, promote inclusive and sustainable industrialization and foster innovation.

Goal 10. Reduce inequality within and among countries.

Goal 11. Make cities and human settlements inclusive, safe, resilient and sustainable.

Goal 12. Ensure sustainable consumption and production patterns.

Goal 13. Take urgent action to combat climate change and its impacts.

Goal 14. Conserve and sustainably use the oceans, seas and marine resources for sustainable development.

Goal 15. Protect, restore and promote sustainable use of terrestrial ecosystems, sustainably manage forests, combat desertification, and halt and reverse land degradation and halt biodiversity loss.

Goal 16. Promote peaceful and inclusive societies for sustainable development, provide access to justice for all and build effective, accountable and inclusive institutions at all levels.

Goal 17. Strengthen the means of implementation and revitalize the Global Partnership for Sustainable Development (UN, 2015).

3.5   Ten Principles of the Global Impact and Supply Chain Sustainability

The ten principles of global impact and supply chain sustainability are divided into 4 main categories namely human rights, labour, environment and anti-corruption (UN Global Compact, 2015).

3.6   Human Rights

Principle 1: Businesses should support and respect the protection of internationally proclaimed human rights.

Principle 2: Make sure that they are not complicit in human rights abuses.

3.7   Labour

Principle 3: Businesses should uphold the freedom of association and the effective recognition of the right to collective bargaining.

Principle 4: Elimination of all forms of forced and compulsory labour.

Principle 5: Effective abolition of child labour.
Principle 6: Elimination of discrimination in respect of employment and occupation.

3.8   Environment

Principle 7: Businesses should support a precautionary approach to environmental
   challenges.
Principle 8: Undertake initiatives to promote greater environmental responsibility.
Principle 9: Encourage the development and diffusion of environmentally friendly
   technologies.

3.9   Anti-corruption

Principle 10: Businesses should work against corruption in all its forms, including
extortion and bribery (UN Global Compact, 2015).

4   Supply Chain Innovation and Barriers

As a result of 'workforce health and safety incidents, labour disputes, geopolitical
conflicts, raw materials shortages, environmental disasters and modern slavery there
is now growing awareness of supply chain risks among consumers, investors,
employees and local communities resulting in the inclusion of 'sustainability'
agenda to the company's procurement and sourcing criteria (EY, 2016). Supply
chains are the engines for global economies, they help to deliver goods and services
and connect businesses and their employees across industrial, cultural, geographical
and regulatory boundaries. 'Supply chains have become global, highly complex and
of significant scale. Building and maintaining resilient supply chains is a key success
factor for business in a globalized and fast-changing world. For some time now,
technical quality, cost effectiveness, speed of delivery and reliability have been the
focus of supply chain and procurement' (EY, 2016). Sustainability has the potential
to assist management as it faces greater pressure to reduce costs in the short term
while building the business foundation for sustainable performance, sustainable
supply chain management and long-term growth (Resaee, 2018). Consequently,
supply chain sustainability focuses on the creation, protection, growth and management of these long-term environmental, social and economic values for all
stakeholders involved in bringing products and services to market and the encour-
agement of effective governance practices, throughout the lifecycles of goods and
services (UN Global Compact, 2015). Through supply chain sustainability,
companies fulfil twin objectives by protecting the long-term viability of their
business and also securing a social license to operate.

   The ability to implement sustainable supply chain results in increased value vis a
vis the firm's core competence, it reduces cost and has positive impact on the

environmental. Sustainable innovation encompasses the introduction of a novel approach to production processes, technologies, product techniques, and organizational systems with a focus on reducing environmental degradation. Various academics identify the importance of sustainable innovative practices in helping firms achieve sustainability goals. For example, Cai and Zhou (2014) note that sustainable innovation strategies can assist in dealing with sustainability issues within manufacturing processes and supply chains. Increasingly, there is evidence to suggest that companies continue to embrace resilience and responsibility in their supply chain management to adapt to externalities such as geopolitical conflicts, changing weather patterns and raw materials shortages, and to improve their impacts on the workforce, local communities and the environment in the places where they procure, source, manufacture and transport their products and services (UN Global Impact, 2015).

This view aligns with the concept of social entrepreneurship, i.e. entrepreneurial activities that pioneer new product concepts which meet social needs using viable business models. The concept of shared value is also of significance here. Shared value involves 'policies and operating practices that enhance the competitiveness of a company while simultaneously advancing the economic and social conditions in the communities in which it operates. Shared value creation focuses on identifying and expanding the connections between societal and economic progress, achieved by reconceiving products and markets, redefining productivity in the value chain, and building supportive industry clusters at the company's locations' (Porter & Kramer, 2011). Furthermore, honest measurement of financial, human, social and environmental indicators whilst often challenging is crucial in assessing short and long-term performances (Porter & Kramer, 2011).

As shown in Fig. 1, this research postulates sustainable supply chain management involves robust organisational systems with access to marketing and networks which has a clear appreciation of the environment and stakeholder influence. Furthermore, adoption of strategies to manage supply chain activities that take into consideration economic and financial, legal regulatory and institutional framework, social and cultural aspects and novel approach to technological development is crucial to success. Successful innovation is to a large extent dependent on skill capabilities, expertise, access to finance, market knowledge, research and development and establishing effective collaboration and cooperation with supply chain partners. While there is support in the academic literature on the potential benefits of sustainability, there is also recognition of the costs and there are ongoing debates as to whether sustainability investments in environmental and social initiatives pay off in terms of customers' satisfaction and perception toward products and services (Bansal & McKnight, 2009; Luchs et al., 2010; Carter & Easton, 2011; Fawcett & Waller, 2011; Rezaee, 2018).

SMEs are often confronted by barriers in their quest to achieve sustainable innovation. Gupta and Barua (2018) classified barriers faced by SMEs into seven main categories: managerial, technological, resource related, financial, regulatory, economic and external partnership-related barriers. They present findings which show that technological, resource-related barriers, financial and economic barriers

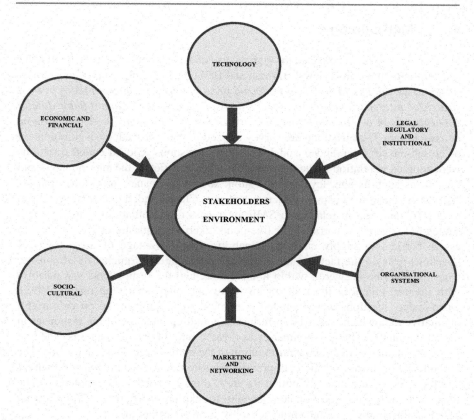

Fig. 1 Framework on sustainability innovation. Source: Author

are the most prominent barriers that militate sustainable innovation amongst the SMEs studied. More recently, Gupta et al. (2020), opines that various barriers hinder sustainable innovations, they cite technological, regulatory and financial barriers, lack of awareness and understanding among organizations about benefits of sustainable innovations, lack of skilled manpower to implement sustainable innovations, lack of customer acceptance and lack of awareness about sustainable products, lack of investment in research and development related to sustainable products as factors which hinder sustainable innovation. Various studies have also arrived at findings consistent with the view that lack of initiatives, lack of pressure from regulatory bodies, non-conductive legal regulatory systems, lack of technology, lack of skill, expertise and training and technological advancement are major impediments to sustainable supply chain innovation, as well as inadequate financial resources to invest and associated cost of technology (Stewart et al., 2016; De Jesus & Mendonça, 2018). It is important to clearly identify these barriers, categorise them in order of significance and address them in order to better promote SME adoption, implementation and upscaling of supply chain sustainability innovations.

5   Methodology

Data for the study was collected through the use of a questionnaire survey instrument and semi-structured interviews. Bryman and Bell (2011) define a survey as *a cross-sectional design in which data is collected predominantly by questionnaire or by a structured interview in order to collect a body of quantitative or quantifiable data in connection with two or more variables, which are then examined to detect patterns of association.* The questionnaire survey utilised in this study adopts a combination of closed-ended, open-ended and Likert scale questions. Closed-ended questions were adopted in relation to the demographic characteristics of the entrepreneurs and their firms such as age, level of education, source of finance and motivation for starting the business, business structure and size. Closed-ended questions were also used to obtain data in relation to SME awareness of sustainability initiatives and strategies to overcome barriers. The five-point Likert scale question style was used to assess barriers to supply chain sustainability. The advantage of these types of questions rests on the fact that they are easier and relatively quicker to administer. Questions can be answered within a short time, and the respondents can perform with greater reliability the task of answering the questions (Oppenhiem, 1992). Open-ended questions were included to allow the entrepreneurs to provide useful qualitative data which was used in further expounding on some of the responses in the questionnaire. The questionnaire was divided into three main sections. The first section contained demographic information of the entrepreneurs and their firms. The second section measured the entrepreneur's awareness of sustainability initiatives and the various sustainability initiatives they adopt in their business practices. The final section of the questionnaire sought to explore barriers to supply chain sustainability innovation and strategies to overcome barriers faced.

Convenient sampling was used to identify the entrepreneurs who participated in the study. Convenience sampling involves the use of readily available participants to participate in the study. The test of sample design is how well the sample structure represents the characteristics of the population it purports to be a representative of (Cooper & Schindler, 2011), bias inevitably occurs where the number of subjects is too small. Babbie (2010) states that the larger a sample the more representative of the population it becomes, and so the more reliable and valid the results based on it will become. This view is supported by Saunders et al. (2015) who note that the larger the sample size, the lower the likely error in generalising to the population. The decision was therefore made to adopt a larger sample size in order to minimise the potential bias, which can often arise through convenience sampling. In total, 500 entrepreneurs in Lagos Nigeria took part in the study. A total of 459 fully completed and usable questionnaires were extracted from the initial 500 questionnaires distributed. Several studies have worked with remarkably fewer samples and have achieved satisfactory results. Lagos State was selected as the apt location for the study as Lagos is the commercial nerve centre of the country and is the most diverse state in the country (Lincoln, 2017).

Semi-structured interviews were conducted to complement the questionnaire survey. A total of 20 semi-structured interviews were carried out with some of the entrepreneurs who took part in the questionnaire survey with a view to elicit further information on the main issues relevant to the overarching research aim. This sample size is of an acceptable limit for an exploratory study of this kind (Burns, 2000; Jamali, 2009). In arriving at an appropriate sample size, various academic perspectives on research methodology were considered. Guest et al. (2006) suggested that conducting twelve interviews is adequate for a research in which the aim is to understand commonalities between fairly homogenous groups. Conversely, Creswell (2007) recommended that twenty-five to thirty interviews should be sought where the sample population is heterogeneous, and the research question is comprehensive. The selection of chosen interviewees was based on the knowledge of these individuals on sustainability issues and their awareness of sustainability development drives in Nigeria. Purposive non-probability sampling was adopted to identify suitable participants. Each interview lasted between 50 and 60 min, the questions were extracted from three main themes in the questionnaire instrument namely (1) awareness of sustainability, (2) main barriers affecting supply chain sustainability in Nigeria and (3) strategies to overcome the barriers. The rationale for adopting semi-structured interviews rests on the premise that the study being exploratory in nature seeks to provide a narrative of real-life occurrences and events from the entrepreneur's own perspective (Burns, 2000). The semi-structured interviews provided the researcher with ample opportunity to follow up ideas from the questionnaire survey carried out with the entrepreneurs, probe responses and investigate motives and feelings on some of the questions listed on the questionnaire (Bell, 1993). Handwritten notes were taken during each interview and information obtained from the semi-structured interviews was transcribed in order to make it more presentable and bolster the results obtained from the questionnaire survey.

In order to ensure validity and reliability, no leading questions were posed to the entrepreneurs. Furthermore, the questionnaire and interview questions were based on stable sources of literature and sound theoretical underpinning derived from the sustainability literature. This finds support from Gill and Johnson (2010) who posit that reliability derives from the clear formulation of questions, which facilitates understanding and hence correct answering. In order to ensure reliability, the interview questions and questionnaire were based on sound theoretical underpinnings derived from the sustainability, supply chain and entrepreneurship literature. The questionnaire was piloted to unearth potential weaknesses so that appropriate corrections were made before the main data collection was carried out. Data obtained from the questionnaire was analysed using the Statistical Package for Social Sciences (SPSS 27 for Mac). Descriptive statistics such as percentages and cross-tabulations were used to analyse the information obtained from the entrepreneurs. Where relevant, verbatim quotes from the open-ended responses extracted from the questionnaire and semi-structure interviewees were used in supporting the discussion of findings thus aiding better understanding of the developing themes.

6   Analysis of Findings

The analysis is divided into two main sections. The first section presents demographic findings obtained from the entrepreneurs in the food and agriculture industry. The second section provides a discussion of the findings obtained in relation to the entrepreneur's awareness of sustainability initiatives, barriers to supply chain sustainability innovation and strategies to overcome barriers faced.

6.1   Demographic Information of SMEs in the Nigerian Food Industry

The analysis in this section focuses on the findings obtained in relation to the demographic information about the entrepreneurs such as gender, marital status, age, education, experience, motivation for starting their business and sources of finance. The latter parts of the analysis present findings on the demographic information about the SME firm such as business structure, firm age and employees. In total 459 entrepreneurs in the food and agriculture industry took part in the study, 396 of the entrepreneurs operated various activities in the food industry ranging from grocery stores, supermarkets, food and drinks services, catering, restaurant, fast food, convenience stores, market stalls, wholesale and retail food distributors. 63 of the sampled businesses operated in the agriculture sector. Table 1 shows an equal percentile of male and female entrepreneurs in the food industry; 198 male entrepreneurs and 198 female entrepreneurs. Of the 63 entrepreneurs sampled in the agriculture sector, 18 (28.6%) were males and 45 (71.4%) were females.

The finding shows that a large proportion of the entrepreneurs were married. The finding also shows that the average entrepreneur in the food industry was 41–50. All of the entrepreneurs in the agriculture sector were in the 41–50 age bracket. Majority of the male and female entrepreneurs were relatively well-educated, the entrepreneurs reported that they had obtained a diploma, degree or postgraduate/professional qualification. In addition, many of the entrepreneurs had either professional or skilled manual work experience.

One hundred fifty-eight entrepreneurs representing 39.9% in the food industry and 24 entrepreneurs representing 38.1% in the agriculture sector stated that they had obtained their sources of finance from family and friends. Eighty six entrepreneurs representing 21.7% in the food industry and 11 entrepreneurs representing 17.5% in the agriculture sector stated that they had utilised their own savings in starting the business. In addition, 83 entrepreneurs representing 21% stated that they had obtained their start-up capital from formal financial institutions. Majority of the entrepreneurs in both the food and the agriculture industry were driven to entrepreneurship as a result of threat of unemployment (43.9% in the food industry and 52.4% in the agriculture sector) and financial motives (29.5% in the food industry and 22.2% in the agriculture sector).

The findings presented in Table 2 show the demographic information of the firm. Many of the businesses were run as sole proprietorship businesses representing

Table 1 Demographic information of the entrepreneurs

Demographic	Food industry	Agriculture
Gender		
Male	198 (50%)	18 (28.6%)
Female	198 (50%)	45 (71.4%)
Marital status		
Single	42 (10.6%)	10 (15.9%)
Married	303 (76.5%)	42 (66.7%)
Divorced	27 (6.8%)	6 (9.5%)
Widowed	24 (6.1%)	5 (7.9%)
Age		
20–30	46 (11.6%)	0 (0%)
31–40	60 (15%)	0 (0%)
41–50	200 (50.5%)	63 (100%)
Above 51	90 (22.7%)	0 (0%)
Education		
Primary	51 (12.9%)	0 (0%)
Secondary	78 (19.7%)	3 (4.8%)
Diploma	150 (37.9%)	27 (42.9%)
Degree	90 (22.7%)	24 (38.1%)
Postgraduate/professional qualification	27 (6.8%)	9 (14.3%)
Experience		
Prior SME experience	59 (14.9%)	8 (12.7%)
Professional	177 (44.7%)	27 (42.9%)
Skilled manual	107 (27.0%)	18 (28.6%)
Unskilled manual	22 (5.6%)	4 (6.3%)
Unemployed	31 (7.8%)	6 (9.5%)
Motivation for business start-up		
Desire to be independent	44 (11.1%)	4 (6.3%)
Threat of unemployment	174 (43.9%)	33 (52.4%)
Financial motives	117 (29.5%)	14 (22.2%)
Job satisfaction/self-fulfilment	28 (7.1%)	2 (3.2%)
Attractive lifestyle	3 (0.8%)	0 (0%)
Helping to create jobs	30 (7.6%)	10 (15.9%)
Source of start-up capital		
Family and friend	158 (39.9%)	24 (38.1%)
Savings	86 (21.7%)	11 (17.5%)
Formal sources of finance	83 (21.0%)	19 (6.6%)
Informal source of finance	69 (17.4%)	9 (14.3%)

70.5% in the food industry and 63.5% in the agriculture sector. The finding vis-a-vis firm age shows a staggering age range a high proportion of the business in the food industry were between 11 and 20 years old (28.8%) and above 21 years (29.3%).

Table 2 Demographic information of the firm

Demographic	Food industry	Agriculture
Business structure		
Sole trader	279 (70.5%)	40 (63.5%)
Company	79 (19.9%)	17 (27.0%)
Partnership	38 (9.6%)	6 (9.5%)
Firm age		
1–5	60 (15.2%)	4 (6.3%)
6–10	106 (26.8%)	31 (49.2%)
11–20	114 (28.8%)	8 (12.7%)
Above 21	116 (29.3%)	20 (31%)
Employees		
Less than 10	42 (10.6%)	2 (3.2%)
11–50	353 (89.1%)	61 (96.8%)
Above 50	1 (0.3%)	0 (0%)

There is also a high representation of relatively young businesses; 1–5 years representing 15.2% and 6–10 years representing 26.8%. The findings obtained from businesses in the agriculture sector show that majority of the businesses were relatively young with 31 of the 63 entrepreneurs surveyed stating that they had set up the business in the last 6–10 years. There is also a high percentage of relatively older agriculture businesses representing 31% of the surveyed businesses. An overwhelming majority of the businesses in both the food industry (89.1%) and the agriculture sector (96.8%) employed between 11 and 50 employees.

6.2 Entrepreneur's Awareness of Sustainability Initiatives

The entrepreneurs were asked their perception on sustainability issues and on national and international sustainability initiatives. The findings obtained show a clear lack of awareness amongst many of the entrepreneurs on supply chain sustainability. Two hundred seventy-three (68.9%) of the 398 entrepreneurs surveyed in the food industry and 51 (81%) of the 63 entrepreneurs in the agriculture sector stated that they were not familiar with national sustainability measures, the United Nations Sustainable Development Goals (SDGs) or the Ten principles of the global impact and supply chain sustainability. This finding raises concerns as it contradicts the current publications which boost of the Nigerian Government's effort where global sustainability drive is concerned. For example, the World Bank (2020) reports that Nigeria has made commitments to reduce food loss and waste at both the global level through the sustainable development goals and their nationally determined contribution under the Paris Climate Agreement, as well as regionally through the Malabo Declaration.

The World Bank (2020) report stated that Nigeria had specifically committed vis-a-vis the SDG 12.3 to halve per capita global food waste at the retail and consumer levels and reduce food losses along production and supply chains, including post-harvest losses by 2030. In addition, in relation to the Malabo Declaration the Nigerian government pledges to halve the current levels of post-harvest losses by the year 2025. The implication of this finding is that in spite of many assurances provided by the Nigerian government on its commitment to global sustainability effort, the practical reality is that there is a clear lack of follow-through. It is clear that there is no real willingness by the Nigerian government to engage and extensively invest in sustainability measures as seen in various Western countries. Many entrepreneurs lack basic awareness of sustainability as a construct let alone measures that are needed to ensure effective engagement with international sustainability agenda within their business practice or the wider supply chain. The inertia of the Nigerian government was confirmed by some of the entrepreneurs who took part in the interviews who stated:

It is all talk, it is all posturing. . .see this country has failed to provide good roads, facilities and infrastructure to its people for the past 60 years, sustainability issues are things rich people focus on. Many Nigerians are poor and just looking to get by on a daily basis. . . It is not that they don't want to do it, they cannot afford to do it even if they are aware of it.

Nigeria has not had constant electricity for 60 years; you think it can engage in any credible discussion on sustainability issues on a global scale.
I am only aware of sustainability measure and try to implement this in my business because I was educated in the UK. It is somewhat challenging, and you only have your passion and drive to push you forward. Many Nigerians do not appreciate the measures as they have zero awareness of it.
Nigeria has a long road ahead of it. It is a complete shame that this country can't seem to focus on formulars for success. They would rather steal the money than invest it in either infrastructure or future survival of the next generation.
There is a lack of understanding among many Nigerians of issues such as corporate social responsibility of sustainability. Some people will think it is a scam or a way to collect more money. . . .
To be honest many businesses in Nigeria are crippled by a lack of resources and know how. It is not a lack of willingness; it is often as a result of a lack of understanding or lack of finance and expertise.
The Nigerian Government fails at every angle to engage or invest in initiatives which make life easier for the average Nigerian. We are plagued with bad roads, no infrastructure, electricity and so on. Asking the average business to invest in sustainability measures is added burden which many cannot afford. The Nigerian Government should roll out the incentives and ensure that they incentivize businesses to engage as many countries around the world have done.

The findings also show that many of the entrepreneur's engagement in sustainability initiatives is often done on an informal and ad hoc basis. Majority of the entrepreneurs stated that they engage in cost-effective ways of doing business, reduce the amount of material used in production and packaging, reduce energy consumption, use natural resources and engage in recycling efforts. In addition, the entrepreneurs commented that they often engage in community and religious support initiatives as well as employee welfare initiatives. Many of the entrepreneurs show a clear understanding of the issues surrounding forced labour and child labour and some of the entrepreneurs who took part in the interviews discussed ongoing issues in the country such as kidnappings and the need to ensure better safety and security for all Nigerians.

6.3   Barriers to Supply Chain Sustainability Innovation in the Food Industry

The entrepreneurs in the food industry were asked to identify the main barriers inhibiting supply chain sustainability innovation in the food industry and agriculture sector (See Table 3). The findings show that many of the entrepreneurs identified economic and financial barriers as the main barriers affecting sustainable supply chain innovation (representing 40.7% of the entrepreneurs). Technological barriers were identified as the second main barrier (representing 21.0%) and legal/regulatory and institutional barriers were identified as the third main barrier (representing 18.9%).

In order to better understand the effect of the three barriers on sustainable supply chain innovation, the entrepreneurs in the food industry were asked to rate their agreement about statements on each barrier on a 5-point Likert Scale (1) Strongly Disagree, (2) Disagree, (3) Neutral, (4) Agree and (5) Strongly Agree.

Four statements were used in gauging the entrepreneur's perception on economic and financial barriers namely; lack of capital to carry out innovation activities; high transaction costs; high initial investment in technology and uncertainty about return on investments (See Table 4). The finding shows that the entrepreneurs perceive the high transaction cost involved in sustainable investments as a significant barrier preventing their engagement with an overwhelming 92.4% (representing 63.4% 'Strongly Agree' and 28% 'Agree'). This is followed closely by a lack of capital to carry out innovation activities with a cumulative representation of 80.3% of the entrepreneurs (representing 54.3% 'Strongly Agree' and 26% 'Agree'). Uncertainty about return on investment and high initial investment in latest technology also received high scores with a cumulative total of 70.4% and 70.2%, respectively.

In relation to technology barriers, five statements were used namely lack of technology to facilitate resource optimisation, lack of technical expertise and training, gap between design and implementation of technologies, Lack of R&D and innovation capabilities and finally lack of waste management and recycling facilities.

Table 3 Barriers to sustainable supply chain innovation in the food industry

Main barriers to sustainable supply chain innovation	Food industry
Economic and financial	161 (40.7%)
Technological	83 (21.0%)
Legal/regulatory and institutional	75 (18.9%)
Social and cultural	24 (6.1%)
Organisational	29 (7.3%)
Marketing	5 (7.9%)

Table 4 Economic and financial barriers (food industry)

	Strongly disagree	Disagree	Neutral	Agree	Strongly agree
Economic and financial barriers					
Lack of capital to carry out innovation	18 (4.5%)	54 (13.6%)	6 (1.5%)	103 (26.0%)	215 (54.3%)
High transaction costs	8 (2.0%)	22 (5.6%)	4 (1.0%)	111 (28.0%)	251 (63.4%)
High initial investment in technology	26 (6.6%)	80 (20.2%)	12 (3.0%)	106 (26.8%)	172 (43.4%)
Uncertainty about return on investment	28 (7.1%)	83 (21.0%)	6 (1.5%)	103 (26.0%)	176 (44.4%)

Table 5 Technology barriers (food industry)

	Strongly disagree	Disagree	Neutral	Agree	Strongly agree
Technology barriers					
Lack of technology to facilitate resource optimization	28 (7.1%)	58 (14.6%)	12 (3.0%)	163 (41.2%)	135 (34.1%)
Lack of technical expertise and training	13 (3.3%)	29 (7.3%)	10 (2.5%)	63 (15.9%)	281 (71.0%)
Gap between design/ implementation of technologies	31 (7.8%)	171 (43.2%)	7 (1.8%)	91 (23.0%)	96 (24.2%)
Lack of R&D and innovation capabilities	24 (6.1%)	41 (10.4%)	5 (1.3%)	60 (15.2%)	266 (67.2%)
Lack of waste management and recycling facilities	16 (4.0%)	29 (7.3%)	13 (3.3%)	215 (54.3%)	123 (31.1%)

The findings presented in Table 5 show that an overwhelming majority of the entrepreneurs stated that they 'Agree' or 'Strongly Agree' that lack of technical expertise and training (86.9% cumulative) and lack of waste management and recycling facilities (85.4% cumulative) were significant barriers to sustainable supply chain innovation. In addition, 82.4% cumulative indicated that they 'Agree' or 'Strongly Agree' that lack of R&D and innovation capabilities were impediments to sustainable supply chain innovation. The finding also shows that 41.2% stated that they 'Agree' and 34.1% stated that they 'Strongly Agree' that lack of technology to facilitate resource optimisation is a barrier to entrepreneurs in the food industry (cumulative percentage of 75.3). A cumulative fifty percent of the entrepreneurs

Table 6 Legal regulatory and institutional barriers (food industry)

	Strongly disagree	Disagree	Neutral	Agree	Strongly agree
Legal regulatory and institutional barriers					
Inadequate institutional framework	24 (6.1%)	27 (6.8%)	10 (2.5%)	122 (30.8%)	213 (53.8%)
Lack of incentives	29 (7.3%)	42 (10.6%)	14 (3.5%)	120 (30.3%)	191 (48.2%)
Lack of pressure and non-conducive legal system	32 (8.1%)	39 (9.8%)	4 (1.0%)	93 (23.5%)	228 (57.6%)
Multiple, complex and changing regulations	23 (5.8%)	21 (5.3%)	31 (7.8%)	122 (30.8%)	199 (50.3%)
Red tape and lengthy documentation process	23 (5.8%)	26 (6.6%)	25 (6.3%)	111 (28.0%)	211 (53.3%)

were not convinced that gaps between design and implementation of technologies represent barriers to sustainable supply chain innovation.

As can be seen from Table 6, five statements were used in gauging the entrepreneur's perception on legal regulatory and institutional barriers namely; inadequate institutional framework; lack of incentives; lack of pressure and non-conducive legal system; multiple, complex and changing regulations; red tape and lengthy documentation process. The findings show that a cumulative majority (84.6%) of the entrepreneurs 'Agree' or 'Strongly Agree' that inadequate institutional framework is a major impediment affecting sustainable supply chain innovation in the food industry. A similar percentage of entrepreneurs also 'Agree' or 'Strongly Agree' that red tape and length of document process (81.3% cumulative); lack of pressure and non-conducive legal environment (81.1% cumulative) and multiple, complex and changing regulations (81.1% cumulative) were significant barriers. A cumulative 78.4% also 'Agree' or 'Strongly Agree' that lack of incentives from the Nigerian Government is a significant barrier to sustainable supply chain innovation in the Nigerian food industry.

6.4 Barriers to Supply Chain Sustainability Innovation in the Agriculture Sector

The same three barriers identified by the entrepreneurs in the food industry were identified by the entrepreneurs in the agriculture sector with slight variations.

Table 7 shows that the entrepreneurs in the agriculture sector identified technological barriers as the main barrier to supply chain sustainability innovation in the sector (representing 33.3%). Legal regulatory and institutional barriers were

Table 7 Barriers to sustainable supply chain innovation in the agriculture sector

Main barriers to sustainable supply chain innovation	Agriculture
Technological barriers	21 (33.3%)
Economic and financial barriers	15 (23.8%)
Legal regulatory and institutional barriers	17 (27.0%)
Social and cultural barriers	3 (4.8%)
Organisational barriers	5 (7.9%)
Marketing and networking barriers	2 (3.2%)

Table 8 Technology barriers (agriculture sector)

	Strongly disagree	Disagree	Neutral	Agree	Strongly agree
Technology barriers					
Lack of technology to facilitate resource optimization	6 (9.5%)	10 (15.9%)	1 (1.6%)	22 (34.9%)	24 (38.1%)
Lack of technical expertise and training	3 (4.8%)	6 (9.5%)	3 (4.8%)	16 (25.4%)	35 (55.6%)
Gap between design/ implementation of technologies	8 (12.7%)	18 (28.6%)	4 (6.3%)	23 (36.5%)	10 (15.9%)
Lack of R&D and innovation capabilities	5 (7.9%)	1 (1.6%)	1 (1.6%)	16 (25.4%)	40 (63.5%)
Lack of waste management and recycling facilities	4 (6.3%)	2 (3.2%)	1 (1.6%)	35 (55.6%)	21 (33.3%)

identified as the second main barrier representing 27% of the businesses. The third barrier identified by the entrepreneurs in the agriculture sector was economic and financial barriers representing 23.8% of the surveyed businesses. In order to better understand the effect of the three barriers on sustainable supply chain innovation, the entrepreneurs in the agriculture sector were asked to rate their agreement about statements on each barrier on a 5-point Likert Scale (1) Strongly Disagree, (2) Disagree, (3) Neutral, (4) Agree and (5) Strongly Agree.

The findings presented in Table 8 show that an overwhelming majority of the entrepreneurs in the agriculture sector indicated that they 'Agree' or 'Strongly Agree' that lack of R&D and innovation capabilities and lack of waste management and recycling facilities were major barriers impeding sustainable supply chain innovation representing a cumulative percentage of 88.9, respectively for both statements. Lack of technical expertise and training received the third-highest representation with a cumulative 81%, followed by lack of technology to facilitate resource optimisation with a cumulative 73%. 52.4% indicated that gaps between design and implementation of technologies represent barriers to sustainable supply chain innovation in the agriculture sector (representing 36.5% 'Agree' and 15.9% 'Strongly Agree').

Table 9 Legal regulatory and institutional barriers (agriculture sector)

	Strongly disagree	Disagree	Neutral	Agree	Strongly agree
Legal regulatory and institutional barriers					
Inadequate institutional framework	4 (6.3%)	5 (7.9%)	3 (4.8%)	19 (30.2%)	32 (50.8%)
Lack of incentives	2 (3.2%)	5 (7.9%)	1 (1.6%)	20 (31.7%)	35 (55.6%)
Lack of pressure and non-conducive legal system	4 (6.3%)	8 (12.7%)	3 (4.8%)	11 (17.5%)	37 (58.7%)
Multiple, complex and changing regulations	1 (1.6%)	6 (9.5%)	1 (1.6%)	18 (28.6%)	37 (58.7%)
Red tape and lengthy documentation process	4 (6.3%)	4 (6.3%)	2 (3.2%)	17 (27.0%)	36 (57.1%)

Table 10 Economic and financial barriers (agriculture sector)

	Strongly disagree	Disagree	Neutral	Agree	Strongly agree
Economic and financial barriers					
Lack of capital to carry out innovation activities	5 (7.9%)	5 (7.9%)	4 (6.3%)	23 (36.5%)	26 (41.3%)
High transaction costs	1 (1.6%)	0 (0.0%)	2 (3.2%)	28 (44.4%)	32 (50.8%)
High initial investment in latest technology	8 (12.7%)	7 (11.1%)	4 (6.3%)	25 (39.7%)	19 (30.2%)
Uncertainty about return on investment	8 (12.7%)	7 (11.1%)	4 (6.3%)	23 (36.5%)	21 (33.3%)

As can be seen from Table 9, a cumulative majority of 87.3%, respectively, indicated that they 'Agree' or 'Strongly Agree' that lack of incentives and multiple, complex and changing regulations were major impediments, which pose as barriers to effective sustainable supply chain innovation in the agriculture sector. 84.1% and 81% also 'Agree' or 'Strongly Agree' that red tape and length of document process and inadequate institutional framework are major impediments, respectively. A cumulative 76.2% indicated that lack of pressure and non-conducive legal environment were significant barriers.

The finding presented in Table 10 shows that an overwhelming cumulative percentage (95.2%) perceive the high transaction cost involved with sustainability activities as a major impediment in engaging with sustainability supply chain innovation. This is followed closely by a lack of capital to carry out innovation activities with a cumulative representation of 77.8% of the entrepreneurs (representing 41.3% 'Strongly Agree' and 36.5% 'Agree'). The high initial investment in latest technology represented a cumulative 69.9% and uncertainty about return on investment had a cumulative 69.8%.

The results obtained from the entrepreneurs find support in extant literature. For example, many researchers have presented findings consistent with the fact that lack of technology which facilitates resource optimisation and lack of skill, expertise and training and technological advancement are major impediments to sustainable supply chain innovation (Stewart et al., 2016; De Jesus & Mendonça, 2018). Technological capability determines a firm's level of innovation. SMEs in Nigeria struggle to adopt technology due in part to their lack of competency and cost implications associated with technology. In some cases, it can be argued that familiarity with traditional technologies or usual customs and practices prevents people from wanting to invest in sustainable products or deviate from the norm (Kirchherr et al., 2018). There is evidence in the academic literature to show remarkable differences between SMEs that embrace innovation and those that do not. The attitude of a business to technology has been shown to determine their level of success and innovation. Innovation helps SMEs increase the quality of their products and maintain their performance levels. SMES operate in a globalised, highly competitive and interconnected world, therefore, their ability to embrace technological advancements and utilise these to be more innovative is crucial to their survival and growth. In addition, the creation and management of recycling facilities are also integral to ensuring success and a lack of such initiative could hamper real progress (Stewart et al., 2016; De Jesus & Mendonça, 2018; Gupta et al., 2020).

Some of the interviewees who took part in the study also stated:

> It requires a lot of capital investment and technology which you will need to import in some cases to ensure good quality. . .You also need to train people and engage in monitoring and so on. There is no benefit for now. We are all trying not to fail.
>
> Lack of money makes it much more difficult to include any sustainable measures.
>
> Sustainability is expensive, it will require acquisition of machinery and also training. It is not that one does not want to engage. It involves a high investment outlay. . . especially when you consider the high cost of living and the high cost of doing business in Nigeria.
>
> It is more work and to be honest, it can be a thankless task as many people are not aware.
>
> It is not as simple as mentioning sustainability without talking about other aspects as well such as increased cost, training of employees on any new technology or initiative that you have implemented and also where this involves customer use, training the customers as well. These are things that we need to consider, where will the money come from.
>
> . . .Entrepreneurs in Nigeria are already suffering and have to cope with obstacles that their counterparts don't have in some of the developed countries that you are coming from. . .these added complexities are what stops Nigerian entrepreneurs from fully engaging with sustainability measures.
>
> Some Nigerian entrepreneurs should be commended; they have adopted sustainable strategies they have acquired from Europe, US and UK and have implemented them in their business practices here in Nigeria. More needs to be done to help them continue to evolve and to pass on some of the wealth of knowledge to other entrepreneurs.

The perception of the Nigerian entrepreneurs vis-a-vis financial, legal regulatory and institutional barriers as a major impediment to sustainable supply chain innovation also find support in academic literature. Sound legal regulatory systems and favourable regulations and policies are necessary for the effective functioning of markets. Regulations and policy initiatives should be designed and implemented in a way to benefit society, enhance shared value and set goals that stimulate innovation. In addition, they should highlight societal objectives, set performance standards, define phase-in periods for meeting standards and put in place performance reporting systems with government investing in the required infrastructure for implementation and monitoring. Thus, creating a level playing field to encourage companies to invest in shared value rather than maximize short-term profit (Porter & Kramer, 2011). Various studies have arrived at findings consistent with the view that lack of initiatives, lack of pressure from regulatory bodies, non-conductive legal regulatory systems as well as inadequate financial resources to invest and associated cost of technology are major impediments to sustainable practices (Stewart et al., 2016; Gupta et al., 2020).

The interviewees also helped further shed light on the findings obtained from the questionnaires. Some of the responses provided include:

> How do you force your suppliers to also engage? It is a two-way street. I think for the most part many businesses in Nigeria try not to do anything that they think might affect the society or the environment negatively. Many of the actions carried out which are not sustainable are either done as a result of lack of awareness... they are not intentional.
>
> The Cost is too high...The Nigerian government needs to provide what is needed, that is their job, often times available funds are stolen by these corrupt bureaucrats.
>
> ...It is a shame, there is too much red tape, lack of credible sustainable framework and weak monitoring of any initiatives.
>
> Nigerian environment is not conducive for entrepreneurs. It stifles growth and advancement. A lot of the entrepreneurs' resources is depleted by unnecessary expenditure...for example the heavy investment required in providing 24-hour access to electricity.
>
> ...Also, some people stick to what they know so they are not really concerned about sustainable technology, due in part to a lack of understanding and added cost implications.
>
> Nigerian government needs to do more to foster the entrepreneur's uptake of sustainability measure in their business practices. Our regulatory bodies are reactive rather than active...many of the people who are supposed to be aware of sustainability issues in government do not have a clue....
>
> There are no sustainability initiatives really, it is serious lip service... There is no one to implement measure. Many of the time, the people who are tasked with creating these initiatives and promoting uptake are the ones who steal them and save it for themselves or their family. They are a bunch of corrupt civil servants who continue to prevent any advancement in this country.
>
> ...It is due largely to weak legal, regulatory and institutional framework; lack of action by the government on sustainability initiatives and endemic bureaucratic practices and red tape...Nigeria requires a complete rethink of everything, it is a disgrace what is allowed to happen in the country.

6.5   Strategies to Overcome Barriers

The entrepreneurs suggested a number of strategies to overcome the barriers (See Table 11). Amongst the entrepreneurs in the food industry, the three main strategies identified include sustainable technology development (26.3%); legal regulatory and environmental strategy (23.5%); networking (22.2%). The entrepreneurs in the agriculture sector however identified four main strategies namely networking

Table 11 Strategies to overcome barriers

Strategies to overcome barriers		
Sustainable technology development	104 (26.3%)	14 (22.2%)
Environment and legal regulatory	93 (23.5%)	14 (22.2%)
Networking/collaboration	88 (22.2%)	15 (23.8%)
R&D	39 (9.8%)	9 (14.3%)
Economic initiatives/incentives	29 (7.3%)	3 (4.8%)
Marketing/promotion	25 (6.3%)	2 (3.2%)
Skill development	18 (4.5%)	6 (9.5%)

(22.8%), legal regulatory and environmental strategy (22.2%), sustainable technology development (22.2%) and R&D (14.3%).

Many of the entrepreneurs interviewed advocated for better networking among Nigerian entrepreneurs. This view finds support in the academic literature, for example Gupta et al. (2020) recognises the importance of networking in order to build collaborative competence of the firm and also between external organizations and institutions through information and technology exchange, training workshops and research and development. This cooperation will enable entrepreneurs in Nigeria to adopt initiatives to promote sustainable development thereby better preparing them to participate on a global scale and allowing them to engage more with international global sustainability initiatives such as the United Nations SDGs (Griggs et al., 2013). Some of the interviewees stated:

Networking is the key if you are looking to ensure Nigeria embraces sustainability. . .it is a misconception to think that Nigerians are not aware of sustainability issues. Many Nigerians who were educated in the West have returned back and are actively transferring technology back to Nigeria. . .but how will you know if there is no avenue to engage with these people.

This is where some of these Non-Governmental Organisations need to focus on, bringing like-minded entrepreneurs together so that there can be easy exchange of information and expertise. . .it can also allow some form of mentoring to occur between entrepreneurs.

Networking is important, but some Nigerians are scared to engage. You can't blame them really, there are security issues and also some might be afraid that others might steal their idea. There has to be a neutral platform, where honesty and trust exist to allow collaborative endeavours to flourish.

There is a need to ensure avenue for information on sustainability is created to spread across various businesses and industry in Nigeria, for example the agriculture sector could benefit a lot from information on new sustainable technologies.

. . .How can we benefit if we don't know?

Available government initiatives must be made public, no nepotism in distribution of such initiatives as is often the case here in Nigeria, it is shared by government officials to their friends and families. . . A hub (network) would be useful to enable effective dissemination of information and collaboration.

The result obtained on legal regulatory and environmental strategy as a way to overcome barriers is to be expected. There is a need to ensure legal regulatory and environmental policies are in place to promote sustainability practices not only in achieving sustainable supply chain innovation amongst the food and agriculture industry but for entrepreneurship as a whole in Nigeria regardless of sector/industry. The policies could take the form of tax exemptions, tax cuts, infrastructure support, process innovation, access to latest green and sustainable technologies (including intellectual property development) and recycling and waste management support (Gupta et al., 2020). One of the entrepreneurs who took part in the interview stated:

> The fact remains Nigeria is in need of policy initiatives that work. . .Such policies could be targeted specifically to businesses in certain sectors/industries e.g. like you are doing focusing on the agriculture sector and food industry. It could also be targeted at specific entrepreneurs e.g. high growth or highly innovative entrepreneurs.

The interviews carried out with the entrepreneurs helped confirm the need for an effective strategy aimed at helping entrepreneurs acquire latest technologies and helping them develop technological skills and competencies which are crucial to sustainable development and innovation. There are incentives available in Nigeria which seeks to encourage entrepreneurship, encourage acquisition of technologies and reward entrepreneurs for recycling and reusing facilities within the business for example:

- Pioneer Status: Tax holiday provided for 7 Years. Business must be included in the list of pioneer industries.
- Investment Tax Credit: New machinery bought to replace an old machinery attracts 15% investment tax credit; a business that purchases locally manufactured plant, machinery for business use is allowed 15% on the fixed asset; companies engaged in research and development for commercialisation are allowed 20% investment tax credit on their expenditure.
- Rural Investment Allowance: Afforded to businesses that set up their business and provide electricity, water, tarred roads telephone service within 20 kilometres.
- Labour Intensive mode of production: Tax concessions for 5 years; graduated rate, e.g. 100 employees = 6%, 200 employees = 7%.
- Export Free Zone: 100% profit of export-oriented business established within an export-free zone exempt from tax for the first 3 years.

Many of the entrepreneurs were not aware of the existence of some of the incentives on offer and as such are unable to benefit from them. Various regulatory bodies and government parastatals should endeavour to promote initiative on offer and encourage businesses to engage. Some of the entrepreneur's interviews stated:

> It is not a lack of initiative, there are a host of initiative on offer, the question is whether people are aware.
> Many Nigerians are intentionally kept in the dark so that they do not benefit from the initiatives on offer. . .many of our leaders are cruel to the people, it is intentionally done to keep people subjugated.
> I am not aware of any of the allowance and incentive's, this is not a good way to treat people, by not giving them adequate information which can help them excel in business you

make it easy for them to fail... The Nigerian Government needs to do more to ensure better promotion of the incentives on offer so people can benefit.

7   Conclusion

The findings show that many of the entrepreneurs were unaware of international sustainability frameworks and many of the sustainability practices adopted were often informal and on an ad hoc basis. The primary barrier affecting supply chain sustainability innovation among the Nigerian entrepreneurs surveyed are economic and financial barriers, technological barriers and legal regulatory and institutional barriers. The study makes empirical contributions to the burgeoning literature in this area thus filling gaps in the literature on barriers to supply chain sustainability in the SME sector in Africa. The potential implication for Nigeria specifically and Africa more generally is that countries on the African continent lag behind vis-a-vis advancement in sustainability measures. Entrepreneurial leadership and skill are significantly important in achieving sustainability innovation within the SME sector. Better education amongst businesses on the continent in order to bring them in line with current thinking in the field is required. The entrepreneurs involved in the study advocate for sustainable technology development, legal regulatory and environmental strategies as well as enhanced networking as ways of overcoming some of the barriers identified. The study makes meaningful contributions to the existing body of literature on sustainability and entrepreneurship in Nigeria by identifying and developing a framework of barriers and strategies to overcome the barriers identified based on extant academic literature and empirical contributions from the entrepreneurs who took part in the research. The study framework presented in Fig. 2 is composed of the three main barriers namely technology, regulatory, economic and financial barriers. A description of the barriers and questions to consider by the entrepreneur in ensuring the business is well equipped to engage in sustainable supply chain innovation is included. Finally, the framework proposes action plans for both the entrepreneur and the government.

There is a need to ensure that entrepreneurs include sustainability goals in the initial stages of their operations. These goals can then be extended as the business grows across the supply chain with a view to develop a collaborative partnership with suppliers to help convey the entrepreneurs' core sustainability value and commitment whilst also emphasising the importance of sustainability agenda across the supply chain. The importance of active participation in industry collaborations cannot be overemphasised as evidenced by the academic literature and responses obtained from the entrepreneurs who took part in the study. This strategy helps in giving the sector/industry and entrepreneurs a common voice and also helps in leveraging their collective influence in the supplier chain. Such collaboration could include access to materials on vital national and international principles and areas of priority as well as sustainability initiatives available to entrepreneurs.

SUPPLY CHAIN SUSTAINABILITY INNOVATION FRAMEWORK			
ELEMENT	DESCRIPTION	INITIAL QUESTIONS TO CONSIDER BY THE ENTREPRENEUR	ACTION PLAN
TECHNOLOGY	- Recognition and acquisition of technology to facilitate resource optimization - Entrepreneur and employee technical expertise and skill/ Training and development - R&D and innovation capabilities	- Is the Entrepreneur aware of current technological developments in the field? - Is the technology relevant to both the business and the supply chain? - Does the integration of the technology into the supply chain take into consideration the triple-bottom-line of profit, people and planet? - Will the technology ensure resilience across the supply chain? - Are measures in place to ensure the entrepreneur and employees possess technical expertise and skills needed? - Is there an internal framework/policy for R&D, training and development? - Is there a policy on production of environmentally friendly products, recycling and waste management? - Are measures in place to assess and manage potential risks? - Is the supply chain transparent enough in order to have a clear understanding of who the suppliers are, their location where they source the products from and their exposure to risk?	**Entrepreneur** - Networking to building collaborative competence of the firm, external organizations/institutions and the society through information and technology exchange, training workshops and R&D - Engagement with stakeholders and local community on sustainability agenda - Dedicate resources to supply chain sustainability and integrate sustainability criteria into procurement and supply chain - Capacity building and knowledge sharing across the business and with employees and suppliers - Monitoring framework to assess performance and reporting performance information **Government** - Incentives which seeks to encourage entrepreneurs acquire technologies, reward them for recycling and reusing facilities within the business - Promote the incentives on offer through the regulatory bodies and NGO's so entrepreneurs are aware
LEGAL REGULATORY AND INSTITUTIONAL	- Conducive legal, regulatory and Institutional framework/systems - Available sustainability Incentives - Consistent and clear regulations - Reduced red tape and streamlined documentation process	- Is the entrepreneur aware of regulations, policies and initiatives? - Are measures in place to ensure the business, employees and consumers comply with regulations and policies?	**Entrepreneur:** - Identify available sustainability initiatives - Secure business allowance and reliefs - Networking with other entrepreneurs to ensure up to date information on regulations and available incentives **Government:** - Adequate institutional framework - Introduction of sustainability incentives which encourages businesses to consider sustainability dimensions - Consistent and clear regulations - Conducive legal system - Pressure from regulatory bodies - Reduced red tape - Streamlined documentation process
ECONOMIC AND FINANCIAL	- Initial and Subsequent Capital for innovative activities	- Is the entrepreneur aware of the initial and subsequent investment cost of the technology? - Has the entrepreneur carried out a projection to ensure he/she is certain of the return on investment? - Are measures in place to reduce transaction costs without compromising product quality or customer satisfaction?	**Entrepreneur:** - Identify and secure available subsidies, grants, loans for sustainability activities - Identify and secure business rate cuts, tax relief and allowance **Government:** - Provision of business rate cuts, tax relief and allowance - Provision of subsidies, grants, loans for sustainability activities

Fig. 2 Supply chain sustainability innovation framework in Nigeria. Source: Author

8 Recommendations

The recent global pandemic has shown a brighter light on sustainability and the need to ensure that 'sustainability' remains at the forefront of global agenda. The pandemic has also uncovered vulnerabilities from both government and industry perspectives across the globe. The recommendations proposed below take into consideration the impact of the current pandemic on the global supply chain in better helping entrepreneurs invest in more sustainable businesses and industries which are resilient to disruptions.

8.1 Entrepreneur

- Dedicate resources to supply chain sustainability and integrate sustainability criteria into procurement and supply chain.
- Capacity building and knowledge sharing across the business and with suppliers.
- Networking with other entrepreneurs to ensure up-to-date information on regulations and available incentives. Networking to ensure collaborative

competence of the firm, external organizations and institutions through information and technology exchange, training workshops and R&D.

- Identify and secure available grants and subsidies, business rates, tax relief and loans for sustainability activities.
- Reassessment of supply chain risks in order to ensure resilience.
- Ensure supply chain transparency in order to have a clear understanding of who the suppliers are, their location where they source the products from and their exposure to risk.

8.2   Government

Governments and NGOs will be most effective if they think in value terms and focus on measuring environmental performance and introducing standards, phase-in periods, and support for technology that would promote innovation, improve the environment, and increase competitiveness simultaneously. They should consider the introduction of:

- Incentives which seek to encourage entrepreneurs to acquire technologies, reward them for recycling and reusing facilities within the business and ensure they promote the incentives on offer through the regulatory bodies and NGOs, so entrepreneurs are aware.
- Grants and subsidies, business rates, tax relief and loans for sustainability activities.
- Adequate legal, regulatory and institutional framework, consistent and clear regulations, reduced red tape and streamlined documentation process.

9   Limitation and Area for Further Studies

Future studies could explore barriers to sustainable supply chain innovation in other sectors so that the issue of variance in the extent of reported barriers can be better comprehended, for example potential influence of the three remaining factors namely social and cultural barriers, organisational barriers and marketing and networking barriers. It would have also been beneficial to have carried out more interviews with the entrepreneurs as their perceptions proved insightful in better understanding some of the findings obtained. A limitation of this study is that it relates to research based on exploring the barriers to supply chain sustainability at a specific time thus may not provide a holistic picture. Especially in light of the pandemic, a clearer picture can be obtained if examined over a prolonged period of time taking into consideration pre- and post-pandemic activities.

References

Babbie, E. (2010). *The basics of social research* (5th ed.). Cengage Learning, Wadsworth Publishing Company.

Bansal, P., & McKnight, B. (2009). Looking forward, pushing back and peering sideways: Analyzing the sustainability of industrial symbiosis. *Journal of Supply Chain Management, 45*, 26–37.

Bell, J. (1993). *Doing your research project*. Open University Press.

Burns, R. B. (2000). *Introduction to research methods* (4th ed.). Pearson Education.

Bryman, A., & Bell, E. (2011). *Business research methods* (3rd ed.). Oxford University Press.

Cai, W. G., & Zhou, X. L. (2014). On the drivers of eco-innovation: Empirical evidence from China. *Journal of Clean Production, 79*, 239–248.

Carter, C. R., & Easton, P. L. (2011). Sustainable supply chain management: Evolution and future directions. *International Journal of Physical Distribution Logistics and Management, 41*, 46–62.

Cooper, D. R., & Schindler, P. S. (2011). *Business research methods* (11th ed.). Irwin Publishers.

Creswell, J. (2007). *Mail and internet surveys: The tailored design method* (7nd ed.). Wiley & Sons.

Cruz, N., & Marques, R. (2014). Scorecards for sustainable local governments. *Cities, 39*, 165–170.

De Jesus, A., & Mendonça, S. (2018). Lost in transition? Drivers and barriers in the eco- innovation road to the circular economy. *Ecological Economics, 145*, 75–89.

Elkington, J. (1994). Towards the sustainable corporation: Win-win-win business strategies for sustainable development. *California Management Review, 36*(2), 90–100.

Elkington, J. (2018). *Harvard business review*. *"25 Years Ago I Coined the Phrase 'Triple Bottom Line.'* Here's Why It's Time to Rethink It." Accessed 01 October 2020.

EU Data Trend. (2019). *Data and trend EU food and drink industry*. Food Drink Europe.EU.

EY. (2016). *The state of sustainable supply chains: Building responsible and resilient supply chains*. The United Nations Global Compact.

Fawcett, S. E., & Waller, M. A. (2011). Cinderella in the C-suite: Conducting influential research to advance the logistics and supply chain disciplines. *Journal of Business Logistics, 32*, 115–121.

Gill, J., & Johnson, P. (2010). *Research methods for manager* (4th ed.). Paul Chapman.

Griggs, D., Rajan, A. J., Stafford-Smith, M., Gaffney, O., Rockström, J., Öhman, M. C., Shyamsundar, P., ... Noble, I. (2013). Policy: Sustainable development goals for people and planet. *Nature, 495*(7441), 305.

Guest, G., Bruce, A., & Johnson, L. (2006). How many interviews are enough? An experiment with data saturation and validity. *Field Methods, 18*(1), 59–82.

Gupta, H., Kusi-Sarpong, S., & Rezaei, J. (2020). Barriers and overcoming strategies to supply chain sustainability innovation. *Resources, Conservation & Recycling, 161*.

Gupta, H., & Barua, M. K. (2018). A framework to overcome barriers to green innovation in SMEs using BWM and fuzzy Topsis. *Science and Total Environment, 633*, 122–139.

Jamali, D. (2009). Constraints and opportunities facing women entrepreneurs developing countries. *Gender Management Journal, 24*(4), 232–251.

Kirchherr, J., Piscicelli, L., Bour, R., Kostense-Smit, E., Muller, J., Huibrechtse-Truijens, A., & Hekkert, M. (2018). Barriers to the circular economy: Evidence from the European Union (EU). *Ecology Economics, 150*, 264–272.

Kuratko, D. F., & Hodgetts, R. M. (2007). *Entrepreneurship: Theory and process, practice* (7th ed.). Thompson South Western Publishing.

Lincoln, A., Adedoyin, O., & Croad, J. (2020). Goal 16 of the UN Sustainable Development Agenda and Its Implications for Effective Governance and Sustainability in the Nigerian Banking Sector. In S. O. Idowu, R. Schmidpeter, & L. Zu (Eds.), *The future of the UN sustainable development goals: Business Perspectives for global development in 2030*. Springer.

Lincoln, A. (2017). Corporate social responsibility in challenging times: A consideration of how small and medium scale enterprises attempt to Deal with CSR challenges in Nigeria. In S. O.

Idowu & Stephen Vertigans Adriana Schiopoiu Burlea (Eds.), *Corporate social responsibility in times of crisis. Practices and case from Europe, Africa and the world.* Springer.

Lincoln. (2018). Nigerian entrepreneurship in the twenty-first century: Corporate social responsibility challenges and coping strategies of Nigerian male and female entrepreneurs. In D. Opoku & E. Sandberg (Eds.), *Challenges to African entrepreneurship in the 21st century* (pp. 117–146). Palgrave Macmillan.

Lincoln, A., Adedoyin, O., & Croad, J. (2016). Fostering Corporate Social Responsibility Among Nigerian Small and Medium Scale Enterprises. In S. O. Idowu (Ed.), *Key initiatives in corporate social responsibility: Global dimension of CSR in corporate entities.*

Luchs, M. G., Naylor, R. W., Irwin, J. R., & Raghunathan, R. (2010). The sustainability liability: Potential negative effects of ethicality on product preference. *Journal of Marketing, 74,* 18–31.

National Bureau of Statistic. (2020). *Nigerian unemployment rate.* Available at: http://www.tradingeconomics.com/nigeria/unemployment-rate

NIPC. (2016, October 22). *Nigerian Investment promotion commission.* Nigeria's food industry worth over $3.2 billion. Available at: https://nipc.gov.ng/2016/10/22/nigerias-food-industry-worth-3-2-billion/

OECD Investment Policy Reviews OECD Investment Policy Reviews: Nigeria. (2015). *OECD publishing.* Accessed Nov, 2020, from https://books.google.co.uk/books?id1/4A_BzCQAAQBAJ&pg1/4PA248&lpg1/4PA248&dq1/4SupervisoryþInterventionþGuidelineþ%282011%29þNigeria&source1/4bl&ots1/4oo_3dSIWPG&sig1/4pVrAinBnKTam4hNLW334daSbJR4&hl1/4en&sa1/4X&ved1/40ahUKEwiT3dDts8rLAhUI0RQKHRpHCa8Q6AEIPDAF#v1/4onepage&q1/4Supervisory%20Intervention%20Guideline%20%282011%29%20Nigeria&f1/4false

Oppenhiem, A. N. (1992). *Questionnaire design and attitude measurement.* Heinemann Publishers.

Pantouvakis, A., Vlachos, I., & Zervopoulos, P. D. (2017). Market orientation for sustainable performance and the inverted-U moderation of firm size: Evidence from the Greek shipping industry. *Journal of Cleaner Production, 165,* 705–720.

Porter, M. E., & Kramer, R., (2011). *The big idea creating shared value.* [pdf] Accessed Oct 10, 2020, from https://files.transtutors.com/cdn/uploadassignments/2703816_3_shared-value-harvard-business-review.pdf

Rezaee. (2018). Supply chain management and business sustainability synergy: A theoretical and integrated perspective. *Journal of Sustainability.* MDPI. Accessed Oct 10, 2020, from file:///Users/adebimpelincoln/Downloads/SSRN-id3148737.pdf

Saunders, M., Lewis, P., & Thornhill, A. (2015). *Research methods for business students* (7th ed.). Pitman.

Statista Research Department. (2020). *US food retail industry–statistics and facts.*

Stewart, R., Bey, N., & Boks, C. (2016). Exploration of the barriers to implementing different types of sustainability approaches. *Procedia CIRP, 48,* 22–27.

Economy Watch. (2010). *Food industry, food sector, food trade.*

UN. (2015). *Food and agriculture in Nigeria: Nigeria at a glance.* Available at: http://www.fao.org/nigeria/fao-in-nigeria/nigeria-at-a-glance/en/

UN Global Compact. (2015). *Supply chain sustainability: Practical guide for continuous improvement.* 2nd Edition.

UK National Statistics. (2020). *Food statistics in your pocket: Food chain.* Department for Environment Food and Rural Affairs. Gov.UK.

Woldie, A., Leighton, P., & Adesua, A. (2008). Factors influencing small and medium enterprises (SME's) in Nigeria: An exploratory study of owner/manager and firm characteristics. *Journal of banks and Bank System, 3.*

World Bank. (2020). *Nigeria: Food smart country diagnostic: 2020 International Bank for Reconstruction and Development/The World Bank.* Available at http://documents1.worldbank.org/curated/en/703791601302657183/pdf/Nigeria-Food-Smart-Country-Diagnostic.pdf

World Commission on Environment Development. (1987). *Our common future.* Oxford University Press.

Adebimpe Lincoln , **PhD**, holds an LLB and an LLM in Commercial Law from Cardiff University and an MBA in International Business and a PhD in Entrepreneurship and Business Development from the University of South Wales. She holds a Postgraduate Diploma in Legal Practice and also holds various teaching qualifications in Higher Education. She is a Fellow of the Higher Education Academy and an Associate Member of the Chartered Institute of Personnel Development. She has 17 years' experience in Higher Education. She has held positions at Cardiff University, University of South Wales, and was a Senior Lecturer in Law at Cardiff Metropolitan University before taking up a position as an Assistant Professor in Saudi Arabia. She currently works with the University of Liverpool online Master's in law Programmes and is an Adjunct Professor with Robert Kennedy College Switzerland.

Adebimpe Lincoln has vast experience supervising research students. She has supervised a wide range of master's level research including MSc, MBA and LLM dissertations. She has also supervised students at Doctoral Level in the area of Entrepreneurship and Corporate Governance. Her research interests lie in the area of Female Entrepreneurship, Small and Medium-Sized Enterprise (SME) sector, Leadership Practices of Nigerian Entrepreneurs, Corporate Governance and Board Diversity, Sustainability and Corporate Social Responsibility among Nigerian SMEs. She has published and presented a number of articles on Leadership, CSR, Governance and Entrepreneurship in Nigeria.

Sustainable Supply Chains in Bolivia: Between Informality and Political Instability

Boris Christian Herbas-Torrico, Björn Frank,
Carlos Alejandro Arandia-Tavera, and Pamela Mirtha Zurita-Lara

Abstract

Nowadays, Bolivia is experiencing big social and economic changes. Due to strong economic growth and deep political changes, the Bolivian society has high expectations for a better future. Multiple factors, most rooted in changing consumer realities, are driving the awakening to the importance of the management and sustainability of supply chains in developed and developing countries. More consumers are also looking for reassurance from manufacturers that their purchased goods are sourced and produced in an environmentally, socially sustainable, and safe fashion. As a consequence, more national regulations are expected to appear in the future to reflect these new consumer demands. Due to social and political instability, attaining supply chain sustainability in Bolivia is far from realistic. However, manufacturers should take several steps to address some of the factors that limit their control over supply chains. Bolivian firms will have to reduce suppliers' complexity, improve supply chain transparency, integrate third parties, and improve material procurement practices. Therefore, our study analyzes sustainable supply chain management practices in the foods and beverage industry of Bolivia. In this chapter, we study the case of Bolivia and attempt to summarize the challenges and opportunities faced by Bolivian firms in the development of sustainable supply chains. In particular, we collected data from nine foods and drink manufacturers of the foods and beverage industry. We find that the implementation of sustainability practices in Bolivian supply chains is still at an early stage. Moreover, due to high political instability and a huge

B. C. Herbas-Torrico (✉) · C. A. Arandia-Tavera · P. M. Zurita-Lara
Universidad Católica Boliviana "San Pablo", Cochabamba, Bolivia
e-mail: bherbas@ucb.edu.bo

B. Frank
Waseda University, Tokyo, Japan

© The Author(s), under exclusive license to Springer Nature Switzerland AG 2022
S. O. Idowu, R. Schmidpeter (eds.), *Case Studies on Sustainability in the Food Industry*, Management for Professionals,
https://doi.org/10.1007/978-3-031-07742-5_3

informal economy, supply chains in Bolivia are highly uncertain and extremely challenging to manage.

Keywords

Sustainable supply chain · Bolivia · Foods and beverage industry · Informality · Political instability

1   Introduction

Globalization has increased the dependency on global supply chains that stretch between developed and developing countries. These countries have different economies, legislations, regulations, and standards, where developing countries supply raw materials to industrialized developed countries. Supply chains in developing countries face problems such as political instability, corruption, labor-intensive industries, deteriorated or inexistent infrastructure, low availability of innovations, child labor, underemployment, and low education of their populations (Akamp & Müller, 2013). As a consequence, developed countries pressure developing countries to solve these problems and develop transparent and sustainable supply chains. Bolivia is one of the least developed countries in the Americas and thus suffer from many of these problems. In this chapter, we study the case of Bolivia and attempt to summarize the challenges and opportunities faced by Bolivian firms in the development of sustainable supply chains. In particular, we collected data from nine manufacturers of the foods and beverage industry. This chapter is organized as follows. In the next section, we present information regarding the Bolivian political and economic context. Unlike in other developing countries, due to constant political instability, supply chains in Bolivia face constant challenges when offering their products to local and international markets. Afterward, we present information regarding the characteristics of the business environment in the country. Subsequently, we present information on the sustainability of nine supply chains of the foods and beverage industry. Finally, we summarize our findings and suggest pathways for improving the sustainability of supply chains in Bolivia.

2   The Political Context of Bolivia

Over the years, Bolivia has been characterized by constant political instability and economic inequality (Herbas-Torrico et al., 2021). The history of political changes and inequality dates to its foundation as a country. However, the modern era is of particular interest (the 1990s to date) because Latin American economies were experiencing a boom in elected liberalist politicians (e.g., Carlos Menem in Argentina and Fernando Collor de Mello in Brazil) and because of the growing importance of global supply chains. In the case of Bolivia, in the 1990s, President Gonzalo Sanchez de Lozada (1993–1997) adopted economic liberalism and gave support to free-market economic policies and to private property as the means of production. During this period, Bolivian state-owned firms were partially privatized (1996) with the state retaining a noncontrolling ownership share in privatized assets (*capitalización*). However, these economic policies were badly implemented and led to unemployment, massive poverty and inequality, and thus political instability. As a consequence, in the year 2005 growing political instability and violent protests led to the resignation of Mr. Sanchez the Lozada and the election of the far-left politician Evo Morales. Among the new economic policies given by Mr. Morales were a new constitution, nationalization of privatized state firms, higher control of interest rates, and open criticism of free trade agreements. This presidency coincided with an increase in the international prices of raw materials and thus higher income for state firms dedicated to the exploitation and trade of raw materials (United Nations, 2005). Consequently, the Bolivian economy started to grow to unprecedented levels with lower unemployment and significant poverty reduction (Herbas-Torrico et al., 2021). Despite positive macroeconomic and microeconomic conditions, political instability rose again in 2019 due to accusations of contested elections. In November 2019, Mr. Morales resigned and the presidency was claimed by a member of the political opposition: Jeanine Añez. Her rise to power coincided with the first wave of the COVID-19 pandemic. In an attempt to control the number of infections, she implemented a strict quarantine. The economy was hit hard and political instability rose again. Subsequently, new elections in December 2020 brought to power the current president: Luis Arce. Mr. Arce is the former Minister of Economics during Morales's presidency and his chosen successor.

As previously suggested, the history of Bolivia is filled with political instability, which influences the firm's supply chain operations. Particularly, in an operational context, political instability leads to increased supply chain uncertainty and excessive amounts of buffer stocks. Moreover, political instability causes underemployment and an increase in the informal economy (Webb et al., 2013; Elbahnasawy et al., 2016). Informal economies, such as in Bolivia, show a lack of transparency and accountability in their supply chains. These supply chains are characterized by labor abuses such as unsafe work conditions, child labor, mandatory overtime work, payment of less than the minimum wage, abusive discipline, sexual harassment, or violation of labor laws and regulations (Huq et al., 2014; Jia et al., 2018). Moreover, Bolivia has the biggest informal economy in the world, which comprises 62.3% of its GDP (Medina & Schneider, 2018). Therefore, we study the degree of sustainability of supply chains in Bolivia as a perennially unstable country with the largest informal economy in the world.

3   The Economic Context of Bolivia

The Bolivian economy is not industrialized and has few large firms and a significant number of small firms (Morales, 2020). In Bolivia, eight out of ten jobs are created in small firms, which contribute to 83% of Bolivian employment (Opinión, 2017). Most of these jobs are in microenterprises and related to self-employment without any social benefits. In other words, workers do not enjoy benefits such as health care rights, retirement, protection against accidents and occupational diseases, maternity leave, and sexual harassment protection. Moreover, many of the formal and informal small, medium, and large firms have a few workers working outside labor laws with low wages (Morales, 2020). Consequently, due to the size of the informal economy and worker defenselessness, the Bolivian economy is vulnerable to persistent political instability and exhibits irregular growth patterns (Hussain, 2014). Furthermore, due to these problems, innovation and entrepreneurship take a backseat precluding any form of competition. For example, to start a formal business in Bolivia, one needs at least 45 days (Del Castillo, 2020). These kinds of disincentives, in the form of regulations and administrative procedures, cause higher economic informality, lower productivity, and higher costs (Strobel, 2010). Therefore, Bolivia has one of the least competitive economies in the world (World Economic Forum, 2019).

From an industrywide perspective, compared to other industries, the foods and beverage industry is the industry that adds the most value to raw materials (Ministerio de Desarrollo Productivo y Economía Plural, 2019). Around 9% of the employment in Bolivia is concentrated in this industry (INE, 2020). Moreover, Morales (2020) suggests that this industry is characterized by strong economic relationships with suppliers from the informal economy. Therefore, for our research, we sampled industries from the foods and beverage industry, which reveals important information about sustainable supply chain management practices in Bolivia.

4   Sustainable Supply Chains in Developing Countries

Developing countries usually play the role of raw material producers or manufacturers, and face issues that influence the efficiency of their supply chains. These countries face different problems such as instability of their governments and policies, corruption, labor-intensive industries, lack of infrastructure, limited access to technology, underemployment, child labor, and a low education level (Akamp & Müller, 2013). By contrast, in developed countries, due to customer pressure and regulations, a primary goal is to achieve the sustainability of supply chains. Sustainable Supply Chain Management (SSCM) is defined as the management of material, information, and capital flows as well as cooperation among companies along the supply chain while integrating goals from all three dimensions of sustainable development. These three dimensions consist of economic, environmental, and social goals, which are derived from customer and stakeholder requirements (Seuring & Müller, 2008). Coordination between supply chain members is important to achieve sustainability goals. All members need to comply with environmental and social priorities to retain their status and place in the supply chain, while competitiveness is accomplished by fulfilling consumer requirements and economic factors. The inability of one stage, or player, in the supply chain would affect the supply chain's overall efficiency and productivity. Compared to developed countries, developing countries face more challenges because their economic gains tend to be dependent on the extraction of natural resources. Specifically, rapid urbanization in developing countries and rising living conditions bring about dilemmas and challenges that are difficult for global supply chains to take into account (Jayanti & Rajeev Gowda, 2014). In general, accelerated economic growth and alleviation of poverty take priority over the conservation of the environment (Ben Brik et al., 2013). Consumers are more concerned with meeting their fundamental needs than with the quality of their purchases because of their socio-economic conditions (Azmat & Ha, 2013). This causes weak consumer demand for sustainable products, which are frequently offered at a premium price and are thus unaffordable for many consumers (Ben Brik et al., 2013). Therefore, the social and environmental effects of manufacturing processes remain largely ignored (Hutchins & Sutherland, 2008). Moreover, the dynamicity and complexity of business conditions and the absence of strong institutions in developing countries make it difficult for supply chains to understand, and take effective action against, factors that hinder the achievement of sustainability goals (Silvestre, 2015). In particular, in developing countries, the informal economy is an increasingly legal avenue for entrepreneurship and self-employment. In these countries, companies located in the formal economy, both domestic and foreign, should treat the informal economy as an essential route for their goods and services to be marketed. As a consequence, in developing countries, the informal economy is inextricably linked to both domestic and international formal economies, with growth in the formal economy often crucial for the growth in the informal economy (Holt & Littlewood, 2014). Therefore, in developing countries, informal suppliers are a core component of a variety of goods and services at different levels of the supply chain.

In the case of Bolivia, as we mentioned above, 8 out of 10 jobs are related to small firms, which constitute 83% of employment. Moreover, most of these small firms are associated with self-employment that lacks any social benefits. Furthermore, Bolivia is one of the poorest countries in the Americas, has the biggest informal economy in the world, and is generally characterized as politically unstable (The Fund for Peace, 2020). Therefore, it is expected that few firms will be able to comply with Bolivian labor laws. Moreover, due to political and economic uncertainty, firms need to work with a limited number of formal and informal available suppliers that hardly have any traceability in their supply chains. The average Bolivian supply chain consists of a variety of formal and informal suppliers, some of which rely on logistics networks connected to the formal economy. Some major formal organizations, such as the foods and beverage industry, rely on rural and informal suppliers who manufacture their products and sell them in the formal economy. These peculiar Bolivian characteristics make it difficult for firms to develop sustainable supply chains. Therefore, we surveyed managers in nine Bolivian firms in the foods and beverage industry. For our research, we surveyed managers because their leadership is a driver for the adoption of sustainability practices in the supply chain. Their leadership reflects both the firm's values and the commitment of top management (van Hoof & Thiell, 2015; Jia et al., 2018). Moreover, we selected this industry due to its strong ties and dependency on the informal economy. Particularly, compared to other countries, the agricultural development in Bolivia has been impeded by an extremely low productivity, a poor geographic distribution of the population across productive land, and a constant lack of transportation facilities (Gigler, 2009; Torres, 2011). Therefore, many firms in the foods and beverage industry work with urban and rural suppliers that hardly have any ties with the formal economy. Consequently, we expect to find below-average SSCM practices for Bolivian firms.

5   Methodology

To examine whether Bolivian firms from the foods and beverage industry comply with globally accepted SSCM practices, we developed a new survey based on the research of Bastian and Zentes (2013) and Zhu et al. (2018). The survey had two sections: the firm's supply chain characteristics and the firm's strategic supply chain planning. Specifically, the first section evaluated supplier complexity, supplier ethical issues, environmental and social requirements, third-party integrations, supply chain transparency, long-term relations, operational performance management, social performance management, and environmental performance management. On the other hand, the second section evaluated the firm's supply chain planning, materials planning, production planning, distribution planning, and supplier quality and reliability. The survey used 7-point multi-item Likert scales and was filled by nine Chief Executive Officers of firms in the foods and beverage industry. For each firm, we report the mean across multiple items in each category of evaluation. The surveyed firms were (1) Gran Oro, a food firm dedicated to wholesale distribution of cereals; (2) Industrias Asín, a food firm dedicated to the production of different types of snacks (e.g., potato chips and roasted peanuts); (3) Golden T, a beverage firm dedicated to the production of tea beverages; (4) Impasa, a food firm dedicated to the production of bakery goods; (5) SaxSay, a food firm dedicated to the production of healthy, gluten-free snacks; (6) Zelada, a food firm dedicated to the production of bakery goods; (7) Vascal, a beverage firm dedicated to the production of juices and soft drinks; (8) Distribuidora de Alimentos—Minuto Verde, a food firm dedicated to the distribution of healthy foods, mainly frozen fruits and vegetables; and (9) La Candelaria, a beverage firm dedicated to the production of craft beer. All sampled firms are small and medium-sized enterprises and both manufacture and distribute their products for the Bolivian market.

In the following two sections, we present the analysis for both survey sections.

6   Results on Firms' Supply Chain Characteristics

We surveyed nine supply chain characteristics: supplier complexity, supplier ethical issues, environmental and social requirements, third-party integration, supply chain transparency, long-term relations, operational performance management, social performance management, and environmental performance management.

(a) *Supplier complexity.* The distribution of products takes place across networks, and the intermediaries are the independent individuals or firms that make the

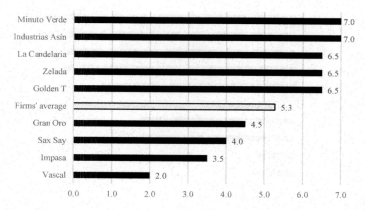

Fig. 1 Supplier complexity

product available for sale within the channel. Every firm has several suppliers, and an increase in the number of suppliers also increases both the channel complexity (Bastian & Zentes, 2013) and product costs (Ulloa & Rojas, 2014). Therefore, we used two items to determine the degree of the supplier's complexity (Bastian & Zentes, 2013). As Fig. 1 shows, on average, surveyed firms consider that their suppliers have an acceptable level of complexity (mean (M) = 5.3). Moreover, more than half of the firms consider that their suppliers do not have complex supply chains. In addition, one firm considers that its suppliers have quite complex supply chains.

(b) *Supplier ethical issues.* Supplier ethical issues occur in the procurement process when a firm's supplier abuses its power or unfairly treats its suppliers through measures such as bribery or abuse of contract terms. Moreover, developing countries generally have complex trade regulations and are frequently characterized by unethical suppliers (Bastian & Zentes, 2013). Therefore, we measured supplier ethical issues using two items (Bastian & Zentes, 2013). Figure 2 suggests that surveyed firms, on average, report that they do not source from countries with ethically doubtful practices (M = 4.5). In addition, most surveyed firms consider that their products or raw materials do not come from ethically compromised suppliers. Furthermore, only two firms report that they have sourced their products from unethical suppliers.

(c) *Environmental and social requirements.* This concept refers to the degree of voluntary standardization and control over the firm's ethical, social, and environmental requirements in the value chain (Elkington, 1998; Herbas Torrico et al., 2018). The fulfillment of these ethical requirements is the basis of product labels and certifications that communicate to consumers the environmental and social value of the firm's products. To measure the fulfillment of these ethical requirements, we used five items (Bastian & Zentes, 2013; Vachon & Klassen, 2006) and found the results shown in Fig. 3. On average, the surveyed firms report a relatively low compliance with environmental and social requirements

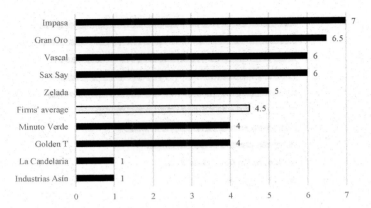

Fig. 2 Supplier ethical issues

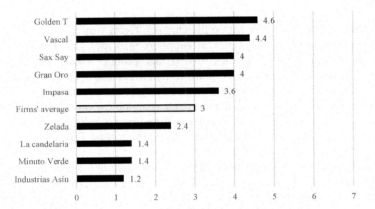

Fig. 3 Environmental and social requirements

(M = 3). An examination of individual firms suggests that most surveyed firms somewhat comply with these requirements, whereas four out of nine firms virtually do not comply with any environmental and social requirements.

(d) *Third-party integration.* This refers to supply chain monitoring by independent agencies and suppliers (Bastian & Zentes, 2013). We used six items to measure the degree of participation of third-party organizations in the monitoring and implementation of a firm's supply chain (Bastian & Zentes, 2013). For the average firm, Fig. 4 shows a moderate degree of integration of independent agencies in the monitoring and implementation of supply chains (M = 3.4). Moreover, only three firms exhibit an above-average integration of third-party firms in their supply chains, whereas more than a third of the firms have not, or virtually not, integrated any third-party agencies for monitoring their supply chains.

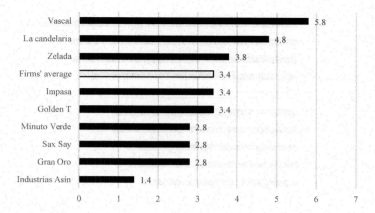

Fig. 4 Third-party integrations

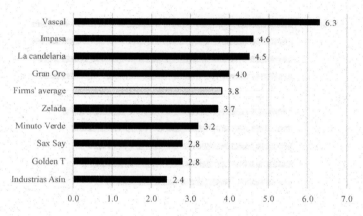

Fig. 5 Supply chain transparency

(e) *Supply chain transparency.* This concept originates from the disclosure of information to consumers, investors, and other stakeholders about the degree of compliance of the firm's supply chain with consumer requirements (Sodhi & Tang, 2019). In this case, we used thirteen items from Bastian and Zentes (2013). The results are shown in Fig. 5. On average, firms have moderately transparent supply chains (M = 3.8). An analysis of individual firms suggests that more than half of these firms have low-to-average supply chain transparency, and only the firm of Vascal has a highly transparent supply chain.

(f) *Long-term relations.* This variable attempts to measure the systematic and strategic coordination among members of the supply chain for long-term improvement and success (Espinal & Montoya, 2009). As a measure, we used six items from Bastian and Zentes (2013). According to the results in Fig. 6, on average, firms tend to develop long-term relations with their suppliers (M = 5.0). Around half of the surveyed firms consider that they cultivate strong

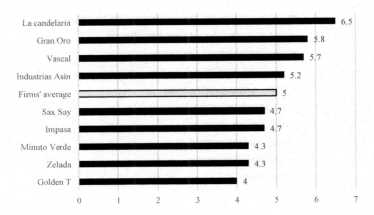

Fig. 6 Long-term relations

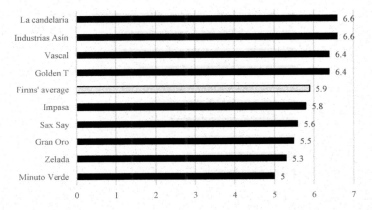

Fig. 7 Operational performance management

long-term supplier relations; and most firms develop at least moderately strong
long-term relationships with their suppliers (Fig. 6).

(g) *Operational performance management.* This variable is related to the coordination of all processes within a firm to ensure that the employees work together to
meet key supply chain objectives (Gardner et al., 2019; Kauppila et al., 2020).
To measure operational management performance, we used eight items from
Bastian and Zentes (2013) and Chen et al. (2004). Figure 7 shows that most of
the surveyed firms exhibit a fairly high operational management performance
(M = 5.9). Around half of the surveyed firms show an even very high operational management performance, and the other half of these firms have a good
operational management performance.

(h) *Social performance management.* This term refers to an approach that positions
societal welfare and the fulfillment of social requirements at the center of
strategic and operational decisions (Köksal et al., 2018). For this variable, we

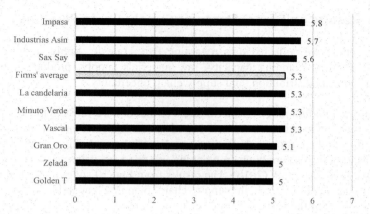

Fig. 8 Social performance management

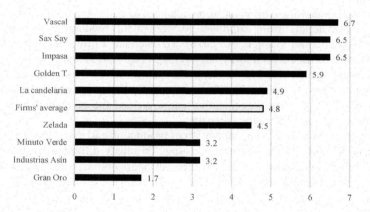

Fig. 9 Environmental performance management

used ten items from Bastian and Zentes (2013). Using this measurement approach, we found the results depicted in Fig. 8. All firms show high social performance management (M = 5.3). Moreover, compared to all other measured variables, there is a relatively high agreement among all surveyed firms regarding the implementation of social performance management practices.

(i) *Environmental performance management.* This is a management tool to help reduce a firm's environmental impact and to fulfill or surpass the environmental standards of a country (Mol, 2015). We used eleven items from Bastian and Zentes (2013) to measure this variable. The results in Fig. 9 suggest that most firms have a medium to high environmental performance management (M = 4.8). More than half of the firms show higher than average environmental performance management, whereas a third of the surveyed firms exhibit a very or relatively low environmental performance management.

7   Results on Firms' Strategic Supply Chain Planning

In this section, we collected data regarding five strategic supply chain planning features: a firm's planning capacity, raw materials supply, production process, finished products distribution, and supply chain uncertainty.

(a) *Supply chain planning.* This term refers to the forward-looking method of asset coordination to optimize the delivery of goods, services, and information from the firm to the customer, balancing supply and demand (Chae et al., 2014). To measure a firm's supply chain planning, we used ten items from Trkman et al. (2010) and Zhu et al. (2018). The results from Fig. 10 suggest that the average firm has medium to high supply chain planning capabilities ($M = 4.5$). Two-thirds of the surveyed firms have low or medium to average supply chain planning capabilities, and merely one firm (Vascal) has very high supply chain planning capabilities.

(b) *Materials planning.* This is the approach used to assess the needs and quantities of raw materials from suppliers to implement production (Stefanovic & Stefanovic, 2009; Wang et al., 2016). To calculate a firm's materials planning capacity, we used four items from Trkman et al. (2010) and Zhu et al. (2018). The results included in Fig. 11 show that the average firm has medium materials planning capacity ($M = 3.7$). Moreover, around half of the surveyed firms have medium to high materials planning capacity, and the other half have low to medium materials planning capacity.

(c) *Production planning.* The preparation of production and assembling modules in a firm is considered production planning. To satisfy different consumers, production planning allocates the resources of operators' activities, materials, and production capacity (Heo et al., 2012). To measure production planning, we used seven items from Trkman et al. (2010) and Zhu et al. (2018). Figure 12 indicates that the average firm has good production planning capabilities

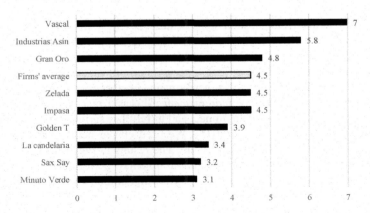

Fig. 10 Supply chain planning

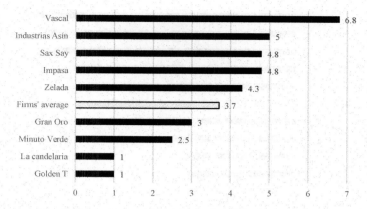

Fig. 11 Materials planning

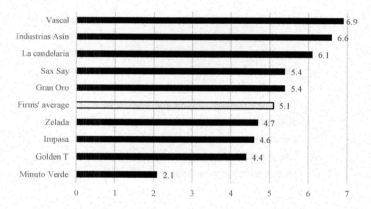

Fig. 12 Production planning

(M = 5.1) and that more than half of the surveyed firms show medium-high production planning capabilities. A third of the surveyed firms show average-medium production planning capabilities, and only one firm (Minuto Verde) has very low production planning capabilities.

(d) *Distribution planning.* This is a structured approach to ensuring that the supply of products to multiple distribution centers is conducted correctly, taking into account the delivery of products in the right quantity, to the correct location, and at the desired time (Trkman et al., 2010). To evaluate distribution planning capabilities, we used six items from Trkman et al. (2010) and Zhu et al. (2018). The results included in Fig. 13 show that the average firm has good distribution planning capabilities (M = 4.8). More than half of the surveyed firms have good to high distribution planning capabilities, and a third of the firms have low distribution planning capabilities.

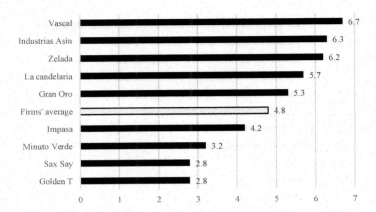

Fig. 13 Distribution planning

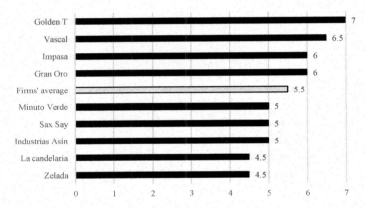

Fig. 14 Supplier quality and reliability

(e) *Supplier quality and reliability.* A supplier's quality practices should be understood as those practices that aim at reducing defects and improving performance features of products, whereas reliability practices are those practices that aim at minimizing the supplier's failure in production and delivery (González-Benito & Dale, 2001). To estimate this variable, we used two items from Trkman et al. (2010) and Zhu et al. (2018). The results listed in Fig. 14 indicate that firms, on average, are satisfied with their suppliers' quality and reliability (M = 5.5). Around half of all firms are satisfied with their suppliers' quality and reliability, and the other half of the surveyed firms show medium-high satisfaction.

Fig. 15 Supply chain characteristics (left) and strategic supply chain planning (right) for all firms

8 Discussion and Conclusions

The results shown in sections 6 and 7 were summarized in Fig. 15. This figure suggest that Bolivian firms from the foods and beverage industry have three main strengths in their supply chains. These are a medium-high operational performance management (M = 5.9), a medium-high social performance management (M = 5.3), and medium-high long-term relations (M = 5.0). On the other hand, the three main supply chain weaknesses of the surveyed firms are a medium-high supplier complexity (M = 5.3), a medium-low supply chain transparency (M = 3.8), and a medium-low integration of third parties (M = 3.4). Such observations indicate that the surveyed firms have efficiency-oriented processes, have a high social commitment, and value long-term relations with their suppliers. However, these firms have complex suppliers, seldom disclose information on their supply chains, and are barely monitored by independent agencies or suppliers. In regards to strategic supply chain characteristics, the two main strengths are medium-high supplier quality and reliability (M = 5.5) and strong production planning capacities (M = 5.1). The main strategic supply chain weakness is materials planning (M = 3.7). As a consequence, these results suggest that firms emphasize the focus on processes, systems, or procedures, and trust their suppliers' quality and reliability, but often experience problems procuring materials for their production processes, due to the characteristics of Bolivia mentioned above.

Our findings can be explained by two main themes in Bolivia: high political instability and a large informal economy. First, Bolivia is a developing country and historically has been characterized as a politically unstable country (Herbas-Torrico et al., 2021). Therefore, while the surveyed firms are socially committed, have efficiency-oriented processes, and value long-term relations with their suppliers, the high political instability makes it difficult for them to effectively manage and achieve sustainable supply chains. Specifically, political instability causes constant material procurement problems due to protest marches and road blockades (Arce & Rice, 2009; Los Tiempos, 2020). For example, the political crisis of 2019 caused 150 firms to declare bankruptcy and close down (Pagina Siete, 2019). In developed countries, this number appears insignificant, but it is large relative to the size of the Bolivian economy. Therefore, to overcome the country's uncertainty due to political instability, as predicted by Goonatilake (1984), firms in developing countries, such as Bolivia, carry excessive amounts of buffer stocks and must carry high inventory costs to meet their production targets.

On the other hand, the huge informal economy in Bolivia, combined with high political instability, causes firms to depend on complex suppliers, to have nontransparent supply chains, and to avoid being monitored by independent agencies. Specifically, firms from the foods and beverage industries must work with rural suppliers that hardly have any relationship with the formal economy, or in extreme cases do business with food smugglers to procure locally unavailable materials for their production processes. These firms can hardly comply with the sustainability and transparency levels required by independent agencies (Morales, 2020). Therefore, as we predicted, Bolivian firms show below-average supply chain sustainability.

Under such conditions, firms should aim to achieve two main objectives: first, to reduce their buffer costs while staying competitive; and second, to convince the Bolivian government to reduce the informal economy and develop a better business infrastructure and economic policy. To achieve the first objective, firms must ensure the efficiency of materials supply by increasing the number of suppliers, working closely with suppliers to ensure supply efficiency, and improving their inventory control techniques constantly. To achieve the second objective, through the commerce and industry chambers, Bolivian firms should work together with the government to develop policies that improve the business infrastructure and economic policy. Moreover, firms should contact workers' unions to work on a common agenda to develop social capital. Historically, Bolivian firms and workers' unions have been characterized by strong and uncompromised ideological positions: firms with neoliberalism and workers' unions with Marxism. Therefore, they should find a middle ground to work together toward developing a formal and competitive economy.

Finally, our study shows that the implementation of SSCM practices in Bolivia is still at an early stage. Due to our limited sample size, more studies are needed to confirm our findings. However, our results are consistent with the Bolivian reality and arrive at similar conclusions as other studies about developing countries (Chen, 2012; Elbahnasawy et al., 2016; Krause et al., 2000; Nguyen et al., 2018; Ogbo et al., 2014; Silvestre, 2015). Therefore, our chapter concludes that due to high political instability and a huge informal economy, sustainable supply chains in developing countries, such as Bolivia, are highly uncertain and extremely challenging to manage.

References

Akamp, M., & Müller, M. (2013). Supplier management in developing countries. *Journal of Cleaner Production, 56*, 54–62. https://doi.org/10.1016/j.jclepro.2011.11.069

Arce, M., & Rice, R. (2009). Societal protest in post-stabilization Bolivia. *Latin American Research Review, 44*(1), 88–101.

Azmat, F., & Ha, H. (2013). Corporate social responsibility, customer trust, and loyalty-perspectives from a developing country. *Thunderbird International Business Review, 55*(3), 253–270. https://doi.org/10.1002/tie.21542

Bastian, J., & Zentes, J. (2013). Supply chain transparency as a key prerequisite for sustainable Agri-food supply chain management. *International Review of Retail, Distribution and Consumer Research, 23*(5), 553–570. https://doi.org/10.1080/09593969.2013.834836

Ben Brik, A., Mellahi, K., & Rettab, B. (2013). Drivers of green supply chain in emerging economies. *Thunderbird International Business Review, 55*(2), 123–136. https://doi.org/10.1002/tie.21531

Chae, B. K., Olson, D., & Sheu, C. (2014). The impact of supply chain analytics on operational performance: A resource-based view. *International Journal of Production Research, 52*(16), 4695–4710. https://doi.org/10.1080/00207543.2013.861616

Chen, A. M. (2012). *The informal economy: Definitions, theories and policies* (No. 1). *Women in informal employment globalizing and organizing.* Retrieved from https://www.wiego.org/sites/default/files/migrated/publications/files/Chen_WIEGO_WP1.pdf

Chen, I. J., Paulraj, A., & Lado, A. A. (2004). Strategic purchasing, supply management, and firm performance. *Journal of Operations Management, 22*(5), 505–523. https://doi.org/10.1016/j.jom.2004.06.002

Del Castillo, J. I. (2020, January). *Doing business in Bolivia: A case study in the Andean regulatory framework* (01). *Development research working paper series.* Institute for advanced development studies. Retrieved from https://ideas.repec.org/p/adv/wpaper/202001.html

Elbahnasawy, N. G., Ellis, M. A., & Adom, A. D. (2016). Political instability and the informal economy. *World Development, 85*, 31–42. https://doi.org/10.1016/j.worlddev.2016.04.009

Elkington, J. (1998). Partnerships from cannibals with forks: The triple bottom line of 21st-century business. *Environmental Quality Management, 8*(1), 37–51. https://doi.org/10.1002/tqem.3310080106

Espinal, A., & Montoya, G. (2009). Information technologies in supply chain management. *Dyna, 76*(157), 37–48.

Gardner, T. A., Benzie, M., Börner, J., Dawkins, E., Fick, S., Garrett, R., . . . Wolvekamp, P. (2019). Transparency and sustainability in global commodity supply chains. *World Development, 121*, 163–177. https://doi.org/10.1016/j.worlddev.2018.05.025

Gigler, B. S. (2009). *Poverty, inequality and human development of indigenous peoples in Bolivia* (Working Paper No. 17). Georgetown University. Retrieved from https://pdba.georgetown.edu/CLAS RESEARCH/Working Papers/WP17.pdf

González-Benito, J., & Dale, B. (2001). Supplier quality and reliability assurance practices in the Spanish auto components industry: A study of implementation issues. *European Journal of Purchasing and Supply Management, 7*(3), 187–196. https://doi.org/10.1016/S0969-7012(00)00030-7

Goonatilake, P. C. L. (1984). Inventory control problems in developing countries. *International Journal of Operations and Production Management, 4*(4), 57–64. https://doi.org/10.1108/eb054726

Heo, E., Kim, J., & Cho, S. (2012). Selecting hydrogen production methods using fuzzy analytic hierarchy process with opportunities, costs, and risks. *International Journal of Hydrogen Energy, 37*(23), 17655–17662. https://doi.org/10.1016/j.ijhydene.2012.09.055

Herbas Torrico, B., Frank, B., & Arandia Tavera, C. (2018). Corporate social responsibility in Bolivia: Meanings and consequences. *International Journal of Corporate Social Responsibility, 3*(7), 1–13. https://doi.org/10.1186/s40991-018-0029-0

Herbas-Torrico, B., Frank, B., & Arandia-Tavera, C. (2021). Corporate social responsibility in Bolivia: Context, policy, and reality. In S. O. Idowu (Ed.), *Current global practices of corporate social responsibility.* Springer.

Holt, D., & Littlewood, D. (2014). The informal economy as a route to market in sub-Saharan Africa–observations amongst Kenyan informal economy entrepreneurs. In S. Nwankwo &

K. Ibeh (Eds.), *The Routledge companion to business in Africa* (1st ed., pp. 198–217). Routledge.

Huq, F. A., Stevenson, M., & Zorzini, M. (2014). Social sustainability in developing country suppliers: An exploratory study in the ready-made garments industry of Bangladesh. *International Journal of Operations and Production Management, 34*(5), 610–638. https://doi.org/10.1108/IJOPM-10-2012-0467

Hussain, Z. (2014). *Can political stability hurt economic growth? World bank.* Retrieved February 14, 2021, from https://blogs.worldbank.org/endpovertyinsouthasia/can-political-stability-hurt-economic-growth#:~:text=The standard definition of political, instability tends to be persistent

Hutchins, M. J., & Sutherland, J. W. (2008). An exploration of measures of social sustainability and their application to supply chain decisions. *Journal of Cleaner Production, 16*(15), 1688–1698. https://doi.org/10.1016/j.jclepro.2008.06.001

INE. (2020). *Boletín estadístico 2020.* La Paz. Retrieved from https://www.ine.gob.bo/index.php/comunicacion/boletines-estadisticos/#399-boletin-estadistico-20-21

Jayanti, R. K., & Rajeev Gowda, M. V. (2014). Sustainability dilemmas in emerging economies. *IIMB Management Review, 26*(2), 130–142. https://doi.org/10.1016/j.iimb.2014.03.004

Jia, F., Zuluaga-Cardona, L., Bailey, A., & Rueda, X. (2018). Sustainable supply chain management in developing countries: An analysis of the literature. *Journal of Cleaner Production, 189.* https://doi.org/10.1016/j.jclepro.2018.03.248

Kauppila, O., Valikangas, K., & Majava, J. (2020). Improving supply chain transparency between a manufacturer and suppliers: A triadic case study. *Management and Production Engineering Review, 11*(3), 84–91. https://doi.org/10.24425/mper.2020.134935

Köksal, D., Strähle, J., & Müller, M. (2018). Social sustainability in apparel supply chains: The role of the sourcing intermediary in a developing country. *Sustainability, 10*(4), 1039. https://doi.org/10.3390/su10041039

Krause, D. R., Scannell, T. V., & Calantone, R. J. (2000). A structural analysis of the effectiveness of buying firms' strategies to improve supplier performance. *Decision Sciences, 31*(1), 33–55. https://doi.org/10.1111/j.1540-5915.2000.tb00923.x

Los Tiempos. (2020, August 27). Bolivia, ¿un país adicto a los bloqueos? *Actualidad.* Retrieved from https://www.lostiempos.com/oh/actualidad/20200827/bolivia-pais-adicto-bloqueos

Medina, L. & Schneider, F. (2018). *Shadow economies around the world: What did we learn over the last 20 years?* IMF Working Paper No. 18/17. International Monetary Fund., https://ssrn.com/abstract=3124402

Ministerio de Desarrollo Productivo y Economía Plural. (2019). *Informe estadístico industrial de Bolivia.* La Paz. Retrieved from https://siip.produccion.gob.bo/noticias/files/BI_050820209ce83_bolivia.pdf

Mol, A. P. J. (2015). Transparency and value chain sustainability. *Journal of Cleaner Production, 107*, 154–161. https://doi.org/10.1016/j.jclepro.2013.11.012

Morales, R. (2020). The resilience of informal workers to COVID19 and to the difficulties of trade. *Sociology International Journal, 4*(4), 92–95. Retrieved from https://www.medcrave.org/index.php/SIJ/article/view/18878

Nguyen, M. H., Phan, A. C., & Matsui, Y. (2018). Supply chain management in developing countries: Empirical evidence from Vietnamese manufacturing companies. *International Journal of Productivity and Quality Management, 24*(4), 566–584. https://doi.org/10.1504/IJPQM.2018.093455

Ogbo, A. I., Victoria, O. I., & Ukpere, W. I. (2014). The impact of effective inventory control management on organizational performance: A study of 7up bottling company Nile mile Enugu, Nigeria. *Mediterranean Journal of Social Sciences, 5*(10), 109–118. https://doi.org/10.5901/mjss.2014.v5n10p109

Opinión. (2017, February 19). Ocho de cada 10 empleos, la mayor parte precarios, se crean en las mypes. *Informe Especial.* Retrieved from https://www.opinion.com.bo/articulo/informe-especial/cada-10-empleos-mayor-parte-precarios-crean-mypes/20170219235300674422.html

Pagina Siete. (2019, December 1). La crisis deja perdidas por $us 72 Millones y 150 Empresas Afectadas. *Empresa.* Retrieved from https://www.paginasiete.bo/inversion/2019/12/1/la-crisis-deja-perdidas-por-us-72-millones-150-empresas-afectadas-238984.html

Seuring, S., & Müller, M. (2008). From a literature review to a conceptual framework for sustainable supply chain management. *Journal of Cleaner Production, 16*(15), 1699–1710. https://doi.org/10.1016/j.jclepro.2008.04.020

Silvestre, B. S. (2015). A hard nut to crack! Implementing supply chain sustainability in an emerging economy. *Journal of Cleaner Production, 96,* 171–181. https://doi.org/10.1016/j.jclepro.2014.01.009

Sodhi, M. M. S., & Tang, C. S. (2019). Research opportunities in supply chain transparency. *Production and Operations Management, 28*(12), 2946–2959. https://doi.org/10.1111/poms.13115

Stefanovic, N., & Stefanovic, D. (2009). Supply chain business intelligence: Technologies, issues and trends. In *Artificial intelligence an international perspective* (pp. 217–245). Springer.

Strobel, T. L. (2010). Entry and exit regulations: The world bank's doing business indicators. *CESifo DICE–Report Journal for Institutional Comparisons, 8*(1), 42–53. Available online: https://www.ifo.de/en/node/29072

The Fund for Peace. (2020). *Fragile states index annual report 2020.* Retrieved from www.fundforpeace.org

Torres, C. (2011). *Agriculture in Bolivia.* Retrieved February 14, 2020, from http://nigthcatd.blogspot.com/2011/03/agriculture-in-bolivia.html

Trkman, P., McCormack, K., De Oliveira, M. P. V., & Ladeira, M. B. (2010). The impact of business analytics on supply chain performance. *Decision Support Systems, 49*(3), 318–327. https://doi.org/10.1016/j.dss.2010.03.007

United Nations. (2005). *Trade and development report.* United Nations. Available online: https://unctad.org/system/files/official-document/tdr2005_en.pdf

Ulloa, I., & Rojas, C. E. (2014). Diagnóstico de la cadena logística de exportación del banano ecuatoriano hacia estados unidos de américa. *Saber, Ciencia y Libertad, 9*(1), 77–90. https://doi.org/10.18041/2382-3240/saber.2014v9n1.1985

Vachon, S., & Klassen, R. D. (2006). Extending green practices across the supply chain: The impact of upstream and downstream integration. *International Journal of Operations and Production Management, 26*(7), 795–821. https://doi.org/10.1108/01443570610672248

van Hoof, B., & Thiell, M. (2015). Anchor company contribution to cleaner production dissemination: Experience from a Mexican sustainable supply programme. *Journal of Cleaner Production, 86,* 245–255. https://doi.org/10.1016/j.jclepro.2014.08.021

Wang, G., Gunasekaran, A., Ngai, E. W. T., & Papadopoulos, T. (2016). Big data analytics in logistics and supply chain management: Certain investigations for research and applications. *International Journal of Production Economics, 176,* 98–110. https://doi.org/10.1016/j.ijpe.2016.03.014

Webb, J. W., Bruton, G. D., Tihanyi, L., & Ireland, R. D. (2013). Research on entrepreneurship in the informal economy: Framing a research agenda. *Journal of Business Venturing, 28*(5), 598–614. https://doi.org/10.1016/j.jbusvent.2012.05.003

World Economic Forum. (2019). *The global competitiveness report 2019.* Retrieved from https://es.weforum.org/reports/global-competitiveness-report-2019

Zhu, S., Song, J., Hazen, B. T., Lee, K., & Cegielski, C. (2018). How supply chain analytics enables operational supply chain transparency: An organizational information processing theory perspective. *International Journal of Physical Distribution and Logistics Management, 48*(1), 47–68. https://doi.org/10.1108/IJPDLM-11-2017-0341

Boris Christian Herbas-Torrico is a Research Professor at the Universidad Católica Boliviana "San Pablo" in Cochabamba, Bolivia. He is a D. Eng. (PhD) and M.Eng. in Industrial Engineering and Management from the Tokyo Institute of Technology (Japan). Most of his research interests are marketing, production, applied statistics, and corporate social responsibility. He currently serves as a researcher at the Exact Sciences and Engineering Research Center (CICEI) from the Bolivian Catholic University.

Björn Frank is an Associate Professor at Waseda University, Japan. He holds a D. Eng. (PhD) from the Tokyo Institute of Technology, an M.Eng. from Ecole Centrale de Lyon, France and an MSc from Technische Universität Darmstadt, Germany. His research interests are in marketing, new product development, and sustainability. He currently serves as an editorial review board member for the *Journal of the Academy of Marketing Science*, as an editorial board member for the *Journal of the Japanese Society for Quality Control*, and as an area editor for the *Journal of the Japan Industrial Management Association*.

Carlos Alejandro Arandia-Tavera holds a bachelor's degree in industrial engineering from the Universidad Católica Boliviana "San Pablo". He had different positions in manufacturing firms in Bolivia. Currently, he works as a production controller in Alicorp Bolivia.

Pamela Mirtha Zurita-Lara holds a bachelor's degree in industrial engineering from the Universidad Católica Boliviana "San Pablo". She has working experience in supply chains and human resources.

Why Chicken? Fileni (Italy): Between Taste, Circular Economy and Attention to the Territory

Mara Del Baldo

Abstract

The agrifood system has a huge impact on several features concerning sustainability, including food safety and food security policies. In this vein, investigating why and how companies' business models incorporate these issues, has been gaining momentum within the scientific, political and managerial agenda. Do agrifood companies only implement strategies addressed to product quality and their characteristics of health, hygiene and sanitation or, in a broader sense, do they incorporate aspects relating to environmental and social sustainability? To answer these questions, this chapter introduces a case study relative to an Italian agrifood company (Fileni Group) that is among the leading European chicken farm and meat producers. The Group has embraced a circular economy putting into practice several strategies of eco-innovation through a regeneration production system based on the use of renewable sources, respect for animals, minimising waste, protecting the environment and generating socio-economic benefits to the territory and the local community. The case analysis allows us to point out the sunken reasons for food safety that rest on an ethical-driven business model, rather than on opportunistic reasons.

Keywords

Food safety · Food security · Sustainability · Circular economy · Case study · Fileni (Italy)

M. Del Baldo (✉)
Department of Economics, Society and Politics, School of Economics, University of Urbino Carlo Bo, Urbino, Italy
e-mail: mara.delbaldo@uniurb.it

© The Author(s), under exclusive license to Springer Nature Switzerland AG 2022
S. O. Idowu, R. Schmidpeter (eds.), *Case Studies on Sustainability in the Food Industry*, Management for Professionals,
https://doi.org/10.1007/978-3-031-07742-5_4

1   Introduction

Circular economy (CE) has been gaining momentum in academia and the managerial and political debate. The awareness of the need for different ways of producing and consuming is gaining ground worldwide. CE assures a different approach to the relations between the economy, environment, and society that, if implemented, can contribute to developing a more harmonic system (Stewart & Niero, 2018; Rizos et al., 2015; Geissdoerfer et al., 2017; Kirchherr et al., 2017). CE is particularly relevant within the agrifood system, which plays a fundamental role on a global scale concerning environmental issues, as well as food security and food safety issues.

In poor countries, food security aims to ensure that all people have sufficient nutrition (FAO, 2019a, b; Godfray et al., 2010; FAO & WHO, 2019). In rich countries, food safety mainly refers to avoiding diseases caused by food insecurity, deriving from the combined effect of the goods attributes capable of influencing consumers' health. Food safety involves all practices used to deliver safe food and ensure consumer health. Governments, businesses, and societies place great emphasis on ensuring firms follow food safety compliance (Escanciano & Santos-Vijande, 2014; Bar & Zheng, 2019; Yadav et al., 2021). Moreover, food safety is linked to the quality of life and the environment, compromised by the negative externalities of industrial agriculture and intensive farming (Rizov, 2020; Fraser & Charlebois, 2016; Bentivoglio & Giampietri Finco, 2016) and by the activities of transformation and modern food distribution, in the context of global supply chains (Filippi & Chapdaniel, 2021).

Agriculture is one of the major causes of climate disruption. In the agricultural sector, natural resources (land and water) are currently exploited at unprecedented annual levels. Together with climate change, this generates tremendous pressure on humanity's ability to feed itself (Flavelle, 2019). The agrifood sector absorbs 70% of the world's freshwater consumption (Roger, 2019) and accounts for over 30% of greenhouse gas emissions (Ahuja, 2019).

Firms in the agrifood sector have attracted the attention of governments, as well as public opinion and NGOs due to the effect of their production on natural resources and the communities in which they operate (Hartmann, 2011; Caracciolo & Lombardi, 2012). The potential vulnerability of agrifood companies to public opinion has driven them to integrate, within corporate strategies, environmental, ethical and human rights issues (Lombardi et al., 2015).

The agrifood sector is the largest manufacturing sector within the EU. It represents one of the main drivers of the EU economy, contributing to both economic output and employment (Finco et al., 2018). As in other countries, in Italy, over the years, the criticism of intensive farming has become increasingly intense, due to the effects produced on human health, animals and the environment (Golini et al., 2017; Battistelli & Campanella, 2020). In this context, organic production is assuming a key role since it provides for the cancellation of the use of pesticides and antibiotics (Tomašević et al., 2013; Abboud, 2019).

Increased consumer awareness and technological advancement have contributed to enhancing the number of food standards, indicating their relevance for research in this area (Mosquera et al., 2013).

In this vein, the empirical analysis is useful to deeply understand why and how companies' business models can incorporate the issue of food safety and food security. Do they (only) implement strategies addressed to the quality of the products and their characteristics of health, hygiene and sanitation, or, in a broader sense, do they incorporate aspects relating to environmental and social sustainability? To answer these questions, this chapter introduces a case study relative to an Italian agrifood company (Fileni Group) that is among the leading European chicken farm and meat producers. This case study has been selected because the Group has embraced a circular economy orientation and is currently putting into practice several strategies of eco-innovation (Cooreman-Algoed et al., 2022; Wang et al., 2022) through a sustainable production system "that is restorative or regenerative by intention and design" (Ellen Mac Arthur Foundation, 2015, p. 2), based on the use of renewable sources, minimising waste, protecting the environment and generating socio-economic benefits to the territory and the local community.

The remainder of the chapter is organised as follows: section 2 presents the theoretical background and sect. 3 introduces the methods and the case study. Section 4 presents and discusses the main findings, and, finally, section 5 draws conclusions in terms of insights and implications deriving from the study.

2 Theoretical Background

Food safety involves all practices used to deliver safe food and ensure consumer health. Governments, businesses and societies have placed great emphasis on firms' food safety compliance (Escanciano & Santos-Vijande, 2014; Bar & Zheng, 2019; Yadav et al., 2021).

Worldwide, each year, more than 600 million people are affected by food-borne illness, with 420,000 deaths. The World Health Organization estimates that over 33 million healthy lives are lost each year. Due to the consumption of unsafe food more than 23 million people, of whom three million are children under the age of 5, fall ill from unsafe food, resulting in 5000 deaths per year (WHO, 2020). As a consequence of food-borne illnesses, food scandals and accidents that occur continuously, consumer interest in food safety have increased (Rafeeque & Sekharan, 2018), as well as the interest of scholars.

As a part of the large interdisciplinary field of agrifood studies (Worosz et al., 2008), research on food safety has gained momentum in the last few decades. Agrifood scholars have been contributing to address a wide range of normative and pragmatic questions concerning the formulation, implementation and assessment of food safety policies. Among the lines of inquiry, they have also addressed the way the socially significant concerns about the agrifood system—equity, transparency, sustainability, responsibility and accountability—can be incorporated into policy, as well as how a food safety policy, that also embodies moral and cultural values, can be created, shaped and negotiated (Thonney & Bisogni, 1991).

Although the food sector is generally considered to be low-tech, innovation-based strategies are becoming increasingly important for agrifood businesses (Capitanio et al., 2010; Lefebvre et al., 2015). The importance of eco-innovation and its contribution to reducing food safety and environmental risks has generated a research stream fuelled by numerous contributions in recent years (Bossle et al., 2016; Cuerva et al., 2020; Triguero et al., 2018; Cuerva, 2019; Triguero, 2019; Corcoles, 2019; Gonzales-Moreno et al., 2019; Wang et al., 2022).

Providing food security and food safety is fundamental for every economy and society; it increases consumer confidence, grants legitimacy for business operations (Bush, 2018; Riganelli & Marchini, 2016) and represents a foundation for sustainable development. Food safety is essential to achieve sustainable development goals (SDGs) and in particular the goals "No Poverty" (SDG 1), "Zero Hunger" (SDG 2), "Good Health and Well-being" (SDG 3) and "Responsible Consumption and Production" (SDG 12) (Kumar & Giri, 2021; UN, 2015).

Among the factors that influence the companies' decision to introduce food safety measures are both regulatory provisions envisaged at the national or supranational level and expected competitive advantages (Hammoudi et al., 2009). The adoption of measures to comply with high standards of food quality and food safety increases the level of barriers to entering the sector and protects against the competition.

Globally, the Food and Agriculture Organization (FAO), the World Health Organization (WHO) and the World Trade Organization (WTO) deal with food safety issues and bring together food safety norms and standards (WHO, 2020).

Standards programmes and certification bodies play a substantial role (Bar & Zheng, 2019) by stating minimum requirements for quality and avoiding fraudulent practices (Khan et al., 2019). Traditionally, public standards have addressed food safety, which is outlined by government agencies and generally mandatory in practice (Hatanaka et al., 2005; Ehrich & Mangelsdorf, 2018).

The scope of food safety standards (FSS) has also expanded over time due to the growth of private standards (Casolani et al., 2018; Russo et al., 2014) that have traditionally focused on safety and quality issues. Lately, FSS have evolved to include other aspects like animal welfare and ethical and environmental concerns (Schuster & Maertens, 2015). Private FSS adoption is voluntary. Firms are not required to comply by law, the adoption being a part of business-to-business agreements between retailers and agrofood firms (Latouche & Chevassus-Lozza, 2015). However, private FSS are more than just product certifications retailers can ask their suppliers for in terms of specific technologies and logistics (Russo et al., 2014).

Among the aforementioned standards, food traceability is a key tool for food safety management (Xiang, 2015; Abad et al., 2009; Folinas et al., 2006; Violino et al., 2019). Several scholars have highlighted that traceability systems, used to demonstrate the origin and final quality of a product, are essential to improve quality assurance and transparency along the food supply chain (Alfian et al., 2017). A mandatory food traceability law has been implemented in the European Union by introducing the General Food Law. The Codex Alimentarius Commission set global food safety standards on which WTO members should base their sanitary and phytosanitary methodologies. The latter assume particular relevance in the production and marketing of meat. The Codex standard—General principles of food hygiene (food security and food safety)—CAC/RCP 1-1969 describes in detail the Hazard Analysis Critical Control Point (HACCP) system and the guidelines for its application. The HACCP principles are integrated into the official regulations of many countries (Rafeeque & Sekharan, 2018).[1]

To achieve optimised interventions to increase sustainability and security, basic elements should be addressed: (1) agricultural and land-use strategy; (2) crop (or livestock) production and harvesting; (3) processing, storage and distribution and (4) retailing and consumption (Harvey, 2016).

Other food safety management systems have been developed by the International Standard Organization, for example ISO 22000—"Food Safety Management Systems—Requirements for any organization in the food chain", which was introduced in 2005. The standard is applicable to all organizations—regardless of size—involved in any aspect of the food chain to deliver safe products (ISO, 2005; Rafeeque & Sekharan, 2018). A further standard on traceability was published in 2007—ISO 22005: 2007, "Traceability in the feed and food chain—General principles and basic requirements for system design and implementation"—as an extension of the ISO 22000 standard on traceability and an important component of food safety. ISO 22005 is intended for organisations that operate or cooperate at any stage of the food chain. It is not a mandatory standard, like ISO 22000; however, once the company intends to certify itself according to this standard, the standard requires the organisation to monitor and assess food safety, thus requiring internal audits and reviews to evaluate the effectiveness of their system (Færgemand, 2008).

In addition to system certifications, which define a basic quality standard level (hygiene and food safety standards, product traceability and traceability), agrifood product certifications—such as Protected Designation of Origin (PDO), Protected Geographical Indication (PGI) or Guaranteed Geographical Specialties (TSG) certifications—can be adopted by companies that aim to deeply incorporate food safety in their business model. As a result, they contribute to boosting the orientation towards sustainability through an extended implementation of food safety and a food security system as the following case study aims to demonstrate.

---

[1] In Italy, HACCP, which is a system that provides for self-control in a rational and organised way is only mandatory for operators in the post-primary sectors (http://www.salute.gov.it/). The UNI 10854 standard is the reference standard document in Italy for the creation of a management system for self-control based on the HACCP method in compliance with the EU regulation. Brussels, 14.9.2018 SWD (2018) 402 final Commission Staff Working Document accompanying the document Report from the Commission to the European Parliament and to the Council on the overall operation of official controls performed in Member States (2014–2016) to ensure the verification of compliance with food and feed law, animal health and welfare rules {COM(2018) 627 final}.

3 Material and Methods

3.1 Case Study Background

The case study below presented has been selected because it concerns a large agrifood company that deals with the production of chicken meat (Fileni Group), whose activity impacts both food safety and sustainability.

Italy is the European country with the largest number of agrifood products with designation of origin and geographical indication recognised by the European Union (299 PDO, PGI and TSG products and 524 wines with Controlled and Guaranteed Designation of Origin (DOCG), Designation of Controlled Origin (DOC) and Typical Geographical Indication (IGT). This corroborates the quality of Italian productions, but above all the strong link that binds Italian agrifood excellence to its territory of origin (CREA, 2020). In fact, the system of Geographical Indications of the European Union favours the production system and the economy of the territory. It protects the environment because the indissoluble link with the territory of origin requires the protection of ecosystems and biodiversity (Carraresi & Banterle, 2015). Moreover, it supports the social cohesion of the entire community. Thanks to the Community certification, consumers are given greater guarantees with a higher level of traceability and safety than other products (Danezis, Tsagkaris, Brusic & Georgiou, 2016).

The quality of Italian food products has always been a competitive driver in foreign markets (De Chiara, 2021). Among the vast number of Italian food excellence and leading production are poultry products. Within Europe, Italy ranks sixth among the poultry-producing countries after Poland, France, the United Kingdom, Germany and Spain (Kök, 2009). The production of chicken meat (with 948,000 tons recorded and an increase of +1.5% in 2019) confirms excellent levels of being overall self-sufficient at 107.3%. Specifically, 104.8% of the chicken consumed is produced in Italy. World Meat production increased from around 70 million tons in 1961 to 336.4 million tons in 2018, and to over 340 million tons in 2021. In 2020, with a production of 124 billion tons, chicken is the most consumed in the world, more than pork (120 billion) or beef (71 billion) (FAO and WHO, 2019).

3.2 Methodology

The research is grounded on a qualitative methodology based on a descriptive case study (Yin, 2013, 2014; Spence & Rutherfoord, 2004; Eisenhardt et al. 2016) relative to Fileni Group (Fileni Alimentare Spa).

The investigation made use of qualitative techniques. The empirical research was based on two steps devoted to data collection and analysis. In the first one, the analysis was carried out drawing from secondary sources: the Internet, the company's sustainability report and press releases. It aimed to identify the business model, the company's values and the food safety policies adopted through the years. The second step was carried out through interviews addressed to employees and was

Table 1 The business's highlights 2020

Turnover	450 M €
Quantities	113,855,954 kg
EBIDTA	33 M €
CAGR TURNOVER	6% Reference period 2014–2020
Employees	1852 employees and over 1200 collaborators
Productions plants	7

Source: Our elaborations of data included in the sustainability report, 2020

complemented by participant observation to better understand the reasons underpinning the food safety orientation, during events (meetings and conferences) organised outside the company (i.e. University workshops).

3.3   Fileni's World: "Italian family agrifood companies, organic and sustainable"

Fileni Alimentare Spa has operated for over 40 years in the breeding and sale of poultry products. It is a leader in the market of high-quality wellness products. The company was founded by Giovanni Fileni in 1978 in Jesi (Italy). Despite the group being headquartered in the Marches Region, its subsidiaries are spread throughout several Italian regions.

Currently, Fileni is the third placed player in the poultry sector on a national scale and the leading producer of organically reared white meat in Italy and Europe (Table 1). Its clients include large and globalised companies, such as Coop Italia, Esselunga, Autogrill, Ikea, Unilever and Plasmon. It has 1834 employees who are joined by external collaborators (interns, agents and administrators) for a total of 2005 people.

Fileni Group serves different distribution channels, including large distribution (GDO) 31%, organised distribution 4%, normal trade, wholesale, food service, foreign, B2B and ready-to-use gastronomic products.

The company's story is intertwined with that of its founder, Giovanni Fileni, born in Monsano in 1940 to a family of tenant farmers in the Vallesina countryside. After he worked as a mechanic when he was 14 years old, he started poultry farming in

Table 2 Fileni circular production chain

Organic	100% Italian organic cereal; Bio animals	Shortened logistics and constant enhancement of by-products and processing waste
Carbon neutrality	100% purchase of renewable electricity Zero t CO_2; carbon neutral production	Cogeneration, photovoltaic panels and the purchase of electricity with a guarantee of origin
Packaging	Recyclable PAPER, EBox secondary packaging, Eco-tray in cardboard, TREY	Continuous research on packaging
Benefit Corporation (Società Benefit)	On 27 April 2021, Fileni became a benefit company	The corporate purpose incorporates the commitment to pursue, in addition to profit, concrete social and environmental benefits for the local community and the community at large

Source: Our elaboration from the Company's presentations

1965 to experiment with one of the first systems for rearing farmyard animals in the countryside near his family home. Then he built their first barn for rearing 5000 chickens, which would be sold on a door-to-door basis in the local area.

In 1967, with the help of his wife, Giovanni Fileni set up his first shop to sell chickens directly to the public. In the following years, the number of shops rose to 48 and the rearing barns to 15, covering an area of 21,000 square metres. In 1968 his first slaughterhouse was built. Direct contact with the consumer was fundamental to understanding the social changes taking place, and in the late 1980s, Giovanni Fileni was the first to predict and interpret the shift from retailing to large-scale distribution. The chain of shops was closed and a factory producing ready-to-use goods was built in Cingoli in 1989. In 1995 a new, larger slaughterhouse was built, paving the way for a complete meat processing chain. In 2008 the production was located to new a premises in Castelplanio.

At the end of the 1990s, Fileni became a pioneer once he started organic production, an unusual way of breeding that was mostly considered utopian. Fileni acted as an envisioner since he recognised a great opportunity that would lead him to receive the "Good Chicken 2011 award as the "main supplier of white meat chosen for environmental sustainability".

The pillars of Fileni's business model—that I representative of a sustainable and circular production—are summarised in Table 2.

3.4   The Organic Choice: A Journey Towards Sustainability

In 2014 Giovanni decided to launch Fileni BIO, the first line of organic meats in Italy. The poultry organic supply chain represents Fileni's core business, characterised by traits of excellence including:

Fig. 1 University partnerships

- Organic animals are fed with 100% Italian organic non-OGM vegetable feeds, mostly from crops conducted directly or under contract and controlled by Fileni since sowing by using regenerative methods are respectful of the soil fertility. Hence, Fileni does not only grant the transformation of raw materials but also controls the cultivation of cereals and legumes that are necessary for the creation of organic feeds.
- Over 97% of organic production took place without the use of antibiotics (Fileni Sustainability report, 2019 and 2020).

In 2019, the group signed two financial transactions for a total of 60 million euros targeted to improve the farming organic processes and issued its first Eurobond, equal to 20 million, listed on the Vienna stock exchange (Fileni Sustainability report, 2019[2]).

In 2020 the group, which includes several cooperatives formed by local producers set up the newco *Fattorie Venete*, through which a diversification strategy for organic animal proteins was undertaken to expand its supply chain to red meat. The partnership—supported by the General Association of Italian Cooperatives—is actively engaged in creating a system of Italian companies that share sustainable-oriented processes. Fileni is also the promoter of the Agrifood Marche Cluster, a network set up in 2015 in collaboration with the Universities located in the Marche Region (Fig. 1), which aims to promote innovation and circular economy-oriented projects. In turn, it adheres to the national agrifood cluster—CL.AN, the active tool made available by the Italian Ministry of Education, University and Research (MIUR), and the main national entrepreneurial associations (Confagricoltura and Confindustria). Moreover, it is a member of the IFOAM EU Group—recognised with the status of advisory body by the United Nations—that represents the leading movements and associations that promote organic agriculture throughout the world.

A distinctive trait of the group is represented by the numerous certifications achieved that include, among others, the "all-around food safety" certification concerning food safety, food security and the integrity of products (HACCP—UNI 10854: 1999), the management and certification system for organic products (ISO 22000) and the management system for food traceability (ISO 22005).

---

[2]The sustainability report is released according to the GRI Standards (Core option) and submitted to external assurance (Deloitte audit).

As one can draw from the sustainability report, the group's practices will be enhanced in the future, helping them become the leading brand of white meat in the well-being and haute cuisine sector and ready-to-use meals, placing great emphasis on health and the enjoyment of taste.

4   Discussion

The key to Fileni Group's success is grounded on its strategic orientation to provide genuine and high-quality products (Golini et al., 2017). Fileni was among the first Italian companies in the sector to obtain food safety and food security certifications and supply chain traceability.

The strategic orientation that moulds the business model rests on the values system of the company that embodies the values promoted by the founder (Giovanni) and his family (Fileni's sustainability Report 2019). Such values (honesty and respect, listening and communication, initiative and innovation, bravery and determination, teamwork and simplification) are grounded on the social capital typical of the local areas where the company is embedded. In this vein, previous studies have addressed other case studies belonging to the same local area that share common traits deriving from the "genius loci", which is the social capital that characterises territorial companies embedded in the local community and is characterised by an authentic and intrinsic CSR and sustainability orientation. Most of these companies are partners of Fileni group and are involved in several actions aimed to benefit the environmental and social contexts they belong to. Their business model consistently reflects the motivations and values of the top management, which places the principles of food safety, human health and social and environmental sustainability at the forefront. The motivations and purposes are related to the business model rooted in the territory (Del Baldo, 2012).

Giovanni Fileni, born in 1940 into a family of sharecroppers takes his first steps from cultural roots inspired by the peasant culture of quality and genuine food, the respect for the environment and the ethics of work. These cultural assumptions and the strong roots of the past made it possible for the Fileni Group to guarantee economic value, social cohesion and environmental protection.

Fileni has developed an effective Agribusiness System, based on the control of the entire production process, from breeding to distribution (Casolani et al., 2018; De Chiara, 2021). The Group achieved the certification of its quality management system in compliance with UNI EN ISO 9001: 2000 and the IFS prestigious European certification.

Drawing from Fileni's experience, several points of excellence emerged. The company makes innovation and quality, animal welfare and environmental sustainability its distinctive features. It merges tradition and creativity. This trait highlights the strength of 'tradinnovation': the company is anchored to tradition but at the same time attentive to the future thanks to a creative and innovative attitude towards eco-innovation (Bush, 2018). People and the environment are at the core of every choice, in accordance with the achievement of several SDGs (Bebbington & Unerman, 2018).

First, Fileni cultivates the raw materials it uses for organic feed directly or through cultivation contracts with local farmers or neighbouring territories. Through a control plan, the company is able to control the crops from the moment they are sowed and ensure that they comply with the provisions of the organic regulations. Furthermore, thanks to their participation in the A.R.C.A. a project (Agriculture for the Controlled Regeneration of the Environment), conceived by Giovanni Fileni with Bruno Garbini and Enrico Loccioni (Del Baldo, 2013), focused on regenerative agriculture techniques and the use of organic fertiliser that allow for the preservation of cultivated land, maintaining its fertility and protecting it from hydrogeological instability (Geissdoerfer,2017; Kirchherr et al., 2017). The farmer is considered a keeper of the local environment, the keeper and makes people aware that with their purchase choices, they can turn from consumers to regenerators, soil regeneration and promote the use of organic fertiliser from animal sources.

A second distinctive aspect concerns the attention to the local community and the connection to the territory (Sustainability report, 2019, p. 96). In this vein, the A.R.C.A. project originated from a pioneering idea launched in the late 1980s and in March 2016, it became a benefit corporation, with the mission: "Good land and healthy food". The project aims to enhance the role of the farmer who is considered a keeper of the local environment, a custodian of the balance between the man and the earth.

In the last 5 years, to further promote environmental sustainability, over 2500 olive trees have been planted and over 5000 metres of laurel hedges have been planted. For the next few years, Fileni is determined to continue on this path, increasing his commitment to the issues of sustainability and always setting more challenging goals. The objective of Fileni's, 2019–2026 business plan is to develop a project of zootechnical verticalization, which provides full control and ownership of the whole supply chain, from feed to breed processing and biological diversification for all animal proteins.

Regarding the chicken farm, respect for animals is another key point. In the breeding centres, subject to rigorous controls, the chickens are raised on the ground and outdoors for at least 81 days and are free to roam on large organic land. The number of animals is controlled: no more than five animals per square metre. Filter systems for the abatement of dust and odours guarantee respect for the environment, landscape and animals. The organic farms are surrounded by plants and trees which, on the one hand, have the task of shading the fields and guaranteeing a comfortable environment for the animals, and on the other, hide the structures, minimising their visual impact. Soil regeneration is a fundamental element of circularity, and farming is closely linked to a series of interventions conducted in a sustainable perspective of agricultural production. The chicken, therefore, is a fundamental protagonist of the circular Economy because it is able to restore reclaimed land, fertile nature and a healthy environment to its territory.

Fileni's goal is to strengthen the dialogue between farmers and agrifood companies in order to promote circular economy and local sustainability. A.R.C.A. is in fact a participatory project, in which each actor is the protagonist and can add value. It is a network of farmers, agronomists, agricultural companies, technology providers, universities and research centres.

A distinctive point is the social orientation of Fileni's strategies that led to carry out several initiatives have been carried out over the years. For instance, the attention to the community manifested in the support provided after the 2016 earthquake, which destroyed many towns in the Marche, and their collaboration with the University of Camerino in setting up an Innovation Lab aimed at constantly improving agrifood production, with projects ranging from research on packaging to cooking methods ("Supporting the University of Camerino is a duty towards our community") (Fileni Sustainability Report, 2020).

Moreover, the attention to people and the local community has been stated and formalised through the creation of the Marco Fileni Foundation, set up in 2016 by the Fileni family in memory of Marco Fileni, the entrepreneur's son (Fondazione Marco Fileni, 2016). The non-profit organisation purposes are to generate utility and promote social solidarity. Among the initiatives carried out, the foundation supports the right to education for young people and to carry out charity activities. For example, it has provided 80 scholarships, with the motto "we believe in young people" to encourage, through economic support, dozens of young people to continue their studies. In 2019, it promoted the initiative "The world that I would like", in which the kids were invited to promote ideas and projects to protect the environment to point out the dreams of the new generations collected in a notebook.[3] The foundation is particularly attentive to young generations and collaborates with several local, national and foreign universities. The projects with universities cover many areas ranging from strategic marketing, agronomy and biology.[4]Since 2017, the foundation provides scholarships for young students. Attending secondary school and universities.

Finally, the group has implemented a transparency path by adopting several accounting tools, such as the ethical code, the adhesion to the discipline of the voluntary labelling of poultry meat and the releasing of the Sustainability Report that has been released according to a selection of the 2016 "GRI Sustainability Reporting Standards" and the 2014 "Food Processing Sector Disclosures", both published by the GRI, since 2018. The standards used to concern the general disclosure (GRI 102), the economic series (GRI 200), the environmental series (GRI 300) and the social series (GRI 400).

---

[3] www.notebook

[4] For instance, we can enumerate the partnership with the University of Gastronomic Sciences of Pollenz (focused on the disclosure of gastronomic knowledge), the Faculty of Agriculture and the Veterinary University of Perugia, and the CREA, Research body supervised by the Italian Minister.

5   Conclusion

The agrifood system has a huge impact on several features concerning sustainability, including food safety and food security policies. Globally, the agrifood supply chain is under significant pressure regarding food security, climate change and consumer demands for affordable and higher-quality food. Food chain decisions affect productivity, finance, insurance, supply chain management, food security, research and development, environmental stewardship and new operating models (Serazetdinova et al., 2019).

The agrifood sector—regarded as a sector with low research intensity—requires more innovation, which is one of the most important factors to enhance competitiveness and sustainability in both national and international markets (EU, 2020). This particularly applies to Italy, which has strong competition, especially from emerging countries (Horton et al., 2016). In this regard, the case study testifies how effective innovation is embedded in ecosystems capable to incorporate food safety and food security (Worosz et al., 2008; Kök, 2009; Godfray et al., 2010; Grazia & Hammoudi, 2012; Russo et al., 2014; Rafeeque & Sekharan, 2018) that interact with stakeholders and elicit partnerships (such as the Cluster Agrifood Marche) that drive competitiveness and sustainability orientation of the local agrifood sector (Horton et al., 2016; Ammirato et al., 2021).

Both regulatory provisions envisaged at the national or supranational level and expected competitive advantages influence the companies' decision to introduce food safety policies (Hammoudi et al., 2009). However, the adoption of measures to comply with high standards of food quality and food safety are not always sufficient to grant a deep and authentic orientation towards sustainability (Tiozzo et al., 2019). By contrast, when the adoption is intrinsically driven, is grounded on the commitment of the entrepreneurial/top management team and incorporated in the business model, it manifests behind the law in several tools, including voluntary certifications and a number projects oriented to achieve sustainability (Lombardi et al., 2015; Golini et al., 2017; Riganelli & Marchini, 2016).

In this vein, the case analysis highlights the roots that drive business models to effectively contribute to the circular economy in many ways (recycling, environmental protection, regeneration etc.). As the experience of Fileni has demonstrated, genuine products, creativity and innovation are devoted to enhancing the circular economy and the sustainability of the local and global community. The guarantee of food quality and safety, the preservation of the environment and the attention to the territory and the community have always been found alongside the Fileni name, which is oriented to become a leading brand of white meat in the well-being and haute cuisine sector, and the point of reference for a ready-to-use meal, placing great emphasis on health and the enjoyment of taste (Cooreman-Algoed et al., 2022).

The case analysis allows us to point out the "sunken" reasons for food safety that rest on an ethical-driven business model, rather than on opportunistic reasons. The main motivational drivers are represented by the persuaded adherence to the principles of sustainability (environmental, social, territorial and generational) and not by the desire to exploit the growing sensitivity of the market and consumers towards the theme of food safety and sustainable consumption in a profit-seeking logic. In this case, we have found that the orientation is not guided only by the need to adapt to the growing selectivity of consumers' food purchases and consumption choices because the company has anticipated and indeed solicited this trend.

The empirical investigation made it possible to verify that the orientation of Fileni, right from the start and even more so with their latest choices, put food safety and human health at the heart of the company's sustainable strategy. Respect for people, animals and the environment is incorporated into the motivational factors that lead to create a business that contributes to sustainability. A clear sign is given by their transformation in April 2021 into a benefit company, certified as B corp, marking the floor with a choice that is not so diffused among large manufacturing enterprises. Moreover, the great emphasis attributed by Fileni to collaboration with external partners (cooperative, universities and other agribusinesses) should be considered a strength since it contributes to contaminating the entrepreneurial agrifood system through the supply chain, considering that the food safety of the products sold in the final market depends on the behaviour of the various actors located in the various stages of the supply chain.

In summary, despite we claim that the insights derived from the analysis may elicit further research that can extend the investigation to other exemplary case-studies (both on a national and international level) to deepen the knowledge of the relevant factors that make food safety real and effective.

References

Abad, E., Palacio, F., Nuin, M., De Zarate, A. G., Juarros, A., Gómez, J. M., & Marco, S. (2009). RFID smart tag for traceability and cold chain monitoring of foods: Demonstration in an intercontinental fresh fish logistic chain. *Journal of Food Engineering, 93*(4), 394–399.

Abboud, L. (2019, July 23). *Can British farmers achieve net zero carbon emissions by 2050?*. https://www.ft.com/content/7d522ad8-abb4-11e9-8030-530adfa879c2

Ahuja, A. (2019, September 25). *A diet that is healthy for you and for the planet..* https://www-ftcom.ezp.lib.cam.ac.uk/content/b9ecaef0-be88-11e9-9381-78bab8a70848.

Alfian, G., Rhee, J., Ahn, H., Lee, J., Farooq, U., Ijaz, M. F., & Syaekhoni, M. A. (2017). Integration of RFID, wireless sensor networks, and data mining in an e-pedigree food traceability system. *Journal of Food Engineering, 212*, 65–75.

Ammirato, S., Felicetti, A. M., Ferrara, M., Raso, C., & Violi, A. (2021). Collaborative organization models for sustainable development in the Agri-food sector. *Sustainability, 13*, 2301. https://doi.org/10.3390/su13042301

Bar, T., & Zheng, Y. (2019). Choosing certifiers: Evidence from the British retail consortium food safety standard. *American Journal of Agricultural Economics, 101*(1), 74–88.

Battistelli, S., & Campanella, P. (2020). Subcontracting chain and working conditions in Italy: Evidence from the food and meat industry. *Studia z Zakresu Prawa Pracy i Polityki Społecznej (Studies on Labour Law and Social Policy), 27*(2), 135–145. https://doi.org/10.4467/25444654SPP.20.013.11951

Bebbington, J., & Unerman, J. (2018). Achieving the United Nations sustainable development goals: An enabling role for accounting research. *Accounting, Auditing & Accountability Journal, 31*(1), 2–24.

Bentivoglio, D., & Giampietri Finco, A. (2016). The new EU innovation policy for farms and SMEs' competitivenss and sustainability: The case of cluster Agrifood Marche in Italy. *Quality–Access to Success, 17*(S1), 56–63. Bucharest PEEC2016.

Bossle, M. B., De Barcellos, M. D., & Vieira, L. M. (2016). Why food companies go green? The determinant factors to adopt eco-innovations. *British Food Journal, 118*(6), 1317–1333.

Bush, S. R. (2018). Understanding the potential of eco-certification in salmon and shrimp aquaculture value chains. *Aquaculture, 493*, 376–383.

Capitanio, F., Coppola, A., & Pascucci, S. (2010). Indications for drivers of innovation in the food sector. *British Food Journal, 111*(8), 820–838.

Caracciolo, F., & Lombardi, P. (2012). A new-institutional framework to explore the trade-off between agriculture, environment and landscape. *Economics and Policy of Energy and the Environment, 3*, 135–154.

Carraresi, L., & Banterle, A. (2015). Agri-food competitive performance in EU countries A fifteene-year retrostpecive. *Internationla Food and Agribusinesses Management Review, 18*(2), 37–62.

Casolani, N., Liberatore, L., & Psomas, E. (2018). Implementation of quality management system with ISO 22000 in food Italian companies. *CAL, 19*(165), 125–131.

Cooreman-Algoed, M., Boone, L., Taelman, S. E., Van Hemelryck, S., Brunson, A., & Dewulf, J. (2022). Impact of consumer behaviour on the environmental sustainability profile of food production and consumption chains–A case study on chicken meat. *Resources, Conservation & Recycling, 178*(106089), 1–11. https://doi.org/10.1016/j.resconrec.2021.106089

Corcoles, D. (2019). The influence of collaboration with competitors on eco and non-eco-innovation: a study for the Spanish food industry. In A. Triguero & A. Gonzales-Moreno (Eds.), *Research on open-innovation strategies and eco-innovation in agro-food industries* (pp. 83–95). Chartridge Oxford Books.

CREA. (2020). *l'agroalimentare italiano settore chiave dell'economia leader in Europa per valore aggiunto agricolo*, 2019, available at CREA: Food drink Europe, data and trends, EU Food & Drink Industry, 2020, Edition.

Cuerva, M. C. (2019). Influence of suppliers and customers knowledge spurces on product and process eco-innovation in the food industry. In A. Triguero & A. Gonzales-Moreno (Eds.), *Research on open-innovation strategies and eco-innovation in agro-food industries* (pp. 73–82). Chartridge Oxford Books.

Cuerva, M. C., Triguero, A., & Sáez-Martinez, F. J. (2020). The way to be gree: Determinants of eco-process innovations in the food sector. In E. Lavere, R. Blacburn, Ben-Hafaïedh, C. Diaz-Garcia, & A. Gonzales-Moreno (Eds.), *Sustainable entrepreneurship and entrepreneurial ecosystem*. Edward Elgar.

Danezis, G., Tsagkaris, A. S., Brusic, V., & Georgiou, C. A. (2016, July). Food authentication: State of the art and prospects. *Current Opinion in Food* Science, 10. https://doi.org/10.1016/j.cofs.2016.07.003.

De Chiara, A. (2021). The "sustainable ingredient" for the quality and international performance of agrifood products. *Micro and Macro Marketing, 30*(2), 435–458. https://doi.org/10.1431/97443

Del Baldo, M. (2012). Corporate social responsibility and corporate governance in Italian Smes: The experience of some "spirited businesses". *Journal of Management and Governance, 16*(1), 1–36. https://doi.org/10.1007/s10997-009-9127-4

Del Baldo, M. (2013). Stakeholders management approach in Italian 'territorial' companies. Loccioni Group and the 'land of values–LOV' project. *European Journal of International Management, 7*(2), 225–246.

Ehrich, M., & Mangelsdorf, A. (2018). The role of private standards for manufactured food exports from developing countries. *World Development, 101*, 16–27.

Eisenhardt, K. M., Graebner, M. E., & Sonenshein, S. (2016). Grand challenges and inductive methods: Rigor without rigor mortis. *Academy of Management Journal, 59*, 1113–1123.

Ellen Mac Arthur Foundation. (2015). An economic and business rationale for an accelerated transition. In *Towards the circular economy* (Vol. 1). The Ellen MacArthur Foundation.

Færgemand, J. (2008). Making it safe to eat–the ISO 22000 series. ISO. *Focus, September*, 19–22.

Flavelle, C. (2019). Warming planet threatens food supply. A U.N. report warns. The New York times international edition. https://www.nytimes.com/2019/08/08/climate/climate-change-food-supply.html.

Escanciano, C., & Santos-Vijande, M. L. (2014). Implementation of ISO-22000 in Spain: Obstacles and key benefits. *British Food Journal, 116*(10), 1581–1599.

FAO. (2019a). *The state of food and agriculture 2019. Moving forward on food loss and waste reduction, Licence: CC BY-NC-SA 3.0 IGO*. Food and Agriculture Organization of the United Nation.

FAO. (2019b). *The state of food security and nutrition in the world 2019: Safeguarding against economic slowdowns and downturns*. Italy.

FAO and WHO. (2019). Sustainable healthy diets–guiding principles. .

Fileni. (2019). *Fileni Sustainability Report*. www.fileni.it

Fileni. (2020). *Fileni Sustainability Report*. www.fileni.it

EU. (2020). *Food drink europe. Data and trends, EU food & drink industry*, 2020 Edition.

Folinas, D., Manikas, I., & Manos, B. (2006). Traceability data management for food chains. *British Food Journal, 108*(8), 622–633.

Fondazione Marco Fileni. (2016). www.fondazionemarcofileni.it

Filippi, M., & Chapdaniel, A. (2021). Sustainable demand-supply chain: An innovative approach for improving sustainability in agrifood chains. *International Food and Agribusiness Management Review, 24*(2). https://doi.org/10.22434/IFAMR2019.0195

Finco, A., Bentivoglio, D., & Bucci, G. (2018). Lessons of innovation in the agrifood sector: Drivers of innovativeness performances. *Economia Agro-Alimentare, 20*(2), 181–192. https://doi.org/10.3280/ECAG2018-002004

Fraser, E., & Charlebois, S. (2016). Automated farming: Good news for food security, bad news for job security? *The guardian*. Accessed Jan 20, 2022, from www.theguardian.com/sustain able-business/2016/feb/18/automated-farming-food-security-rural-jobs-unemploymenttechnology?CMP=oth_b-aplnews_d-2

Geissdoerfer, M., Savaget, P., Bocken, N. M., & Hultink, E. J. (2017). The circular economy–A new sustainability paradigm? *Journal of Cleaner Production, 143*, 757–768. https://doi.org/10.1016/j.jclepro.2016.12.048

Godfray, C., Beddington, C., Haddad, L., Lawrence, D., Muir, J. F., Pretty, N. J., Robinson, S., Thomas, S. M., & Toulmin, C. (2010). Food security: The challenge of feeding 9 billion people. *Science, 327*(5967), 812–818. https://doi.org/10.1126/science.1185383

Golini, R., Moretto, A., Caniato, F., Caridi, M., & Kalchschmidt, M. (2017). Developing sustainability in the Italian meat supply chain: An empirical investigation. *International Journal of Production Research, 55*(4), 1183–1209. https://doi.org/10.1080/00207543.2016.1234724

Gonzales-Moreno, A., Triguero, A., & Sáez-Martinez, F. J. (2019). Many or trusted partners for eco-innovation? The influence of breath and depth of firms' knowledge network in the food sector. *Technological FORECASTING AND SOCIAL Change, 147*, 51–62.

Grazia, C., & Hammoudi, A. (2012). Food safety management by private actors: Rationale and impact on supply chain stakeholders. *Rivista di Studi sulla Sostenibilità, 2*, 111–143.

Hammoudi, A., Hoffmann, R., & Surry, Y. (2009). Food safety standards and agri-food supply chains: An introductory overview. *European Review of Agricultural Economics, 36*(4), 469–478.

Hartmann, M. (2011). Corporate social responsibility in the food sector. *European Review of Agricultural Economics, 38*(3), 297–324.

Harvey, S. J., Jr. (Ed.). (2016). *Handbook on the human impact of agriculture*. Edward Elgar Publishing.

Hatanaka, M., Bain, C., & Busch, L. (2005). Third-party certification in the global agrifood system. *Food Pol., 30*(3), 354–369.

Horton, P., Koh, L., & Shi Guang, V. (2016). An integrated theoretical framework to enhance resource efficiency, sustainability and human health in Agri-food systems. *Journal of Cleaner Production, 120*, 164–169.

ISO. (2005). *ISO 22000, food safety management systems -requirements for any organization in the food chain*. International Organization for Standardization.

Khan, M. I., Khan, S., & Haleem, A. (2019). Using integrated weighted IRP-fuzzy TISM approach towards evaluation of initiatives to harmonise halal standards. *Benchmark Int. J., 26*(2), 434–451.

Kirchherr, J., Reike, D., & Hekkert, M. (2017). Conceptualizing the circular economy: An analysis of 114 definitions. *Resources, Conservation and Recycling, 127*, 221–232. https://doi.org/10.1016/j.resconrec.2017.09.005

Kök, M. S. (2009). Application of food safety management systems (ISO 22000/HACCP) in the Turkish poultry industry: A comparison based on enterprise size. *J. Food Protect., 72*(10), 2221–2225.

Kumar, S., & Giri, T. K. (2021). The role of sustainability and policy measures in export performance. India studies in business and economics. In R. S. Ratna, S. K. Sharma, R. Kumar, & A. Dobhal (Eds.), *Indian agriculture under the shadows of WTO and FTAs* (pp. 177–192). Springer.

Latouche, K., & Chevassus-Lozza, E. (2015). Retailer supply chain and market access: Evidence from French Agri-food firms certified with private standards. *World Economics, 38*(8), 1312–1334.

Lefebvre, V. M., De Steur, H., & Gellynck, X. (2015). External sources for innovation in food SMEs. *British Food Journal, 117*(1), 412–430.

Lombardi, A., Caracciolo, F., Cembalo, L., Lerro, M., & Lombardi, P. (2015). How does corporate social responsibility in the food industry matter? *New Medit, 3*, 1–9.

Mosquera, M., Evans, E., Walters, L., & Spreen, T. (2013). The US food safety modernization act: Implications for caribbean exporters. *Soc. Econ. Stud, 62*(1&2), 151–176.

Rafeeque, K. T., & Sekharan, N. (2018). Multiple food safety management systems in food industry: A case study. *International Journal of Food Science and Nutrition, 3*(1), 37–44.

Riganelli, C., & Marchini, A. (2016). The strategy of voluntary certification in Italian olive oil industry: Who and why? Recent patents on food. *Nutrition & Agriculture, 8*(1), 9–18.

Rizos, V., Behrens, A., Kafyeke, T., Hirschnitz-Garbers, M., & Ioannou, A. (2015). *The circular economy: Barriers and opportunities for SMEs*. CEPS Working Documents.

Rizov, M. (2020). A vision of the farming sector's future: What is in there for farmers in the time of the second machine age? *Local Economy, 35*(8), 717–722. https://doi.org/10.1177/02690942211010151

Roger, S. (2019, September 1-2). *6 enquetes surl'alimentation de demain, Le Monde*. https://www.lemonde.fr/planete/article/2019/08/31/six-enquetes-sur-l-alimentation-dedemain_5504905_3244.html

Spence, L. J., & Rutherfoord, R. (2004). Social responsibility, profit-maximisation and the small firm owner-manager. In L. J. Spence, A. Habisch, & R. Schmidpeter (Eds.), *Responsibility and social capital*. Palgrave Macmillan. https://doi.org/10.1007/978-1-349-72886-2_4

Russo, C., Perito, M. A., & Di Fonzo, A. (2014). Using private food safety standards to manage complexity: A moral hazard perspective. *Agricultural Economic Review, 15*(2), 1–15. https://doi.org/10.22004/ag.econ.253686

Schuster, M., & Maertens, M. (2015). The impact of private food standards on developing countries' export performance: An analysis of asparagus firms in Peru. *World Development, 66*, 208–221.

Serazetdinova, L., Garratt, J., Baylis, A., Stergiadis, S., Collisone, M., & Davis, S. (2019). How should we turn data into decisions in AgriFood? *Journal of the Science of Food and Agriculture, 99*(7), 3213–3219. https://doi.org/10.1002/jsfa.9545

Stewart, R., & Niero, M. (2018). Circular economy in corporate sustainability strategies: A review of corporate sustainability reports in the fast-moving consumer goods sector. *Business Strategy and the Environment, 27*(7), 1005–1022. https://doi.org/10.1002/bse.2048

Tiozzo, B., Pinto, A., Mascarello, G., Mantovani, C., & Ravarotto, L. (2019). Which food safety information sources do Italian consumers prefer? Suggestions for the development of effective food risk communication. *Journal of Risk Research, 22*(8), 1062–1077.

Thonney, P. F., & Bisogni, C. A. (1991). Food safety policy: A shared responsibility. *Human Ecology Forum, 19*, 13–15.

Tomašević, I., Šmigić, N., Đekić, I., Zarić, V., Tomić, N., & Rajković, A. (2013). Serbian meat industry: A survey on food safety management systems implementation. *Food Control, 32*(1), 25–30. https://doi.org/10.1016/j.foodcont.2012.11.046

Triguero, A. (2019). Does the sector matter? The role of open innovation in the adoption of eco-innovations in food versus non-food industries. In A. Triguero & A. Gonzales-Moreno (Eds.), *Research on open-innovation strategies and eco-innovation in agro-food industries* (pp. 61–72). Chartridge Oxford Books.

Triguero, A., Fernandez, S., & Sáez-Martinez, F. J. (2018). Inbound open innovative strategies and eco-innovation in the Spanish food and beverage industry. *Sustainable Production and Consumption, 15*, 49–64.

UN. (2015). *Transforming our world: The 2030 agenda for sustainable development.* Accessed Oct 20, 2017, from https://sustainabledevelopment.un.org/post2015/transformingourworld

Violino, S., Antonucci, F., Pallottino, F., Cecchini, C., Figorilli, S., & Costa, C. (2019). Food traceability: A term map analysis basic review. *European Food Research and Technology, 245*(10), 2089–2099.

Wang, G., Shi, R., Mi, L., & Hu, J. (2022). Agricultural eco-efficiency: Challenges and progress. *Sustainability, 14*(1051), 1–23. https://doi.org/10.3390/su14031051

World Health Organization (WHO). (2020). *Estimating the burden of foodborne diseases.* Disponibile su: Accessed Feb 3, 2021, from https://www.who.int/activities/estimating-the-burden-of-foodborne-diseases

Worosz, M. R., Knight, A. J., & Harris, C. K. (2008). Resilience in the US red meat industry: The roles of food safety policy. *Agriculture and Human Values, 25*(2), 187–191. https://doi.org/10.1007/s10460-008-9127-z

Xiang, B. (2015). Study on safety management of food traceability based on food supply chain. *Journal of Food Science and Technology, 8*(6), 394–397.

Yadav, D., Dutta, G., & Kumar, S. (2021). Food safety standards adoption and its impact on firms' export performance: A systematic literature review. *Journal of Cleaner Production, 329*, 1297.

Yin, R. K. (2013). Validity and generalization in future case study evaluations. *Evaluation, 19*(3), 321–332.

Yin, R. K. (2014). In Kindle (Ed.), *Case study research: Design and methods.* Sage Publications.

Mara Del Baldo, is an Associate Professor of: Business Administration; Economics of Sustainability & Accountability in the Department of Economics, Society and Politics at the University of Urbino (Italy). She has extensive experience in collaborative research, in conjunction with universities and private industry. She visited several European Universities, including the University of Vigo (Spain), the New Bulgarian University of Sofia (Bulgaria), the Corvinus University in Budapest (Hungary) and the University of Craiova (Romania). She is a member of the European Council for Small Business, the Centre for Social and Environmental Accounting Research (CSEAR), the SPES Institute and the European Business Ethics Network (EBEN) Italia, as well of Italian scientific associations. She is part of the editorial board and she serves as a reviewer of several scientific journals, including the *International Journal of Corporate Social Responsibility* (Springer). Her main research interests include, among others: entrepreneurship and SMEs; Corporate Social Responsibility; sustainability and entrepreneurial business ethics; financial reporting; social accountability; integrated reporting and accounting and gender. She published in both Italian and foreign journals as well as in national and international conference proceedings and books.

Development and Planning of the Strategy against Food Waste in the Spanish Region of Cantabria

Elisa Baraibar-Diez, María D. Odriozola, Ladislao Luna, Ignacio Llorente, Antonio Martín, José Luis Fernández, Ángel Cobo, José Manuel Fernández, and Manuel Luna

Abstract

The need to establish actions to reduce food waste is part of the European Union's comprehensive approach to efficient use of resources so that economic development must derive from a linear economy to a circular economy. One of the most important aspects to be developed in this policy framework is the fight against food waste.

At a regional level, the Government of Cantabria raised strategic lines and transversal measures within the *Social Emergency Plan 2016–2017* to achieve practical proposals aimed at reducing food waste, coordinating efforts that derive from a greater articulation of the agrifood value chain. The research group Economic Management for Sustainable Development of the Primary Sector, to which the authors of this book chapter belong, collaborated with the government of Cantabria to define the *Strategy against food waste in Cantabria*.

This chapter aims to expose the collaborative process of the development and planning of this strategy. The first stage involved searching for the main initiatives carried out by public and private agents in the fight against food waste. As a result, we obtained a catalogue with more than 100 national and international initiatives. The second stage involved meetings with the Cantabria agrifood value chain agents, intending to assess the suggested proposals' suitability. Having collected their feedback, the third stage included the selection of the final recommendations. Finally, the fourth stage included the definition of the final strategy, as well as disclosing issues.

This book chapter emphasizes the need to publicize development processes prior to transversal strategies. It can serve as an example for other regions and communities that have concerns regarding food waste.

E. Baraibar-Diez (✉) · M. D. Odriozola · L. Luna · I. Llorente · A. Martín · J. L. Fernández · Á. Cobo · J. M. Fernández · M. Luna
Universidad de Cantabria, Santander, Spain
e-mail: elisa.baraibar@unican.es

© The Author(s), under exclusive license to Springer Nature Switzerland AG 2022
S. O. Idowu, R. Schmidpeter (eds.), *Case Studies on Sustainability in the Food Industry*, Management for Professionals,
https://doi.org/10.1007/978-3-031-07742-5_5

Keywords

Food waste · Food · Waste · Biowaste · Waste material · Waste matter ·
Agriculture · Agri-food · Agronomy · Agronomics · Farming · Cantabria · Spain ·
Efficiency · Economy · Organization · Enterprise · Business · Industry ·
Hospitality industry · Circular economy · Sustainability · Growth · Development ·
Management · Generation · Strategy · Plan · Social emergency plan · Social ·
Approach · Planning · Planning · Government · Governance

1   Introduction

It is estimated that EU-28 (European Union—28) wastes 88 million tonnes of food
along the supply chain: primary production, processing, wholesale and retail, food
service, and households, meaning that we waste 20% of the total food produced
(Stenmarck et al., 2016). In this sense, technological gears have been set in motion to
investigate how to reduce food waste in all the value chain stages. Thus, valorization
of resources, circular economy, recycling, etc. are concepts that are beginning to be
integrated into organizations and society. Besides, several institutions and awareness
campaigns try to educate at all levels toward a more sensible and responsible
consumption. In addition to the creation in 2016 of the European Union
(EU) Platform on Food Losses and Food Waste, we can highlight the following
campaigns: I Love Leftovers (sustainability.vic.gov.au/campaigns/love-food-hate-
waste) in Australia, Love Food Hate Waste (lovefoodhatewaste.com) in the UK,
Save the Food (savethefood.com) in the USA, or Think.Eat.Save (thinkeatsave.org)
and Feeding the 5000 (feedbackglobal.org) at an international level.

The Sustainable Development Goals (SDGs) help make this problem increasingly
known and recognized. Specifically, SDG 12 is "Ensure sustainable consumption and
production patterns," and target 12.3 reads "By 2030, halve per capita global food
waste at the retail and consumer levels and reduce food losses along production and
supply chains, including post-harvest losses" (United Nations, 2020). Simultaneously,
the alarming volume of food waste coexists with another equally worrying problem,
millions of people worldwide suffering from hunger. This unfair and unbalanced
situation has been previously raised by Tristram Stuart, an entrepreneur and change
maker interested in shifting society's attitude toward food waste (Stuart, 2009).

Given the economic, social, and environmental impacts generated by food waste,
diverse attention is observed in the literature, with multidisciplinary approaches.
Main contributions are related to the quantification of food waste (Barco et al., 2019;
Beretta et al., 2013; Buzby & Hyman, 2012; Hall et al., 2009), focusing on the
environmental impact (Hall et al., 2009; Hamilton et al., 2015) or the nature and
origin of food waste (Aschemann-Witzel et al., 2015). Those issues are treated in the
UK (Mena et al., 2011; Quested et al., 2011; Ventour, 2008), Australia (Pearson
et al., 2013), France (Mourad, 2016), Spain (Gracia & Gómez, 2020; Mena et al.,
2011), the USA (Love et al., 2015; Mourad, 2016), or developing countries (Hodges
et al., 2011; Thi et al., 2015). The effect of financial incentives and other policies to

prevent food waste has also been explored (Edwards et al., 2018; Schanes et al., 2018). Holistic and multi-stakeholder processes against food waste have also been described in several places such as Britain (Caswell, 2008) or Denmark (Halloran et al., 2014). Those impacts justify governments' concern to develop transversal plans that combine efforts and guarantee results. However, the description of government initiatives approached from a holistic perspective is less numerous. Thus, initiatives like the one in Denmark serve as a model to be applied in other territories. In this sense, this paper uses the case of the autonomous community of Cantabria in Spain to illustrate the process of creating a strategy against food waste, approached from a transversal perspective and focusing on standard proposals applied to the entire value chain. Even though the trip is not finished, this contribution aims to raise awareness and promote a change of attitude on the part of governments toward food waste.

2   Background

Cantabria is a single-province Autonomous Region in northern Spain with 580 thousand inhabitants. On January 18th, 2016, the Government of Cantabria approved the Cantabrian Social Emergency Plan 2016–2017 with an investment of 86 million euros within the framework of the Europe 2020 Strategy. This plan refers to "the adoption and implementation of the necessary measures to reduce the number of people at risk of poverty or social exclusion by reinforcing active policies targeted to the labour market, to increase the employability, and improving the objective, adequacy, efficiency, and effectiveness of support measures, including quality services to families." It also considers the three strategic objectives of the National Action Plan for Social Inclusion of the Kingdom of Spain 2013–2016 concerning social and labor inclusion and the systems of economic benefits and essential services. The Cantabrian Social Emergency Plan 2016–2017 was structured in five strategic lines (a) to guarantee a minimum income, (b) to ensure access to basic supplies, (c) to offer measures that impact and improve labor and community inclusion programs, (d) to guarantee access to health services, and (e) to guarantee mobility), containing 104 actions.

Within the framework of the Cantabrian Social Emergency Plan and considering the commitment to achieve Sustainable Development Goals (especially 12.3), the General Office of Social Policy, together with the research group Economic Management for Sustainable Development of the Primary Sector, initiates the process to outline the Strategy against Food Waste in the region. This initiative also responds to the so-called ambitious objectives of the "More Food, Less Waste" Strategy launched by the Spanish Ministry of Agriculture, Fisheries and Food in accordance with the mandate of the European Union, which established in 2016 the EU Platform on Food Losses and Food Waste (FLW). To assess the various alternatives to implement actions aimed at using food in Cantabria, a long-term work is established, which sets out strategic lines and transversal measures in which all the Public Administration is involved.

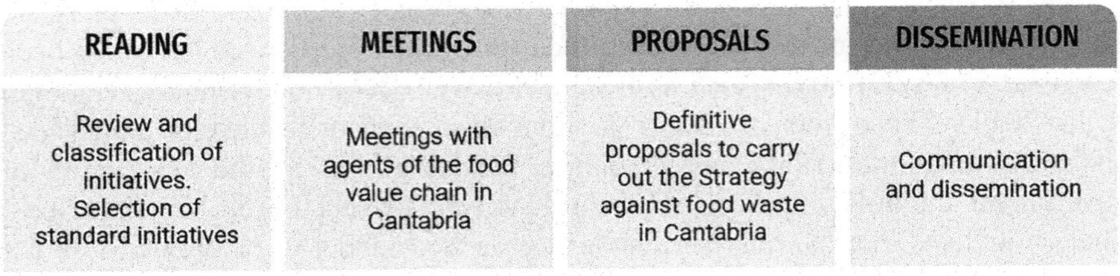

READING	MEETINGS	PROPOSALS	DISSEMINATION
Review and classification of initiatives. Selection of standard initiatives	Meetings with agents of the food value chain in Cantabria	Definitive proposals to carry out the Strategy against food waste in Cantabria	Communication and dissemination

Fig. 1 Stages of the project. Source: Authors

2.1 Procedure for the Elaboration of a Global Strategy

The collaboration agreement signed between the General Office for Social Policy of the Government of Cantabria and the research group proposed four different phases for the implementation of a global strategy: (1) search and analysis of the main initiatives carried out by public and private agents in the fight against food loss and food waste, (2) meetings with agents in the Cantabrian value chain, (3) definite proposals to carry out the plan against food waste in Cantabria, and (4) communication and dissemination (Fig. 1).

The first phase of the project consisted of reviewing as many initiatives as possible to know which types of initiatives were being carried out and at which value chain stage. In this sense, we build on previous work published by several sources such as the Spanish Ministry of Agriculture, Fisheries and Food and its "More Food, Less Waste" Strategy, a program "aimed to reduce food loss and waste and maximize the value of discarded food" (Ministry of Agriculture, 2018). The European Commission categorized practices against food waste into four main groups: (1) Innovation and Research; (2) Awareness, information, and education; (3) Policies, awards, and certification; and (4) Redistribution of food. Based on this classification, we reviewed more than 170 initiatives that we reclassified into subcategories. In this sense, we separated the following categories:

- *Innovation and research* into initiatives based on improving productive efficiency (12 initiatives), technology (11 initiatives), and waste management (5 initiatives).
- *Awareness, information, and education* in awareness campaigns (19 initiatives); events (6 initiatives); dissemination of information between donors and receivers (13 initiatives); educational (9 initiatives); and recommendations (16 initiatives).
- *Policies, awards, and certification* into the empowerment of social exclusion groups (1 initiative) and regulations (3 initiatives).
- *Food redistribution* into initiatives based on the transformation of surpluses (4 initiatives); the distribution of surpluses (30 initiatives); and the prevention of the appearance of surpluses (3 initiatives).

- Other initiatives were categorized as dual (21 initiatives with a twofold purpose), multiple (9 initiatives with various purposes), cross-cutting (3 transversal initiatives), or network initiatives (16 initiatives).

This disaggregation allowed us to establish leaner typologies of initiatives for each stage in the value chain, in line with the recommendations proposed later by the EU Platform on Food Losses and Food Waste (EU Platform on Losses and Food Waste, 2019): primary production, manufacturing stage, retail stage, hospitality/food services, consumer level, food donation; and then to propose to the agents in each stage.

The second phase of the project consisted of identifying the food value chain's primary agents in Cantabria (producers, food processors, wholesalers and retailers, foodservice, and consumers). We separate the agrifood value chain in Cantabria into various stages to classify all the agents along the chain. This was key to detect in which stages there is more waste and, therefore, require more immediate measures. Consumers' perception is beginning to change, and they are now aware of their role in preventing food waste. Not surprisingly, the flash Eurobarometer held in 2015 detailed that 76% of respondents thought that consumers had a role to play in preventing food waste, followed by shops and retailers (62%), hospitality and food service (62%), food manufacturers (52%), public authorities (49%), and farmers (30%). In Spain, these results were slightly different, since the responsibility was held on hospitality and food service (80%), public authorities (77%), consumers (75%), shops and retailers (70%), food manufacturers (70%), and farmers (39%) (European Commission, 2015).

We carried out a first telephone approach with ten organizations in the primary sector (producers of potatoes, vegetables, tomato, dairy products, and meat) to assess food waste behavior. EU statistics do not usually include the primary sector in food waste percentages. The causes of food loss and waste at the farm or production level are diverse (Buzby & Hyman, 2012) but contacted producers usually do not have food surpluses (either because of some planning system, because they have assured their sales, or because they have greater demand than supply). If they do have surpluses, they do compost, reuse them or produce secondary products, or give the produce away to various institutions. These destinations of waste coincide with those exposed by Mena et al. (2011). In the later stages of the value chain, the stage with the highest percentage of food waste is consumers (42%), followed by processing/manufacturing (39%), hospitality/foodservice (14%), and distribution (5%) (Monier et al., 2011).

After that, three meetings were held with various groups: representatives of the primary sector, representatives of wholesalers and retailers, and representatives of consumers and foodservice. Their comments and evaluation on the first set of initiatives obtained from the review allowed us to define the proposals that had the most significant possibility of implementation in the region. The structure of each meeting was similar. First, we set out the meeting's objectives; then, we informed the attendees of the proposed set of initiatives related to its stage in the value chain.

Finally, we collected the perceptions and evaluations of each of the attending representatives.

The first meeting (primary sector) was attended by representatives of the General Office of Livestock, the Food Quality Office in Cantabria, representatives of agricultural cooperatives, producers, Local Development Agencies, and representatives of unions. We proposed four initiatives related to the donation of surpluses, valorization, and sale of secondary products, compost policies, and direct distribution channels. The second meeting (wholesalers and retail) was attended by representatives of the General Office of Environment, representatives of the Trade Service, representatives of the Center of Information and Food Training (CIFA in Spanish), and representatives of four big wholesalers operating at a national level. We proposed four initiatives related to food donation with no commercial value, the creation of a social distribution channel, promotion of happy hours, and direct distribution channels. The third meeting (consumers and foodservice) was attended by representatives of the General Office of Environment, the Trade Service, the General Office for Education, the Consumers Union, the Cantabrian Regional Society for Tourism Promotion (whose acronym in Spanish reads CANTUR), the Hospitality Business Association, three catering services, and the Altamira School of Tourism. In this case, we proposed five initiatives related to responsible sourcing, menu planning and adjustment, maps/routes against food waste, courses and workshops, and money equivalents.

3   A Transversal Strategy against Food Waste

The comments and results obtained after the meetings with the value chain agents in Cantabria led to selecting of eight proposals (see Fig. 2).

Fig. 2 Summary of proposals for the strategy against food waste in cantabria. Source: Authors

3.1 Proposal 1. Promotion of the Use of Food Surpluses Through the Creation of a Label

The creation of a label to promote the use of surpluses (i.e., "I donate," "I do not waste," and "My establishment does not waste"), similar to the anti-waste labeling proposed by (del Giudice et al., 2016), is targeted at producers, industry, retailers, and foodservice. It aims to promote the use of surplus or ugly food, enhancing voluntary workforce or, where appropriate, the creation of work integration social enterprises (WISEs).

The advantages of creating this anti-waste label would be avoiding food losses and food waste; the improvement of corporate social responsibility (CSR), reputation, and visibility of companies with the distinctive; the empowerment of groups of social exclusion; and the generation of synergies. However, this could generate additional costs for the organization (additional transport, storage), and it would require a strong commitment on their part.

The indicators related to this proposal would be (1) Number of companies with the anti-waste label and (2) Kilograms of donated/distributed food.

This proposal is based on other initiatives such as *Huertos de Soria*[1] (Vegetable patches in Soria, Spain), *Es Imperfect* (Flawed, Spain), *La alimentación no tiene desperdicio* (Food has no waste, Spain), *Rubies in the Rubble* (United Kingdom), *Culinary Misfits* (Germany), *Last Minute Market* (Italy), *Ni un pez por la borda* (No fish overboard, EU), *Ampleharvest.org* (USA), or *SecondBite* (Australia).

3.2 Proposal 2. Promotion of the Knowledge of Direct Distribution Channels Through the Dissemination of Information and the Organization of farmer's Markets Where Local Produce Can Be Found

Encouraging the promotion of slow food fairs and farmer's markets would suppose a new commercialization way for producers, avoiding intermediaries and improving visibility for consumers, in line with the cooperation proposed by (Göbel et al., 2015).

This proposal's advantages would be favoring the concept of "from the field to the table" or short food supply chains; the avoidance of intermediaries; the belonging to a network, taking advantage of synergies; one-click information; reaching out to other consumer segments (organic, online, proximity). However, producers need to implement additional physical distribution channels for their products, with the cost of preparation and travel to fairs and markets, and adaptation to online platforms.

The indicators related to this proposal would be (1) Number of fairs and farmer's markets (attendees, number of stands, kilograms of food sold, and redistribution plans).

---

[1] The name of the initiatives has been kept with the original language (translation in brackets).

This proposal is based on other initiatives such as *directodelcampo.com (Straight from the field)*, *Hermeneus*, *Disfruta & Verdura (Enjoy Fruit & Vegetables)*, *demipueblo.es (from my village)*, all of them in Spain.

3.3   Proposal 3. Creation of a Food Distribution Network

The analysis of food waste in alternative food networks has already been explored in the literature (Poças Ribeiro et al., 2019; Seyfang, 2008), but the truth is that one of the main drawbacks identified in producers, industry, and foodservice when donating certain products is the distribution to organizations and social institutions. The creation of food distribution microgrids, integrated within the regional Social Emergency Network, will facilitate the local distribution of such foods with no commercial value or close to their expiration date. The local distribution would minimize the environmental impact.

This proposal's advantages would be the design of optimal routes to contact donors and recipients; donor savings by not having to transport the produce; belonging to a network; connection with the regional Social Emergency Network.

The indicators related to this proposal, targeted to producers, industry, retailers, foodservice, schools, and even consumers, would be (1) Kg of donated/distributed food, (2) Number of beneficiaries, and (3) Km travelled.

This proposal is based on initiatives such as *Leket Israel* (Israel), *Food Cloud* (Ireland), *Approved Food* (the UK), *dc Central Kitchen* (the USA), or *Too Good To Go* (Europe).

3.4   Proposal 4. Supply and Recycling Standards for Hospitality/Foodservice

This proposal aims to target producers, industry, retailers, foodservice, schools, and consumers to convey regulations to avoid the purchase of unrequired production, which may lead to food waste due to mismanagement in transformation, storage, or distribution of food. This proposal is transversal to the entire value chain since it can be applied at any link: producers, industry, distribution, foodservice, and consumers. Some aspects that these responsible sourcing standards would cover would be: responsible and adjusted-to-demand purchases, search for suppliers that can regularly supply, avoidance of marketing strategies that encourage the purchase of large quantities of perishable products in the short term, and lower esthetic standards for fruits and vegetables.

This proposal's advantages would be the improvement of CSR in the supply chain; the possibility of certifying in responsible standards; equivalence with an ethical supply code (not binding, but decisive); improvement in supplying, warehousing, logistics; collaboration with suppliers. However, this could generate changes in the conditions of contracts, penalties, and additional costs.

The indicators related to this proposal would be (1) Number of companies with supply and recycling standards and (2) Kilograms of food waste.

This proposal is based on other initiatives such as *LIFE Zero Cabin Waste* (Spain), *Ecoembes* (Spain), *Nestlé* (Spain), *Grupo Lactalis* (Spain), *Every Crumb Counts* (Europe), *ReFood* (Europe), *Food Waste Reduction Alliance* (USA), or *Slow Food Youth Network* (Global).

3.5 Proposal 5. Analysis and Adjustment of Portions (Foodservice, Schools, and Consumers)

On many occasions, consumers have to waste food because it is cheaper to buy food in larger formats (the so-called economy pack or value pack), which have a large volume of products, and which may not be needed at home (Aschemann-Witzel et al., 2015; Wang et al., 2017). In other areas, it is often the portions that are not appropriate to the consumer's age that generates waste (large kids' menus), or sometimes, the impossibility of being able to request another portion format (for example, half portions). Meal planning and portion size can enhance food waste (Mirosa et al., 2016), so having a better understanding of the menu's composition in a restaurant (portion size and side dishes) will facilitate the non-waste of food. Initiatives to raise awareness or reducing plate waste, such as the "Clean dish, clean conscience!" held in the School of Agriculture canteen at the University of Lisbon (Pinto et al., 2018), proved to be effective. Another way to avoid food waste in foodservice is the use of doggy bags, through which consumers can take home their leftovers.

This proposal's advantages would be the understanding of portions and nutritional information; the avoidance of food waste; the improvement in the management of raw materials. However, portion analysis according to age and type of organization is needed, in addition to the cost of bags or containers in the case of doggy bags.

The indicators related to this proposal would be (1) Number of companies have carried out portion analysis, (2) Number of containers used (doggy bag) for leftovers, (3) Consumer response rate (containers offered/containers accepted), and (4) Evolution of food waste in foodservice.

This proposal is based on other initiatives such as EUROPEN or those implemented by TESCO.

3.6 Proposal 6. Workshops and Awareness Campaigns against Food Waste that Include Sector Recommendations (Benchmarking)

The most significant proportion of food waste along the supply chain is originated in final consumption. The causes that could explain this amount of waste among consumers are the low awareness they have of both the expense that food represents within family budgets and the lack of awareness of the amounts of food waste and the needs it could cover. Awareness courses or workshops could act on the family consumption habits and reduce food losses.

This proposal's main advantage would be an increase in individual awareness, which has proved to be a determinant in the fight against food waste (Chalak et al., 2016). Besides, it would improve learning and acquaintance with new consumption habits, enhance familiar budgets, and promote networks and networking among participants. To do this, it would be necessary to create and coordinate the development of courses and seminars.

The indicators related to this proposal would be (1) Number of courses or workshops developed, (2) Awareness level (before and after), and (3) Amount of food waste (before and after).

This proposal is based on initiatives such as *Sin desperdicios: aprovecha la comida* (No waste: make the most of food, Spain), *Albal reta a Albal* (Albal challenges Albal, Spain), *Slow Food Youth Network* (Global), *Taste before you waste* (Holland), *Cozinha Brasil* (Brasil), *Save food from the fridge* (Global), *Appetite for action* (the UK), *Generation Awake* (Europe), *Io non spreco: adotta un nonno a pranzo* (I do not waste: adopt a grandfather for lunch, Italy), *Io non spreco: snack-saver bag* (I do not waste: snack-saver bag, Italy), and *School Waste Heroes* (the UK).

3.7   Proposal 7. CO_2/Euro Equivalences of Food Waste at Home

Generating food waste in the home is a result "from the interaction of multiple behaviors relating to planning, shopping, storage, preparation and consumption of food" (Quested et al., 2011). Not having the habit of translating the food that is wasted into money means that consumers cannot internalize its value. In this way, monetizing the food waste that an average family can generate would allow greater awareness and could send the message that if they avoid food waste, they could allocate that potential expense to another type of activity (leisure, training, etc.).

It is an effortless and attractive way to raise awareness, adapting equivalences to various publics ("If you avoid food waste, you could buy a toy or go to the circus (for children), you could go to such festival (for teenagers), or you could buy such gadget or go to such trip (for adults)").

The indicator related to this proposal would be the level of awareness.

This proposal is based on initiatives such as *Food Rescue* (the UK), *Garbage for energy* (Spain), *Tu basura vale un huevo* (Your garbage cost the earth, Spain), or *En busca de heroes contra el desperdicio de alimentos* (In search of heroes against food waste, Spain).

3.8   Proposal 8. Map/Route against Food Waste

Sometimes, it is misinformation or lack of visibility that generates food waste. Knowing what is being done in a particular environment can be an incentive and inspire others. These ant-food waste maps may resemble a pub crawl, showing local initiatives that combat food waste. Also, they can be incorporated into an online map of good practices, being available to all those who want to consult them.

The win-win approach is straightforward, improving reputation, visibility, and CSR of those establishments or individuals that join the route, forging a network of good practices.

The indicators related to this proposal would be (1) Number of initiatives adhered to the map/route against food waste and (2) Number of visits registered on the web.

This proposal is based on *La cuina furtiva* (The clandestine kitchen, Spain).

4   Publication of the Strategy in the Official Bulletin of Cantabria

The Strategy against Food Waste in the Autonomous Region of Cantabria was released in the Official Gazette of Cantabria under decree 56/2019 in April 2019. The eight-set of proposals were broken down into 52 initiatives divided into seven lines: (1) Action plan (commitments setting), (2) Actions with producers (9 initiatives), (3) Actions with industry and distribution (14 initiatives), (4) Actions with hospitality (14 initiatives), (5) Actions with educational organizations (2 initiatives), (6) Actions with final consumers (6 initiatives), and (7) Transversal initiatives (7 initiatives).

Each of the initiatives includes the competent office to carry it out as well as the collaborating office. Besides, each of the initiatives details the indicators to be considered (see Fig. 3).

5   Discussion

Although the Strategy against Food Waste has already been released, it is an ongoing project. A new organization and restructuring of Offices within the Government of Cantabria implemented the Strategy against Food Waste be transferred to the Office of Rural Development, Livestock, Fisheries, Food, and Environment. In this sense, the Government of Cantabria is developing during 2020 the awareness campaign "*Re Aprovecha*" (it could be translated as *Re-Use*) through the public company Environment, Water, Waste and Energy of Cantabria, S.A. (MARE, in Spanish) (Fig. 4). This campaign (www.reaprovechacantabria.es) aims to deploy coordinated awareness raising actions within the framework of the Strategy against food waste, highlighting the importance of small actions among citizens. Also, the aim is to reach as many people as possible through actions in various contexts (restaurants, markets, schools, etc.), reporting on the socioeconomic and environmental benefits of curbing food waste.

GOBIERNO
de
CANTABRIA

LINE 5. ACTIONS WITH EDUCATIONAL CENTERS

Goals:

Sensitization and awareness

Which initiatives will be carried out?

Initiatives	Competent department	Collaborating Department	Indicators
Initiative 38. Awareness raising. Information and disclosure campaigns aimed at students, teachers and people in educational centers and school canteens on the value of food and agricultural products, on the causes and consequences of food waste and ways to reduce them and the promotion of responsible consumption and cooking avoiding waste	General Directorate of Innovation and Educational Centers. General Directorate for the environment	General Directorate of Social Policy MARE	Number of campaigns Number of workshops Number of participating students Teachers trained
Initiative 39. Dissemination of good practice guidelines among students and teachers	General Directorate of Innovation and Educational Centers. General Directorate for the environment	General Directorate of Social Policy MARE	Number of workshops Number of participating students Teachers trained

Fig. 3 English translation of an extract of the Strategy against Food Waste. Disaggregation of initiatives within line 5: Actions with educational centers. Source: Cantabrian Official Gazette (04/25/2019), page 23

Fig. 4 Logo of the Strategy against Food Waste (B/W). Source: www. reaprovechacantabria.es

ESTRATEGIA CONTRA EL
DESPILFARRO
DEALIMENTOS

The very recent nature of this process of defining and implementing the Strategy implies not having, at the moment, any measurement with which it can be validated whether the Strategy has the expected results. In the II Social Emergency Plan 2018–2020, the implementation of a pilot experience in a municipality of Cantabria that develops measures of the Autonomous Plan against food waste is proposed. This will allow policymakers to limit and validate the proposed initiatives.

Facing the severe challenge of food waste implies a transversal and multistakeholder approach. This chapter has shown the outline of the Strategy against Food Waste in the Autonomous Region of Cantabria, responding to national and European mandates.

Only from a holistic perspective that considers all the agents of the value chain, is it going to be able to reduce food waste. Besides, in the case of Cantabria, this strategy is coordinated with the Social Emergency Plan, which further closes the concept of circular economy and extends its usefulness to people at risk of exclusion.

References

Aschemann-Witzel, J., de Hooge, I., Amani, P., Bech-Larsen, T., & Oostindjer, M. (2015). Consumer-related food waste: Causes and potential for action. *Sustainability (Switzerland), 7*(6), 6457–6477). MDPI AG. https://doi.org/10.3390/su7066457

Barco, H., Oribe-Garcia, I., Vargas-Viedma, M. V., Borges, C. E., Martín, C., & Alonso-Vicario, A. (2019). New methodology for facilitating food wastage quantification. Identifying gaps and data inconsistencies. *Journal of Environmental Management, 234*, 512–524. https://doi.org/10.1016/j.jenvman.2018.11.037

Beretta, C., Stoessel, F., Baier, U., & Hellweg, S. (2013). Quantifying food losses and the potential for reduction in Switzerland. *Waste Management, 33*(3), 764–773. https://doi.org/10.1016/j.wasman.2012.11.007

Buzby, J. C., & Hyman, J. (2012). Total and per capita value of food loss in the United States. *Food Policy, 37*(5), 561–570. https://doi.org/10.1016/j.foodpol.2012.06.002

Caswell, H. (2008). *Facts behind the headlines Britain's battle against food waste The war on waste.* www.statistics.gov.uk/statbase/product.asp?vlnk=867

Chalak, A., Abou-Daher, C., Chaaban, J., & Abiad, M. G. (2016). The global economic and regulatory determinants of household food waste generation: A cross-country analysis. *Waste Management, 48*, 418–422. https://doi.org/10.1016/j.wasman.2015.11.040

del Giudice, T., la Barbera, F., Vecchio, R., & Verneau, F. (2016). Anti-waste labeling and consumer willingness to pay. *Journal of International Food and Agribusiness Marketing, 28*(2), 149–163. https://doi.org/10.1080/08974438.2015.1054057

Edwards, J., Burn, S., Crossin, E., & Othman, M. (2018). Life cycle costing of municipal food waste management systems: The effect of environmental externalities and transfer costs using local government case studies. *Resources, Conservation and Recycling, 138*, 118–129. https://doi.org/10.1016/j.resconrec.2018.06.018

EU Platform on Losses and Food Waste. (2019). *Recommendations for action in food waste prevention developed by the EU platform on food losses and food waste.* https://eplca.jrc.ec.europa.eu/FoodSystem.html

European Commission. (2015). *Food waste and date marking.*

Göbel, C., Langen, N., Blumenthal, A., Teitscheid, P., & Ritter, G. (2015). Cutting food waste through cooperation along the food supply chain. *Sustainability (Switzerland), 7*(2), 1429–1445. https://doi.org/10.3390/su7021429

Gracia, A., & Gómez, M. I. (2020). Food sustainability and waste reduction in Spain: Consumer preferences for local, suboptimal, and/or unwashed fresh food products. *Sustainability, 12*(10), 4148. https://doi.org/10.3390/su12104148

Hall, K. D., Guo, J., Dore, M., & Chow, C. C. (2009). The progressive increase of food waste in America and its environmental impact. *PLoS One, 4*(11). https://doi.org/10.1371/journal.pone.0007940

Halloran, A., Clement, J., Kornum, N., Bucatariu, C., & Magid, J. (2014). Addressing food waste reduction in Denmark. *Food Policy, 49*(P1), 294–301. https://doi.org/10.1016/j.foodpol.2014.09.005

Hamilton, H. A., Peverill, M. S., Müller, D. B., & Brattebø, H. (2015). Assessment of food waste prevention and recycling strategies using a multilayer systems approach. *Environmental Science and Technology, 49*(24), 13937–13945. https://doi.org/10.1021/acs.est.5b03781

Hodges, R. J., Buzby, J. C., & Bennett, B. (2011). Postharvest losses and waste in developed and less developed countries: Opportunities to improve resource use. *Journal of Agricultural Science, 149*(S1), 37–45. https://doi.org/10.1017/S0021859610000936

Love, D. C., Fry, J. P., Milli, M. C., & Neff, R. A. (2015). Wasted seafood in the United States: Quantifying loss from production to consumption and moving toward solutions. *Global Environmental Change, 35*, 116–124. https://doi.org/10.1016/j.gloenvcha.2015.08.013

Mena, C., Adenso-Diaz, B., & Yurt, O. (2011). The causes of food waste in the supplier-retailer interface: Evidences from the UK and Spain. *Resources, Conservation and Recycling, 55*(6), 648–658. https://doi.org/10.1016/j.resconrec.2010.09.006

Ministry of Agriculture, F. and F. (2018). *More food, less waste strategy 2017-2020.* https://www.menosdesperdicio.es/sites/default/files/documentos/relacionados/estrategia_2017-2020_en.pdf

Mirosa, M., Munro, H., Mangan-Walker, E., & Pearson, D. (2016). Reducing waste of food left on plates: Interventions based on means-end chain analysis of customers in foodservice sector. *British Food Journal, 118*(9), 2326–2343. https://doi.org/10.1108/BFJ-12-2015-0460

Monier, V., Escalon, V., & O'Connor, C. (2011). Preparatory study on food waste across EU 27. *European commission.* https://ec.europa.eu/environment/eussd/pdf/bio_foodwaste_report.pdf.

Mourad, M. (2016). Recycling, recovering and preventing "food waste": Competing solutions for food systems sustainability in the United States and France. *Journal of Cleaner Production, 126*, 461–477. https://doi.org/10.1016/j.jclepro.2016.03.084

Pearson, D., Minehan, M., & Wakefield-Rann, R. (2013). Food waste in australian households: Why does it occur? *The Australasian-Pacific Journal of Regional Food Studies Number, 3.*

Pinto, R. S., Pinto, R. M. D. S., Melo, F. F. S., Campos, S. S., & Cordovil, C. M.-D.-S. (2018). A simple awareness campaign to promote food waste reduction in a university canteen. *Waste Management, 76*, 28–38. https://doi.org/10.1016/j.wasman.2018.02.044

Poças Ribeiro, A., Rok, J., Harmsen, R., Rosales Carreón, J., & Worrell, E. (2019). Food waste in an alternative food network–a case-study. *Resources, Conservation and Recycling, 149*, 210–219. https://doi.org/10.1016/j.resconrec.2019.05.029

Quested, T. E., Parry, A. D., Easteal, S., & Swannell, R. (2011). Food and drink waste from households in the UK. *Nutrition Bulletin, 36*(4), 460–467. https://doi.org/10.1111/j.1467-3010.2011.01924.x

Schanes, K., Dobernig, K., & Gözet, B. (2018). Food waste matters–a systematic review of household food waste practices and their policy implications. *Journal of Cleaner Production, 182*, 978–991. https://doi.org/10.1016/j.jclepro.2018.02.030

Seyfang, G. (2008). Avoiding Asda? Exploring consumer motivations in local organic food networks. *Local Environment, 13*(3), 187–201. https://doi.org/10.1080/13549830701669112

Stenmarck, A., Jensen, C., Quested, T., Moates, G., Buksti, M., Cseh, B., Juul, S., Parry, A., Politano, A., Redlingshofer, B., Scherhaufer, S., Silvennoinen, K., Soethoudt, H., Zübert, C., & Östergren, K. (2016). *Estimates of European food waste levels.*

Stuart, T. (2009). *Waste : Uncovering the global food scandal.* W.W. Norton & Co.

Thi, N. B. D., Kumar, G., & Lin, C. Y. (2015). An overview of food waste management in developing countries: Current status and future perspective. *Journal of Environmental Management, 157*, 220–229). Academic Press. https://doi.org/10.1016/j.jenvman.2015.04.022

United Nations. (2020). *Sustainable development goals: Sustainable development knowledge platform.* https://sustainabledevelopment.un.org/?menu=1300

Ventour, L. (2008). The food we waste. In *Food waste report V2* (Vol. 2, Issue July). WRAP. http://library.wur.nl/WebQuery/clc/1944512

Wang, L., Liu, G., Liu, X., Liu, Y., Gao, J., Zhou, B., Gao, S., & Cheng, S. (2017). The weight of unfinished plate: A survey based characterization of restaurant food waste in Chinese cities. *Waste Management, 66*, 3–12. https://doi.org/10.1016/j.wasman.2017.04.007

Elisa Baraibar-Diez . She is a Lecturer teaching (both in Spanish and in English) in the fields of business administration, entrepreneurship, international business, and simulation in business administration. Her research interests focus on corporate transparency, corporate social responsibility, reputation, and corporate governance. She has taken part in several national and international conferences and held research stays at Institut für Management belonging to Humboldt Universität (Berlin), at Sun Yat-sen University (Guangzhou, China), and at La Trobe University (Melbourne, Australia). Now, she coordinates the Official MBA in the University of Cantabria. She has over 8 years of experience in seafood markets in Spain and Latin America.

María D. Odriozola . She is a Lecturer in the Department of Business Administration at the University of Cantabria, becoming a PhD in 2015 (awarded in 2016 by the CSR Santander Chair at the University of Málaga). She lectures in the area of human resources management (human resources management—HRM, management skills, and new management models) and entrepreneurship. Her main lines of research are focused on aspects related to Corporate and Labor Social Responsibility, labor practices, corporate reputation, and the dissemination of social information. Her research has been part of several papers, book chapters, or proceedings at national and international conferences. Now she is the academic coordinator of the Master in Human Resources: the value of people. She has over 8 years of experience in seafood markets in Spain and Latin America.

Ladislao Luna . He is the Director of the research group "Economic Management for the Sustainable Development of Primary Sector" since 1997. Graduated in Economics and Business Studies at the University of Oviedo in 1983, he became a PhD at the same university in 1993. He is an Associate Professor in the Business Administration Department at the University of Cantabria since 1995. He has over 15 years of experience in seafood markets in Spain and Latin America. Participating as coordinator in ministerial research projects and expert teams, he has authored many works presented in reputed national and international industrial journals and scientific conferences. He has also taken part in the Globefish Research Program (FAO).

Ignacio Llorente . Lecturer in the Department of Business Administration at the University of Cantabria, becoming an awarded PhD in 2013 (by the International Association of Aquaculture Economics and Management and by the Research Awards "Juan María Pares"). He lectures in the fields of business administration and entrepreneurship. He is a specialist in Environmental Impact Assessment at the Universidad Politécnica de Catalunya (2010). He has over 10 years of experience in seafood markets in Spain and worldwide. As an expert in economic management of fisheries and aquaculture, he has participated in the Globefish Research Program (FAO) and in national and international research projects and author of works published in reputed international scientific and industrial journals. Participation in the STECF expert meetings.

Antonio Martín . He is an expert in Human Resources and Management Skills. He has been part of the University of Cantabria since 1983, becoming PhD in Economics and Business Studies in 1992. He is an Associate Professor of Business Organization in the Business Administration Department at the University of Cantabria since 1994. He is the Internship Coordinator of that Faculty since 2014. He has been one of the promoters of the implementation of the Official Master in Business Administration (MBA) at the University of Cantabria, of which he has been a coordinator in the first six promotions (2007–2008 to 2012–2013) and of which he continues as a lecturer. Now, he is the director of the Master in Human Resources Management: the value of people, at the University of Cantabria.

José Luis Fernández . Economist and Associate Professor in the Department of Business Administration at the University of Cantabria, he is PhD in Business Administration from the University of Cantabria in 2009. He has a Master of Arts (MA) degree in Economics from Queens

College of the City University of New York (CUNY) and a Master's Degree in Marketing and Commercial Management from ESIC. His research lines are mainly two: a line of applied research related to the economic management and sustainability of the primary sector, and a second line, more academic, focused on the economic analysis of firms and industries, as well as the study of corporate social responsibility.

Ángel Cobo . PhD in Mathematical Sciences and Associate Professor since 1993, he became a Full Professor in the area of Applied Mathematics in 2019. He is the University Ombudsman of the University of Cantabria since 2013. He has extensive experience in developing computer applications and he is the author of several books on Programming. He has lectured postgraduate courses at more than a dozen Latin American universities. He also has extensive experience in developing e-learning environments and automated analysis of user activities in these environments. As part of the Department of Applied Mathematics and Computing Sciences, he assists the team in big data processing and modelling.

José Manuel Fernández -Polanco. Expert in Marketing Research, he has been lecturing at the University of Cantabria since 1995, becoming PhD in Economics and Business Studies in 2001. He has over 15 years of experience in seafood markets in Spain and worldwide, having contributed to several funded research projects in fisheries economics. He has been a consultant at Globefish Research Programs and at the Fisheries and the Products, Trade and Marketing Branch (FIPMO) in FAO, among other research projects and expert meetings on value chain analysis and seafood consumption. He is a regular participant at STECF economic meetings (aquaculture and processing). Program chair in several WAS conferences in the last years.

Manuel Luna . He is a Researcher Expert in Data Science, becoming PhD in the Department of Business Administration at the University of Cantabria in 2020. He lectures in the area of business administration and entrepreneurship. He is a data scientist, having completed a Master's Degree in Banking and Financial Markets (University of Cantabria and Santander) in 2015, a Master's Degree in Strategic Planning (University Rey Juan Carlos and ESERP Business School) in 2016, and a Master's Degree in Data Science and Big Data (AFI) in 2017.

Food Security in South Africa: Lessons from the COVID-19 Pandemic on Creating Sustainable Value Chains Through Corporate Social Responsibility (CSR)

Mikovhe Maphiri

Abstract

For more than 25 years, South Africans have been living in a pandemic-like environment where access to adequate healthcare, food security, adequate housing and necessities that make people human is unevenly distributed. This position has remained the same even post-democracy, as ineffective methods were used to eradicate inequality and food insecurity. Prior to the pandemic, the hunger of millions of South African households has been a constant presence, where families do not have access to food. The COVID-19 pandemic has brought to light the country's overlooked crisis of food security by exposing with great concern the harsh reality of the lack of access to food for millions of South Africans. According to the Global Food Security Index, South Africa is the most food-secure country on the African continent. However, millions of South Africans go hungry every day. This position is reflected in a lack of access to food in a country where local food production and export are taking place on a massive scale. This paper seeks to provide a new perspective on re-envisioning the food security system in South Africa. The paper will discuss how corporate social responsibility (CSR) could provide avenues for food security and play a role in food equality as a key driver of food security after the pandemic and after the lockdown restrictions have been lifted. Drawing on the strengths of the CSR framework, the paper argues that CSR can enhance an understanding of South Africa's position in ensuring sustainable food security. This will also include the areas in which government and other social partners can intervene in the food security system in a way that promotes human dignity, equality, and the freedom of all persons, as espoused in the Constitution.

M. Maphiri (✉)
University of Cape Town, Cape Town, South Africa
e-mail: mikovhe.maphiri@uct.ac.za

© The Author(s), under exclusive license to Springer Nature Switzerland AG 2022
S. O. Idowu, R. Schmidpeter (eds.), *Case Studies on Sustainability in the Food Industry*, Management for Professionals,
https://doi.org/10.1007/978-3-031-07742-5_6

Keywords

Food security · South Africa · CSR · COVID-19 · Land reform · Democracy ·
NPO · Solidarity Fund · King Codes of Good Corporate Governance · Food
sovereignty

1   Introduction

On 31 December 2019, the World Health Organization (WHO) reported a cluster of
pneumonia cases in Wuhan City, China (WHO, 2020). "Severe Acute Respiratory
Syndrome Coronavirus 2" (SARS-CoV-2) was confirmed as the contributory agent
to what we now know as the novel "Coronavirus Disease 2019" (COVID-19, n.d.
(WHO, 2020). Since then, the virus has spread to more than a hundred countries,
including South Africa, with the WHO declaring COVID-19 a global pandemic
(Cucinotta & Vanelli, 2020). In response to the global pandemic, on 23 March 2020,
the President of the Republic of South Africa, Cyril Ramaphosa, announced a new
measure to combat the spread of COVID-19: a three-week nationwide lockdown
with severe restrictions on travel and movement, supported by the South African
National Defence Force (Disaster Management Act, 2002: Amendment of
Regulations issued in terms of section 27(2)). The first national lockdown took
place from midnight on Thursday, 26 March 2020, to midnight on Thursday,
16 April 2020 (South African Government, 2021).

The President emphasised the need to impose radical measures to avoid what he
described as "an enormous catastrophe" among the population (Disaster Management Act, 2002: Amendment of Regulations issued in terms of section 27(2)). This
meant that people would be allowed to leave their homes only to buy food, to seek
medical help, or in other extreme circumstances. The lockdown was followed by
government regulations that limited public gatherings and travel from high-risk
countries; the borders were closed to reduce the possibility of infection from those
travelling into the country from other countries.

A quarantine period was also applied to inbound travellers and returning citizens.
However, the number of confirmed cases and deaths continue to rise, and the
lockdown was extended to a period of more than 12 months, with eased restrictions
at each level of the lockdown phase. However, more cases are reported each day
(Sacoronavirus, 2020). More and more South Africans have succumbed to the virus
as countries across the globe raced to distribute the vaccine and achieve herd
immunity against the virus. At the time of writing this paper, the country has been
in lockdown for more than 12 months, but with certain restrictions on travel and
movement eased to allow for the recovery of the economy. One of the challenges
that resulted from the pandemic and the lockdown restrictions imposed by the
government was the impact on the economy, and the exacerbation of
socio-economic inequality and food insecurity. The pandemic has posed a threat to
the health, human rights and overall nutrition of people across the world. Globally,
millions of households were already suffering from hunger and malnutrition prior to

the pandemic (Global Hunger Index, 2020). South Africa is a food-secure nation, producing enough food with enough caloric and nutritional quality to feed its population (Teka Tsegay et al., 2014). However, this is not the reality at household level for the millions of zero-, low- and middle-income households that go hungry on a regular basis (Teka Tsegay et al., 2014). The 2020 South African National Health and Nutrition Examination Survey (SANHANES) revealed that 45.6% of the population were food-secure, 28.3% were at risk of hunger (score 1–4) and 26% experienced food insecurity (SANHANES, 2013).

The extent of food insecurity in South Africa is influenced by race (Hanoman, 2017). This racial aspect of the food security crisis stems from the impact of the apartheid regime on socio-economic inequalities. The South African National Health and Nutrition Examination Survey reveals that, prior to the pandemic, the largest percentage of participants who experienced food insecurity were in urban informal areas (32.4%) and in rural formal areas (37.0%) (SANHANES, 2013). The highest prevalence of people being at risk of hunger was in urban informal areas (36.1%) and rural informal areas (32.8%). The lowest prevalence of food insecurity was in urban formal areas (19.0%) (SANHANES, 2013).

The black African race group had the highest prevalence of food insecurity (30.3%), followed by the coloured population (13.1%) (SANHANES, 2013). Moreover, 30.3% of the black African population and 25.1% of the coloured population were at risk of being food insecure. A large percentage of the Indian population (28.5%) was also at risk of being food insecure. The majority of the white race group (89.3%) was food secure, which represents a significantly higher percentage than the other racial groups.

This paper conducts a desktop qualitative study of the Statistics South Africa (Statistics South Africa, 2020) time series data on food access in South Africa and the COVID-19 pandemic. The paper reviews the literature consisting of the primary and secondary sources of law as it seeks to find a new perspective on re-envisioning the food security system in South Africa. The goal is to find new ways to eradicate hunger at household level through corporate social responsibility (CSR). The paper reviews the reports and literature on food security and CSR by investigating the causes of the food security crisis in South Africa. The paper provides an overview of the impact of COVID-19 on food security from an access context and makes recommendations on how CSR could provide avenues for food security by promoting food equality as a key driver of food security during the pandemic and after the lockdown restrictions have been lifted.

2　　Methodology

The paper conducts desktop research on food security in South Africa during COVID-19 pandemic. It reviews historical data on food security prior to and during the pandemic. The rationale is to provide an overview of the food security landscape by exposing the challenges of food security both before and during the pandemic. The paper explores the opportunities for ameliorating food insecurity through the

CSR framework. Due to social distancing regulations, face-to-face interviews with people experiencing food insecurity during the pandemic could not be conducted. Most of the people who could have participated live-in communities where virtual communication is not possible due to a lack of access to computers and Internet services. Hence the research was conducted on a desktop basis.

3   Food Security and Food Sovereignty

At its core, food security refers to the availability of food (is enough food being produced to meet all needs?); access to food (can everyone who needs food obtain that food, and are their claims to a share of available food socially recognised?); and the use of food (is the food nutritionally diverse, socially appropriate and of the requisite quality?) (Labadarios, 2011). These dimensions are interconnected as access to, use of and quality of food determine the extent of the security or insecurity of food. A household is considered food insecure if it fails in any one or all of these dimensions (Teka Tsegay et al., 2014).

Furthermore, food sovereignty goes hand in hand with food security as defined. It is not in binary opposition to the discourse on food security (Desmarais, 2003). It is a rejoinder to the food security discourse adopted by global institutions and international policy analysts in addressing the global food crisis (Windfuhr & Jonsén, 2005). Food sovereignty rejects the understanding of food as a regular commodity by reducing the value of small-scale agricultural producers. It vests the right to control the food systems of nations in people (Desmarais, 2003), including access, production and distribution (Beuchelt & Virchow, 2012). Hence, food security cannot be discussed outside the discussion about the place of food sovereignty in South Africa (Ngcoya & Kumarakulasingam, 2017). It should be noted that the movement on food sovereignty has not yet taken root in South Africa, hence uniting the discussion on food security with the discussion on food sovereignty is essential.

3.1   Challenges to Food Security and Food Sovereignty in South Africa

South Africa has been a continental leader in producing enough food to feed its population (Labadarios, 2011). According to the Global Food Security Index, South Africa ranks first among sub-Saharan countries, based on a 2020 food security score of 57.8, followed by Botswana with a score of 55.5 (Global Food Security Index, 2020). In world rankings, South Africa takes the 48th position in the ranking of the most food-secure countries in the world (). The country is ranked first on the African continent as the most food-secure country. South Africa produces enough food to feed its population of 59,62 million people (StatsSA, 2020), yet household food insecurity and malnutrition are extremely severe. The food security crisis is not primarily based on the availability of food but rather on access to food. The latter is a legacy of apartheid which has had a significant impact on socio-economic inequality and poverty. This lack of access to food is felt particularly by children, with most children in rural communities experiencing long periods of hunger and malnutrition (Seekings et al., 2016).

The apartheid regime was a well-known system of oppression used by the South African government to discriminate against people based on race and gender (Feinstein, 2005). The decades-long system of apartheid resulted in inequality and poverty among racial groups and communities previously discriminated against by apartheid laws and practices. According to the World Bank, South Africa is the most unequal country in the world, as evidenced by the higher Gini coefficient of 63% (World Bank, 2014; WID, 2020). The Gini coefficient measures the distribution of income across the population in a country: the higher the value, the greater the inequality, where high-income individuals receive the greater proportion of the nation's income.

The apartheid system was used to ensure that the resources of the country were distributed in a manner that benefited the white minority. Thus, inhuman systems of control were put in place by passing laws aimed at oppressing black persons (Dubow, 1989). This included the dispossession of land owned by black South Africans as well as the exclusion of blacks from political decision-making processes. Thus, the pattern of land ownership, the geographic distribution of the population, the organisation of industry, the distribution of the country's wealth, access to basic education and food, including the quality of the food, was determined based on race. Although the country transitioned to democracy 28 years ago and is led by the African National Congress (ANC) government, which radically dismantled the system of apartheid, not much has changed in terms of food sovereignty and food security. The discourse on food sovereignty is interlinked with the debate on land reform. Much of the land seized by the white apartheid government has still not yet been returned to its rightful owners (Cousins, 2016).

3.2   Food Insecurity in South Africa during the COVID-19 Global Pandemic

According to the National Income Dynamic Study Coronavirus Rapid Mobile Survey (NIDS-CRAM), a nationally representative survey of the impacts of the COVID-19 pandemic in South Africa, a significant number of households have reported food insecurity. The survey was carried out in three waves. In the first wave, 47% of the surveyed households indicated that they had run out of money to buy food in April (Wills et al., 2020), a month after the government imposed a hard lockdown on 26 March 2020. In comparison, the 2020 General Household Survey reported that 25% of the households had run out of money to buy food. In May and June, after lockdown had been eased, about 21% of the households reported that someone in the household had gone hungry in the past seven days from the date of the survey (Wills et al., 2020). Additionally, 15% of the households indicated that a child had gone hungry in the past seven days from the date of the survey.

In the second wave of the NIDS-CRAM survey, data was collected between 13 July and 13 August 2020. There was a decline in the number of households reporting food insecurity, but the problem persisted despite the government's efforts to roll out relief programmes. The percentage of households that ran out of money to buy food decreased by 10 percentage points to 37%, whilst 16% indicated that someone in the household had gone hungry (Bridgman et al., 2020). The number of children going hungry declined by 4 percentage points to 11%. With regard to adult hunger, where someone in the household had gone hungry, by population group, there was a decline in the percentage of black Africans who went hungry between the first wave (26%) and the second wave (19%). Indians and Coloureds had similar percentages (around 7%) of those who went hungry, whilst the least was among the white population group. Based on the small sample sizes among non-African population groups, it remains difficult to understand the exact percentages of those who went hungry. However, the disparities can be easily revealed. Bridgman et al. (2020) show that child hunger was the highest among black African children compared to other races. This is not surprising given the level of inequality and the historical disparities in South Africa.

During the third wave of collection, which ran between November and December 2020, when the lockdown was further eased, the food security situation seemed to have worsened relative to the second wave. Households that ran out of money to buy food increased from 37% in the second wave to 41% in the third wave (Van der Berg et al., 2020). Household hunger increased to 18%, whilst the percentage of households reporting child hunger increased again to 16%, a worrying finding.

3.3   Efforts to Prevent Insecurity During the Pandemic

To address food insecurity during the pandemic, social relief and social protection programmes were undertaken by the government, intermediary organisations, non-governmental organisations (NGOs), community-based organisations (CBOs), faith-based organisations (FBOs), humanitarian organisations and philanthropic initiatives. The social welfare system in South Africa is strong. The SASSA grant is provided to the most vulnerable to poverty in the community, including the elderly and those living with disabilities. Social grants help poor households to cover the costs of the food and non-food needs of household members and not only grant recipients (Sekhampu & Grobler, 2011). The grant system has expanded to households that did not receive any form of grant, unemployment insurance or NSFAS bursaries through COVID-19 social relief distress (SRD).

NIDS-CRAM data indicates that 5.7% of households received support for food and shelter from NGOs, churches and associations, whilst 8.4% received support from the government. A localised social relief effort has been instrumental in helping communities during the lockdown. Food parcels, valued between R400 and R700 for families with an average household size of 4.5 members, were distributed monthly. The Solidarity Fund and the government provided instrumental information regarding the geographical distribution of food during lockdown. The additional measures implemented included the "provision of a temporary supplementary social assistance benefit in the form of a 'top-up' to existing social grants paid to different categories of beneficiaries with variable amounts and over different time periods" (Bridgman et al., 2020) and the Temporary Employee/Employer Relief Scheme (TERS). However, these efforts to address food insecurity are ad hoc and unsustainable. The relief is temporary and does not empower those who are experiencing food insecurity. Channels for long-term food security through both commercial and subsistence farming must be created. This then leads to the debate on land reform, which goes beyond the current discussion.

3.4   Statistical History of Food Insecurity in South Africa

According to the General Household Survey (GHS, 2020), an annual survey conducted by StatsSA since 2002, there has been a significant drop in food insecurity in South Africa since 1994. The survey measures hunger and food access using a shortened version of the Household Food Insecurity Access Scale: the respondent is asked whether the household has experienced hunger in the past 12 months. Figure 1 shows the percentages of people living below the poverty line in South Africa. This percentage was highest between 2008 and 2009, much of which could be attributed to the global financial crisis.

Figure 2 shows the percentage distribution of households by level of adequacy in food access between 2010 and 2017. Most of the households (more than 75%) annually reported that they had adequate access to food, and this has increased from 76.4% in 2010 to 78.7% in 2017. Between 14% and 16% had inadequate access. Generally, the inadequacy has decreased in the same manner, since severe inadequacy showed a decline from 7.4% in 2010 to 5.5% in 2017.

The above distribution has changed due to the COVID-19 pandemic. The COVID-19 pandemic has shown that the apartheid system continues to have an impact on the country's food security system. Most zero- to low-income households living in the rural areas still live on land which is incapable of sustaining them in terms of food sovereignty or food security. Hence, productivity among black farmers is low, and they are still producing one crop, namely maize. Although the government has attempted to invest in new methods of production by providing capital for irrigation, fences, fertilisers and agricultural equipment, in general, these attempts have been ineffective in creating sustainable food production avenues for people living in the former homelands. On the other hand, white farmers are at the top of the food chain in both commercial and subsistence farming; most food production and exportation companies are owned by white farmers. This has over the years fuelled the debate on land reform and inequality. Thus, land reform as a method of addressing inequality in the food production industry has been a key aspect of creating sustainable methods of diversified food production (Fig. 3).

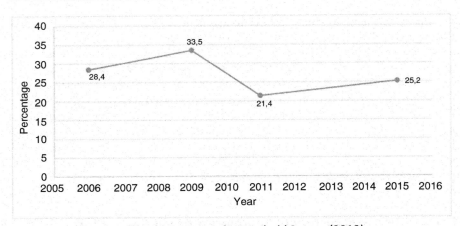

Fig. 1 Proportion of people living below the poverty line in South Africa from 2006 to 2015. Source: Statistics South Africa: General Household Survey (2019)

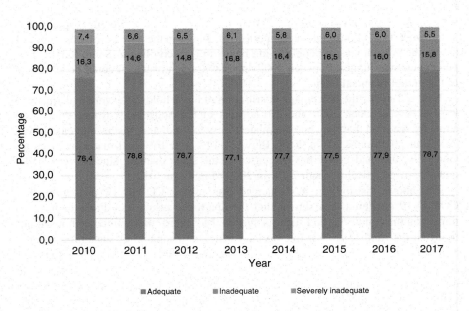

Fig. 2 Percentage distribution of households by level of adequacy in food access between 2010 and 2017. Source: Statistics South Africa: General Household Survey (2019)

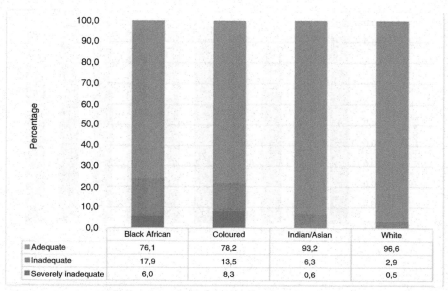

	Black African	Coloured	Indian/Asian	White
■ Adequate	76,1	78,2	93,2	96,6
■ Inadequate	17,9	13,5	6,3	2,9
■ Severely inadequate	6,0	8,3	0,6	0,5

Source: Statistics South Africa: General Household Survey (2017)

Fig. 3 Food access by population group of the household head in 2017. Source: Statistics South Africa: General Household Survey (2017)

Only 2.9% of the white population group had inadequate food access compared to 17.9% of the black African group. 76% of the black African group had adequate food access compared to 96.6% of the white-headed households. A critical influence on food insecurity in South Africa is the immense socio-economic inequality in South Africa, largely caused by the apartheid regime. This impact is reflected in the disparity in the distribution of wealth and income caused by the system of apartheid. Most black South Africans are food insecure because of this inequality, which is largely reflected in the educational, social and living conditions of the black, coloured, Indian and white population groups (Fig. 4).

A further aspect of the food security problem is embedded in the high rate of unemployment. The rate of unemployment in South Africa has been significantly high, both before and during the COVID-19 pandemic (StatsSA, 2020). According to Statistics South Africa, the unemployment rate stood at 30,1% in the first quarter of 2020 and increased by 2,3% in the second quarter of 2020. This increase was also influenced by the impact of COVID-19 pandemic as most businesses downsized, with some closing down due to the pandemic (StatsSA, 2020). An increase in unemployment has an impact on food insecurity as access to food is based on the availability of jobs that will allow people to earn money to buy food. To ensure sustainable food security, the challenges of inequality and unemployment need to be adequately addressed as the problem of food security is not primarily based on the availability of food but on access thereto.

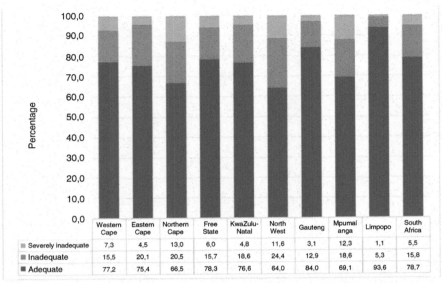

	Western Cape	Eastern Cape	Northern Cape	Free State	KwaZulu-Natal	North West	Gauteng	Mpumalanga	Limpopo	South Africa
Severely inadequate	7,3	4,5	13,0	6,0	4,8	11,6	3,1	12,3	1,1	5,5
Inadequate	15,5	20,1	20,5	15,7	18,6	24,4	12,9	18,6	5,3	15,8
Adequate	77,2	75,4	66,5	78,3	76,6	64,0	84,0	69,1	93,6	78,7

Source: Statistics South Africa: General Household Survey (2017)

Fig. 4 Household food access by province in 2017. Source: Statistics South Africa: General Household Survey (2017)

3.5 The Constitution and the Right to Food

The Constitution is the highest law in the land; every law that conflicts with the rights entrenched in the Constitution, including the right to food, is unconstitutional (The Constitution of the Republic of South Africa Act, 1996). However, this is subject to certain limitations. Section 27(1)(*b*) of the Constitution guarantees the right of access to enough food. The right to food embodies the right of human beings to have access to food and to create food and feed themselves. This right is interconnected with the right to life and human dignity, as the deprivation of a person's right to food fundamentally impacts the right to life and human dignity. Thus, the right to food requires that food be available and accessible without discrimination. The argument is that the right of access to food must be realised by the national government, which must increase food production and make social welfare available to those who do not have access to food (De Visser, 2019). The government must take reasonable legislative and other measures within its available resources to achieve the progressive realisation of each of these rights in a manner that promotes the human dignity and equality of all persons.

3.6 Government Initiatives to Address Food Security in South Africa

In a bid to address food security, South Africa's Department of Social Development and the South African Social Security Agency (SASSA) have in the past made resources available to address the food crisis. SASSA derives its core mandate from section 29 of the Constitution, which provides for "the right of access to appropriate social assistance to those unable to support themselves and their dependants" (The Constitution of the Republic of South Africa Act 1996, s 29). SASSA, which was formerly known as the Department of Welfare, seeks to promote social transformation and social justice as contained in the Constitution. SASSA was established to ameliorate some of the harsh impacts of the apartheid regime by reducing poverty and promoting social integration through the provision of social protection and social welfare to all who live in South Africa.

This approach is based on the commitment to promoting solidarity and humanity by empowering communities and creating conditions for self-sustaining livelihoods (Kaseke, 2010). In a bid to achieve the above, SASSA issues monthly care dependency grants, which include inter alia child support grants, disability grants, older person grants, and social relief of distress grants (Vally, 2016). However, the efficacy as well as the integrity of SASSA in promoting social security were questioned prior to COVID-19 pandemic, as the grants that are crucial in ensuring food security in households were not distributed on time (Vally, 2016). After the government placed the country on lockdown in March 2020, it introduced the Special-COVID-19 Social Relief of Distress grant which was administered by SASSA in which successful applicants would receive an amount of R350 per month (Department of Social Development, 2020). Other government departments, the Solidarity Fund, corporations, and non-profit organisations had to fill in this gap to promote food security. As a result, millions of zero- to low-income households were without food—and some still are without food. This would not have been the case if it were not for the corruption that occurred in the appointment of service providers to distribute social relief of distress grants (SRDs). These grants are used by millions of South African households to buy food and ensure a decent meal. SRDs are food parcels and food vouchers supplied to people experiencing unanticipated hardship or indigent families who have lost a family breadwinner or are victims of a disaster such as COVID-19. From November 2019 to the end of March 2020, SASSA delivered zero food parcels in six provinces, with a third of the population experiencing severe food insecurity, which means that most low-income households ran out of food or in some cases went for days without eating a decent meal. The SRDs issued by SASSA are not sustainable in ensuring long-term food security; innovative and sustainable measures of redress are needed to promote food security.

The Solidarity Fund was established during the pandemic (in March 2020) to support the national health response and to contribute to humanitarian relief efforts to fight COVID-19. The fund works closely with, but independently of, business and government in managing the donations to fund various initiatives to address the impact of the pandemic. However, the effectiveness of the Solidarity Fund is limited by bureaucratic corporate red tape that hampers the delivery of essential services, such as food, to those who need it. Moreover, the value of the Solidarity Fund in addressing the impact of the pandemic on society during the COVID-19 pandemic will require a proper audit; at the time of writing this paper, this had not yet been done.

Local community actions as well as the work done by local NGOs have been critical in supplementing the shortcomings of the government in terms of food security during the COVID-19 pandemic. More than 200,000 NGOs as well as local CBOs are registered with the Department of Social Development. The NGOs perform crucial humanitarian and advocacy functions that help to address the socio-economic issues affecting society. They play an integral role in promoting food security by partnering with local communities, corporate entities and government to deliver food parcels and cater to communities that do not have access to food and are affected by the pandemic. During the pandemic, the Department of Social Development and local NGOs established measures to address the food security crisis by setting up food parcel initiatives for local communities (Parliamentary Monitoring Group, 2020).

However, NGOs are dependent on grant funding and individual donations to sustain their operations. The COVID-19 pandemic has placed local and international businesses in financial distress and NGO donors have had to reduce their spending and retrench workers as a result. The global economic downturn caused by the pandemic means that less funding from both local and international donors will be available to support NGOs.

However, the pandemic has provided an opportunity to review the efficacy and overall sustainability of the structures put in place to promote food security. What is evident is that renewed efforts for self-sustaining food security are needed. A system that will allow South Africans to self-sustain and ensure food security through avenues that promote food sovereignty is pivotal in promoting food security for the most vulnerable. Communities need projects that will be centred on creating sustainable food value chains so that they are not merely passive recipients of government food parcels and social grants.

The process of producing and ensuring food distribution needs to be local and decentralised as opposed to the top-down approach currently used. This also speaks to the debate on land reform and how CSR can play a role in land reform (Kloppers, 2013). What the pandemic has done to South Africa is to expose the progress, if any, which has been made by the ANC government in addressing inequality. The evidence speaks for itself as the extreme income disparity and socio-economic inequality continue to grow, and innovative measures of redress are required. The development of such an economy is dependent on the provision of necessities, such as access to food, and ensuring food security through sustainable food creation processes. Corporations, governments and social partners can play a meaningful role in ensuring the realisation of food security through recalibrated efforts towards corporate social responsibility in food security.

4   CSR in South Africa

4.1   CSR Explained

The concept of corporate social responsibility (CSR) has been used and developed in a broad range of academic disciplines across the globe, resulting in a plurality of definitions and applications (Dahlsrud, 2008). The result is that there is a lack of consensus about what CSR really means and how it should be applied (Dahlsrud, 2008). Here, I follow Johnson (1971) in choosing a traditional definition for CSR in the context of the discussion of food security in South Africa. Traditionally, CSR denotes the voluntary actions of organisations and companies that go beyond the legal requirements to ensure that companies act in a socially responsible manner (Johnson, 1971). In the traditional sense, CSR entails the process of corporations performing altruistic acts through some form of corporate philanthropy where companies address the challenges facing society (De Bakker et al., 2005). This has customarily been through providing food parcels, pro bono health services, medical supplies and other necessities to local communities. It is best defined as a corporate governance and management model used to promote economic prosperity for businesses and sustainable growth (Amodu, 2020).

The theoretical foundation upon which the doctrine of CSR is applied in the South African context is based on the stakeholder value approach to doing business. In terms of this theory, the general understanding is that companies do not operate in a vacuum and that companies affect a broader body of corporate stakeholders than the shareholders of the company (Freeman, 2018). For this reason, it is argued that companies have a social responsibility to stakeholders (Mansell, 2013). The engagement of companies in societal issues is voluntary (Gamerschlag et al., 2011), with some scholars arguing that CSR, more especially in developing economies, such should be made a legal requirement as companies have a social responsibility to the communities in which they do business (Hamann & Kapelus, 2004). This argument brings to the fore one of the most controversial questions in corporate governance. In whose interests should companies operate and what is the fundamental purpose of the corporation? (Dodd, 1932). Scholars from the Anglo-American jurisdictions have advanced arguments to support a shareholder-centric response to this question for decades, with the earliest discussion being by Adam Smith (1822). The highly cited work of Berle and Means (1932) argues that the interest of the company is to ensure maximum returns for shareholder value. Thus, the role of the company is to act in the interests of the shareholders of the company and therefore the social responsibility of corporate managers is to the company (Friedman, 2007). This argument forms part of the shareholder primacy theory, which is the main theory of corporate law in capitalist jurisdictions such as the United States of America and the United Kingdom (Keay, 2007). Under the shareholder primacy theory, the main purpose of a corporation is to ensure maximum returns for shareholders by operating the company in the exclusive best interests of the shareholders (Keay, 2010). However, the efficacy of the shareholder primacy approach to corporate governance undermines the local realities in which companies do business and as such does not present itself as the appropriate lens through which corporate governance and the overall management of companies should be approached in developing economies such as South Africa (Mongalo, 2010).

The approach which has been adopted to promote CSR is the stakeholder-inclusive approach to corporate governance (Esser & Du Plessis, 2007). Under the stakeholder-inclusive approach, a company director must consider the interests of all relevant and identifiable stakeholders (Freeman, 2018). The interests of the environment, the employees, the local community and the consumers must be considered, provided they have a direct interest in the success or failure of the company (Kneale, 2012). In 2001, the Department of Trade and Industry (DTI) published a policy document setting out the guidelines for corporate law reform. The process entailed the modernisation of company law, which included a complete overhaul of company law (Mongalo, 2010). The company law, which was in place at the time, the Companies Act of 1973, was not in line with the new democratic dispensation. Moreover, there was a need to promote global competitiveness as the 1973 Act was not aligned with international corporate law (Mongalo, 2010). The theoretical foundation that was thus adopted by the DTI meant moving away from the shareholder primacy norm where companies operate in the best interests of the shareholders (Andreasson, 2011) and embracing a stakeholder-inclusive approach to doing business. However, the DTI implemented the middle ground of these two concepts, by adopting a United Kingdom-based concept of the enlightened shareholder value theory (Esser & Delport, 2017). Under this theory, company directors must consider the interests of the stakeholders, provided this is in the long-term best interests of the shareholders (Keay, 2007). Critics of the enlightened shareholder value approach argue that this theoretical foundation of the new Companies Act does not in itself present a significant departure from the shareholder primacy norm, which places significant limitations on the effective implementation of CSR by companies in a company law context (Esser & Delport, 2017). Under the enlightened shareholder value theory, the shareholder-centric aspect of corporate governance remains, significantly limiting effective stakeholder accountability and CSR in company law.

The position regarding CSR in South Africa is quasi-judicial as structures are put in place to promote CSR through voluntary codes of conduct (Rossouw et al., 2002) and CSR is implicitly acknowledged in law under the provisions of the Companies Act 71 of 2008 (hereafter "the Companies Act 2008") (Amodu, 2020). CSR under the Companies Act 2008 embraces traditional approaches to CSR which are implied in the provisions of the section 72(4) social and ethics committee, read together with regulation 43 of the Companies Regulations 2011 (Kloppers, 2014). Regulation 43(5)(a)(i)(aa) and (bb) sets out the mandate of the social and ethics committee: to report to the shareholders of the company on the company's social responsibility footprint by monitoring the company's social and economic development impact, including the company's standing in terms of the goals and purposes set out in the ten principles of the United Nations Global Compact, and compliance with laws such as the Employment Equity Act 55 of 1998 and the Broad-Based Black

Economic Empowerment Act 53 of 2003, to name a couple of the laws put in place to promote good corporate governance under the Act (Botha, 2016).

What is more pertinent to the current discussion is that regulation 43 of the Companies Regulations requires the committee to monitor the company's corporate citizenship, including the company's "contribution to the development of communities in which its activities are predominantly conducted or within which its products or services are predominantly marketed and record sponsorship, donations and charitable giving" (Kloppers, 2014). Regulation 43 implies that companies doing business in South Africa must conduct some form of corporate philanthropy, which is generally understood as CSR (Rahman, 2011). However, the effective implementation of the philanthropic dimensions of CSR is yet to be seen. This is because CSR initiatives that promote food security are often sporadic and inconsistent, with a marketing undertone that is used to attract more sales.

The "tick box" approach to CSR is further exacerbated by the implicit reference to CSR in the new Companies Act 2008, and the fact that there is no exclusive framework governing corporate responsibility. The COVID-19 pandemic has not only provided a mirror to reflect on where the country stands since the 1994 democratic elections, it has also shown that the country has an inequality problem, which is worsening the access aspect of the food security problem. For the country to address the inequalities that influence the food security crisis in a sustainable manner, innovative measures of redress need to be carefully considered. CSR can play a role in ameliorating the food crisis. Companies can partner with government in addressing the challenge of access to food. However, for food security to be effectively realised through CSR, the latter must be made a real requirement in law.

Domestic and foreign companies can address the food security crisis through CSR. One of the key challenges in the food crisis is that communities in crisis do not have the resources to embark on small-scale commercial farming that will enable them to invest in the community by planting seasonal fruits and vegetables and thus address the lack of access to food caused by poverty and unemployment. Through CSR companies can provide skills training in food processing and agricultural farming for local communities. The lack of funding from government as well as the highly debated and controversial issue of land reform (Kloppers, 2013) makes the discussion of food security in the land reform context a complex issue. CSR can provide avenues for food security and food sovereignty whilst government oversees the long and highly contentious but crucial discussions on land reform and food security.

4.2   An Overview of the Legal and Regulatory Framework for CSR

Owing to the central role that companies play in developing the economy through their contribution to the country's Gross Domestic Product (GDP), companies need to be seen as good corporate citizens who contribute to the development of the economy and society (Dorasamy, 2013). As corporate citizens, companies need to adopt practices that build their social corporate identity and place (Dorasamy, 2013).

The overarching idea of CSR is, therefore, to ensure that companies have an obligation to meet the needs of a range of stakeholders as part of a commitment to contribute to sustainable economic development, working with employees and local communities (Hamann, 2009). A good corporate citizen is defined as "one that has comprehensive policies and practices in place throughout the business, that enable it to make decisions and conduct its operations ethically, meet legal requirements, as well as show consideration for society, communities and the environment" (Fig, 2005). Some of the measures put in place to promote good corporate citizenship as a significant component of CSR are discussed below.

4.2.1 The Johannesburg Stock Exchange's Socially Responsible Investment Index

A key driver of CSR in South Africa is the Johannesburg Stock Exchange (JSE), which sets standards for listed companies to engage in CSR projects through the socially responsible investment (SRI) index (Kabir et al., 2015). Simply defined, SRI is an investment strategy that is aimed at creating a balance between financial and social objectives (Du Toit & Lekoloane, 2018). First implemented in May 2004, the SRI index seeks to identify companies listed on the JSE that integrate the principles of the triple-bottom line approach (Slaper & Hall, 2011) as well as to provide a tool for a broad holistic assessment of company policies and practices against globally aligned and locally relevant corporate responsibility standards (JSE SRI Index, 2014). The SRI index measures companies' policies, management systems, performance and reporting in relation to the three pillars of the triple-bottom line, namely, environmental sustainability, social and economic sustainability and corporate governance practice (Kabir et al., 2015).

From the investors' perspective, SRI entails investing according to one's ethical considerations, which include the Environmental, Social and Governance (ESG) considerations (Demetriades & Auret, 2014). This means that CSR, as a corporate practice, makes it easier for socially and environmentally conscious investors to determine which companies are worth investing in and which should be avoided if ethical investing is what they desire (Demetriades & Auret, 2014). Investors play a crucial role in the global effort to achieve sustainable development goals by ensuring that capital is appropriately raised and allocated (Widyawati, 2020). From the perspective of food security, this means that investors who wish to invest in companies which seek to promote their corporate citizenship status by advancing food security as a form of CSR get to invest in companies which are aligned with their approach to social responsibility.

4.2.2 The King Codes on Corporate Governance

The King Codes of Corporate Governance can be described as South Africa's take on modern corporate governance. The JSE requires all listed companies to incorporate the King Codes of Corporate Governance principles, to explain why they have not applied the principles, as well as to provide independent CSR assurance (Ackers & Eccles, 2015). Initially, the implementation of the King Code on Governance (King I) in 1992 provided a groundbreaking corporate governance initiative which extended beyond the financial and regulatory dimensions conventionally associated with corporate governance (Ackers & Eccles, 2015). The second King Code (King II) extended this shift towards inclusive corporate governance by incorporating the principles of good social, ethical and environmental practice (Ackers & Eccles, 2015). In 2009, the third King Code (King III) took corporate stakeholder account-ability to another level by requiring companies to integrate CSR, risks and opportunities into their core business strategies (Ackers & Eccles, 2015).

Both King II and King III cater for the need to adopt a "triple-bottom line" approach. Even though the King Codes are not mandatory, they take a "comply or explain" approach in King III which technically seeks to encourage organisations to apply CSR programmes or to justify why they have not adopted them (Nedelchev, 2013). King III defines the concept of CSR as the responsibility of the company in relation to the impacts of its decisions and activities on society and the environment, through transparent and ethical behaviour that, inter alia, contributes to sustainable development, including the health and welfare of society; takes account of the legitimate interests and expectations of stakeholders; complies with applicable law and is consistent with international norms of behaviour; and is integrated throughout the company and practised in its relationships (IoD King Report on Governance, 2009). King IV, which was released in 2016, requires companies to take the level of accountability even further by following the comply and explain approach (IoD King Report on Governance, 2016).

In particular, the self-regulatory aspect of the King Codes specifically opts for the inclusive stakeholder value approach which entails an interdependent relationship between the organisation and its stakeholders (Esser & Delport, 2017). Therefore, an organisation becomes attuned to the opportunities and challenges posed by the triple-bottom line context in which it operates by having regard to the needs, interests and expectations of material stakeholders (IoD King IV, 2016). Accord-ingly, the various interests of different stakeholders should be determined on a case-by-case basis (Esser & Delport, 2017). The King Codes are voluntary in nature and have no legal effect outside of JSE-listed requirements, thus providing significant limitations for the realisation of food security through CSR.

4.2.3 The Companies Act 71 of 2008

The Companies Act 2008 contains significant provisions for companies to consider the needs of wider stakeholder groups. Although the Act does not explicitly refer to the social responsibility of the company, section 7 of the Companies Act as well as the provisions of the section 72(4) social and ethics committee allow companies to report on the social responsibility of the company. The debate is continuing on whether the Companies Act should contain explicit provisions detailing the social responsibility of the company as a legal requirement. Some scholars argue that the inclusion of the committee in the Act does not represent a significant development towards the recognition of CSR under law as it does not explicitly refer to the concept of CSR or legally mandate companies to implement CSR (Botha, 2016).

5   Re-envisioning the Food Production System Through CSR: Opportunities and Limitations

It is important to note from the onset that the food crisis is not solely the outcome of a lack of food. The food system can supply enough food through either production or trade. A key challenge is the affordability aspect of the food security system as most zero- to low-income households cannot afford to purchase the food they need. The efforts to improve food security have focused on increasing the income of zero- to low-income households through social welfare grants, such as those provided by the Department of Social Development through the social development agency, SASSA, or by increasing incomes, employment and entrepreneurship.

Food security is an issue of national priority, which requires innovative measures of redress to successfully realise the right to food. However, the process of increasing income through SASSA is unsustainable, with corruption affecting the administration of the funds, which has led to millions of zero- to low-income households experiencing food security problems, both before and during COVID-19 pandemic. This is not to say that the social grants do not provide much-needed financial assistance to zero- and low-income households, but rather to say that the SASSA grants need to be supplemented with much more sustainable measures of food production and renewed measures to facilitate access to food within local communities. At present, the food security system is technocratic, with centralised solutions to production that obscure the broader dimensions of food production and consumption. This process has serious implications for the domestic economy which is currently plagued by high levels of unemployment and inequality. A more localised approach to food production and overall consumption needs to be followed, one that is cognisant of the social, cultural, political and socio-economic climate of the society. South Africa needs to urgently review the problem of access to food as most of the food produced is exported and not locally consumed. The gap between the excess food that is available and the means of South Africans to have access to food is a key challenge in food security. Companies can solve the critical challenges of our time by promoting inclusive corporate governance through social responsibility. Such inclusive governance entails the process of investing with purpose and empowering the local communities.

The advantage of using CSR in addressing the food security crisis will be the creation of sustainable avenues for food security. This can be done in the context of long-term projects centred on creating sustainable projects on access to food, employing the youth and addressing inequality and unemployment by financing projects that are centred on creating platforms for food sovereignty. This is not to say that companies must give charitable donations to local communities as required under the Companies Regulations 2011, but rather to say that the CSR investment of companies must be focused on creating long-term CSR projects instead of one-off sporadic donations that are unsustainable and poorly managed.

The value for companies beyond addressing the food security problem is that companies can win the public trust by strengthening the agenda of socio-economic transformation for all segments of society. This can also aid in solving some of the inequalities and disparities that South Africans face that go beyond food security. Addressing the food security problem extends beyond the provision of food for nutritional benefits. It has the potential to solve other socio-economic problems that result from food insecurity. Empowering local communities through CSR has the potential to reduce the income inequality gap which is primarily embedded along racial lines. Companies can play a role beyond the COVID-19 food parcel projects by creating sustainable value chains for investing in local food production and distribution through CSR.

However, notwithstanding the opportunities available for food security through CSR beyond the COVID-19 pandemic, CSR has some limitations in its application and form. First, the concept of CSR has been marred by being interpreted by multinational companies as a concept that promotes brand marketing, and for many years it has been used as a "tick box" initiative (Bondy, 2008). Most companies engage in CSR as a one-off initiative with a charity to show undertone, because the legal and regulatory frameworks promoting CSR do not ensure compliance. The King Codes of Good Corporate Governance, which are a key driver of CSR and good corporate governance, are voluntary and non-binding. Although JSE-listed companies are required to implement the King Codes as a listing requirement, not all companies are listed companies. Fundamentally, given the history of the country and the challenges of socio-economic inequality, the new Companies Act had the opportunity to embrace a radical approach to corporate responsibility by making CSR a legal requirement through company law. For now, the approach remains voluntary, which presents significant limitations for promoting food security through CSR.

6   The Way Forward

The future of food security lies in identifying the importance of access to food security and creating sustainable value chains for promoting food security through CSR in more sustainable ways. The recommended approach is the promotion of more localised food economies supporting small-scale farmers. Food security can be effectively realised when government, companies and other social partners invest in small-scale farmers, local community distribution agents, small food manufacturers and agro-processors, street traders and spaza shops (informal convenience shops). This will not only address the food security crisis but also ameliorate the high rates of unemployment and poverty (StatsSA, 2020). If companies support these processes with private sector corporate finance as a part of their CSR agendas, the transition to localised food systems for achieving sustainable food security systems after the COVID-19 pandemic can be realised.

7   Conclusion

This paper provides an overview of the landscape for food security and CSR by reviewing the impact of the COVID-19 pandemic on food security in South Africa. The paper revealed that the challenge to food security is in the lack of access to food created by the socio-economic problems of poverty, unemployment and inequality. The paper provides a theoretical understanding of CSR and food security, describes the multi-dimensional nature of CSR, identifies the barriers to food security through CSR and, finally, discusses the ways in which the food security system can be recalibrated through CSR.

COVID-19 has accentuated the food security crisis in South Africa. Social inequality is a serious feature in the structure and precarious ordering of society. Income-related inequality associated with the ownership of capital and other assets, lack of access to food and services, and individualised inequality hamper the administration of food security. Children, youth, women and people in the rural areas are most at risk as access to food is the key factor in food security. This inequality is on the rise; despite the vast changes brought about by democracy and the eradication of apartheid policies, wealth and basic resources are still accruing to the wealthy.

In the food security system, the COVID-19 pandemic has shown the critical weaknesses and inequalities which reinforce the challenges in securing access to food. The current food and social crisis is a wake-up call for South Africa to address the food security crisis insofar as access to food is concerned. Government, companies and other social partners can play a role in prioritising co-ordinated efforts to achieve food security in all its dimensions. Multi-stakeholder engagements are crucial—with companies and the representatives of those who are food insecure. These engagements should be the basis upon which a road map is formalised for immediate, medium- and longer-term initiatives to achieve food security. Companies, governments and other social partners can contribute to the immediate and long-term view of facilitating food security through CSR. However, in order to do so, CSR must be made a real requirement in law as the voluntary approach to CSR has had minimal impact in creating sustainable value chains for CSR beyond the food security agenda.

The approach to CSR by companies has been typical of most CSR perspectives across the globe. CSR has been voluntary in nature, with companies having a nonchalant approach to how they can play a role in ameliorating the challenges faced in the food security system of South Africa. The approach by companies is more self-serving, with CSR being used as corporate advertising, since most corporations are far from making CSR a real requirement. This perspective is further emphasised by the role of corporate law and the voluntary codes of good corporate governance, the King Codes, in promoting CSR as a real requirement in company law and in corporate South Africa. For South Africa to fully realise the value of CSR in promoting food security, CSR must be made a real requirement in law. This is because government initiatives to promote food security have been insufficient as a result of corruption, maladministration of state funds and the limitations embedded in government and other social partners such as NGOs. CSR in its legal form can assist government's efforts to promote food security. However, to effectively achieve food security through CSR, policy reforms are needed in the legal and regulatory frameworks supporting CSR in South Africa.

References

Ackers, B., & Eccles, N. (2015). Mandatory corporate social responsibility assurance practices: The case of king III in South Africa. *Accounting, Auditing, & Accountability, 28*(4), 515–550. https://doi.org/10.1108/AAAJ-12-2013-1554

Amodu, N. A. (2020). *Corporate social responsibility and law in Africa: Theories, issues and practices*. Routledge.

Andreasson, S. (2011). Understanding corporate governance reform in South Africa: Anglo-American divergence, the King Reports, and hybridization. *Business & Society, 50*(4), 647–673.

Berle, A., & Means, G. (1932). *Private property and the modern corporation*. Macmillan.

Beuchelt, T. D., & Virchow, D. (2012). Food sovereignty or the human right to adequate food: Which concept serves better as international development policy for global hunger and poverty reduction? *Agriculture and Human Values, 29*, 259–273.

Bondy, K. (2008). The paradox of power in CSR: A case study on implementation. *Journal of Business Ethics, 82*(2), 307–323.

Botha, M. M. (2016). Evaluating the social and ethics committee: Is labour the missing link? (1). *Tydskrif vir hedendaagse Romeins-Hollandse Reg, 79*(4), 580.

Bridgman, G., Van der Berg, S., & Patel, L. (2020). *Hunger in South Africa during 2020: Results from wave 2 of NIDS-CRAM*. University of Stellenbosch.

Cousins, B. (2016). *Land reform in South Africa is sinking. Can it be saved*. PLAAS, University of the Western Cape.

COVID-19. (n.d.). *Corona virus South African resource portal*. Accessed May 15, 2021, from https://sacoronavirus.co.za/

Cucinotta, D., & Vanelli, M. (2020). WHO declares COVID-19 a pandemic. *Acta Bio Medica: Atenei Parmensis, 91*(1), 157.

Dahlsrud, A. (2008). How corporate social responsibility is defined: An analysis of 37 definitions. *Corporate Social Responsibility and Environmental Management, 15*(1), 1–13.

De Bakker, F. G., Groenewegen, P., & Den Hond, F. (2005). A bibliometric analysis of 30 years of research and theory on corporate social responsibility and corporate social performance. *Business & Society, 44*(3), 283–317.

De Visser, J. (2019). *Multilevel government, municipalities and food security*.

Demetriades, K., & Auret, C. J. (2014). Corporate social responsibility and firm performance in South Africa. *South African Journal of Business Management, 45*(1), 1–12.

Desmarais, A. A. (2003). The via campesina: Peasant women at the frontiers of food sovereignty. *Canadian Woman Studies, 23*(1).

Disaster Management Act. (2002). *Amendment of Regulations issued in terms of section 27(2)*. Accessed Dec 4, 2020, from 43725gon999.pdf (www.gov.za)

Dodd, E. (1932). E. Merrick. "For whom are corporate managers trustees". *Harv. L. Rev., 45*, 1145.

Dorasamy, N. (2013). Corporate social responsibility and ethical banking for developing economies. *Journal of Economics and Behavioral Studies*.

Du Toit, E., & Lekoloane, K. (2018). Corporate social responsibility and financial performance: Evidence from the Johannesburg stock Exchange, South Africa. *South African Journal of Economic and Management Sciences, 21*(1), 1–11.

Dubow, S. (1989). *Racial segregation and the origins of apartheid in South Africa, 1919–36*. Springer.

Esser, I. M., & Delport, P. (2017). The protection of stakeholders: The South African social and ethics committee and the United Kingdom's enlightened shareholder value approach: Part 1. *De Jure, 50*(1), 97–110.

Esser, I. M., & Du Plessis, J. J. (2007). The stakeholder debate and directors' fiduciary duties. *S. Afr. Mercantile LJ, 19*, 346.

Feinstein, C. H. (2005). *An economic history of South Africa: Conquest, discrimination, and development.* Cambridge University Press.

General Household Survey. (2020). Statistic South Africa Statistical Release General Household Survey 2020.

Freeman, R. E., Harrison, J. S., & Zyglidopoulos, S. (2018). *Stakeholder theory: Concepts and strategies.* Cambridge University Press.

Friedman, M. (2007). The social responsibility of business is to increase its profits. In *Corporate ethics and corporate governance* (pp. 173–178). Springer.

Gamerschlag, R., Möller, K., & Verbeeten, F. (2011). Determinants of voluntary CSR disclosure: Empirical evidence from Germany. *Review of Managerial Science, 5*(2–3), 233–262.

Global Food Security Index. (2019). Available *at South Africa food security* (eiu.com).

Hamann, R. (2009). South Africa: The role of history, government, and local context. In *Global practices of corporate social responsibility* (pp. 435–460). Springer.

Hamann, R., & Kapelus, P. (2004). Corporate social responsibility in mining in southern Africa: Fair accountability or just greenwash? *Development, 47*(3), 85–92.

Hanoman, J. (2017). *Hunger and poverty in South Africa: The hidden faces of food insecurity.* Taylor & Francis.

IOD. (2009). *The king code of governance for South Africa (2009) and king report on governance for South Africa* (2009) (King III).

IOD. (2016). *The king code of governance for South Africa (2016) and king report on governance for South Africa* (2016) (King IV).

Johnson, H. L. (1971). *Business in contemporary society: Framework and issues.* Wadsworth Pub.

Kabir, M. H., Mukuddem-Petersen, J., & Petersen, M. A. (2015). Corporate social responsibility evolution in South Africa. *Problems and Perspectives in Management, 13*(4), 281–289.

Kaseke, E. (2010). The role of social security in South Africa. *International Social Work, 53*(2), 159–168.

Keay, A. (2007). Tackling the issue of the corporate objective: An analysis of the United Kingdom's enlightened shareholder value approach. *Sydney Law Review, 29*, 577.

Keay, A. (2010). Shareholder primacy in corporate law: Can it survive? Should it survive? *European Company and Financial Law Review, 7*(3), 369–413.

Kloppers, H. J. (2013). Driving corporate social responsibility (CSR) through the companies act: An overview of the role of the social and ethics committee. *Potchefstroom Electronic Law Journal/Potchefstroomse Elektroniese Regsblad, 16*(1), 165–199.

Kloppers, H. (2014). Driving corporate social responsibility through black economic empowerment. *Law, Democracy & Development, 18*(1), 58–79. https://doi.org/10.4314/ldd.v18i1.4

Kneale, C. (2012). *Corporate governance in southern Africa.* Chartered Secretaries Southern Africa.

Labadarios, M. (2011). Food security in South Africa: A review of national surveys. *Bulletin of the World Health Organization, 89*(12), 891–899. https://doi.org/10.1590/S0042-96862011001200012

Mansell, S. F. (2013). *Capitalism, corporations and the social contract: A critique of stakeholder theory.* Cambridge University Press.

Mongalo, T. H. (2010). An overview of company law reform in South Africa: From the guidelines to the companies act 2008. *Acta Juridica,* xiii–xxv.

Nedelchev, M. (2013). Good practices in corporate governance: One-size-fits-all vs. comply-or-explain. *International Journal of Business Administration, 4*(6), 75–81.

Ngcoya, K., & Kumarakulasingam, N. (2017). The lived experience of food sovereignty: Gender, indigenous crops and small-scale farming in Mtubatuba, South Africa. *Journal of Agrarian Change, 17*(3), 480–496. https://doi.org/10.1111/joac.12170

Parliamentary Monitoring Group. (2020). *COVID-19 NGO food distribution; SRD R350 grant; with Ministry.* https://pmg.org.za/committee-meeting/30339/

Rahman, S. (2011). Evaluation of definitions: Ten dimensions of corporate social responsibility. *World Review of Business Research, 1*(1), 166–176.

Rossouw, G. J., Van der Watt, A., & Rossouw, D. M. (2002). Corporate governance in South Africa. *Journal of Business Ethics, 37*(3), 289–302.

SANHANES. (2013). *The south African National Health and nutrition examination survey: SANHANES-1.* HSRC Press.

Seekings, J., Nattrass, N., & Seekings, J. (2016). *Poverty, politics & policy in South Africa: why has poverty persisted after apartheid.* Jacana Media.

Sekhampu, T. J., & Grobler, W. (2011). The expenditure patterns of households receiving the State's old-age pension (SOAP) grant in Kwakwatsi township. *The Journal for Transdisciplinary Research in Southern Africa, 7*(1), 41–52. https://doi.org/10.10520/EJC111925

Slaper, T. F., & Hall, T. J. (2011). The triple bottom line: What is it and how does it work. *Indiana Business Review, 86*(1), 4–8.

Smith, A. (1822). *The theory of moral sentiments* (Vol. 1). J. Richardson.

South African Government. (2021). *COVID-19/Novel Coronavirus|South African Government.* https://www.gov.za/Coronavirus

Statistics South AfricaSouth Africa. (2020). *SA economy sheds 2,2 million jobs in Q2 but unemployment levels drop|Statistics South Africa.* http://www.statssa.gov.za/?p=13633

Statistics South Africa. (2017). *General household survey: 2017: Statistical release (P0318).* Department Statistics South Africa Republic of South Africa. https://www.statssa.gov.za/publications/P0318/P03182017.pdf

Statistics South Africa. (2019). *General household survey: 2019: Statistical release (P0318).* Department Statistics South Africa Republic of South Africa. https://www.statssa.gov.za/publications/P0318/P03182019.pdf

Statistics South Africa. (2020). *General household survey: 2020: Statistical release.* Department Statistics South Africa Republic of South Africa. http://www.statssa.gov.za/publications/P0318/GHS%202020%20Presentation%202-Dec-21.pdf

Teka Tsegay, Y., Rusare, M., & Mistry, R. (2014). Hidden Hunger in South Africa: The faces of hunger and malnutrition in a food-secure nation.

The ConstitutionConstitution of the Republic of South AfricaSouth Africa (1996).

Vally, N. T. (2016). Insecurity in South African social security: An examination of social grant deductions, cancellations, and waiting. *Journal of Southern African Studies, 42*(5), 965–982.

Van der Berg, S., Zuze, L., & Bridgman, G. (2020). *The impact of the coronavirus and lockdown on children's welfare in South Africa: Evidence from NIDS-CRAM wave 1.* Department of Economics, University of Stellenbosch.

Widyawati, L. (2020). A systematic literature review of socially responsible investment and environmental social governance metrics. *Business Strategy and the Environment, 29*(2), 619–637.

Wills, G., Patel, L., van der Berg, S., & Mpeta, B. (2020). *Household resource flows and food poverty during South Africa's lockdown: Short-term policy implications for three channels of social protection* https://cramsurvey.org/wpcontent/uploads/2020/07/Wills-household-resource-flows-and-food-poverty-during-South-Africa%E2%80%99s-lockdown-2.pdf.

Windfuhr, M., & Jonsén, J. (2005). *Food sovereignty. Towards democracy in localized food systems.* ITDG Publishing.

World Bank. (2014). https://data.worldbank.org/indicator/SI.POV.GINI?locations=ZA

World HealthHealth Organization. (2020). Accessed from https://www.who.int/emergencies/diseases/novel-coronavirus-2019/events-as-they-happen

World InequalityInequalities Database. (2020). Accessed from https://wid.world/country/south-africa/

Mikovhe Maphiri Attorney of the High Court South Africa, Academic (Lecturer) in the Department of Commercial Law at the University of Cape Town. LLB (University of Limpopo) LLM (University of Cape Town) PhD Candidate (University of Cape Town).

Mikovhe is an early career researcher whose research interests are in corporate governance CSR and sustainability in South Africa. She is a lecturer in the Department of Commercial law at the University of Cape Town where teaches law to both law and non-law students.

Reducing Negative Environmental Impacts in Conventional Agriculture, but Not the Amount of Harvest: A Multi-stakeholder Joint Project in Conventional Citrus Production in Spain

Marina Beermann, Patrick Freund, and Felipe Fuentelsaz

Abstract

An ecological footprint reduction of Germany's retailer EDEKA and its private brand products has led to a multi-stakeholder project in Spanish citrus production. Developing and applying partly participatory action research and Agroecology principles has led to a considerable reduction of negative environmental impacts in citrus cultivation.

Keywords

Agroecology · Sustainable supply chain management · Multi-stakeholder project · Intersectoral partnership · Sustainable citrus production

1   Introduction

Agriculture around the world is contributing to the loss of biodiversity and exacerbates global warming, at the same time agriculture itself is also highly vulnerable to the impacts of climate change and biodiversity loss (WWF, 2020; IPBES, 2019; IPCC, 2018; Steffen et al., 2015). Conventional production of food in

M. Beermann (✉)
WWF/EDEKA Strategic Partnership, WWF Germany, Berlin, Germany
e-mail: marina.beermann@gmx.de

P. Freund
Sustainable Supply Chains, Berlin, Germany
e-mail: patrick.freund@wwf.de

F. Fuentelsaz
Proyectos Agricultura y Agua, WWF España, Madrid, Spain
e-mail: agrofelipe@wwf.es

© The Author(s), under exclusive license to Springer Nature Switzerland AG 2022   159
S. O. Idowu, R. Schmidpeter (eds.), *Case Studies on Sustainability in the Food Industry*, Management for Professionals,
https://doi.org/10.1007/978-3-031-07742-5_7

resource-intensive monocultures, a growing world population and an increasing demand for animal protein, intensify several worldwide societal challenges which are mainly summarized in the sustainable development goals (SDGs). A transition of food systems toward sustainability is the main goal of the strategic partnership of EDEKA and WWF which originally started in 2009. This partnership aims to reduce the environmental footprint of the retailer's private brand products and to promote sustainable consumption of its daily more than 12 million customers in Germany.

Current EU agricultural initiatives like the "Farm to Fork Strategy" show the urgent need for holistic, interrelated, the whole agrifood system encompassing transformations instead of change in subsystems only. Existing intersectoral partnerships already fulfill some relevant preconditions for supply chain overall transformation. The EDEKA/WWF partnership itself as intersectoral partnership and the initiated joint multi-stakeholder project in Spain, which directly include several citrus growers, environmentalists, agricultural experts, trader and retailer, demonstrates this. The participatory-driven project approach enables the integration of diverse but necessary perspectives regarding, e.g., Spanish environmental legislation, specific locally environmental as well as the consideration of retail requirements and shopping habits of German consumers.

2   Conceptional Foundations

This chapter is based on a still running and 2015 initiated joint agricultural multi-stakeholder project. Starting from the idea to integrate conservation goals into concrete agricultural practice for conventional agriculture, the joint project in citrus production in Spain has its origin as one of currently three agricultural projects within the strategic partnership between WWF Germany and EDEKA. Taking this into consideration, several links to ongoing academic research streams exist. In particular to research on the role and functions of intersectoral partnerships, sustainable supply chain management (SSCM), and Agroecology.

2.1   Intersectoral Partnerships

In relation to an increasing relevance of sustainable development in certain domains like states, markets and civil society in general and in particular since the report of the World Summit on Sustainable Development in 2002, intersectoral partnerships were declared as tool of progress for sustainable development. Furthermore, in 2015 the Sustainable Development Goals (SDGs) officially came into force and defined partnerships as original goal, SDG 17.

A common definition defines "intersectoral partnerships as collaborative arrangements in which actors from two or more spheres of society (state, market and society) are involved in a non-hierarchical process, and through which these actors strive for a sustainability goal" (Van Huijstee et al., 2007, 75; Glasbergen, 2007).

In literature, different types of partnerships are differentiated and also in practice a huge and increasing variety of partnership constellations exists (Sperfeld et al., 2018). Van Huijstee et al. (2007, 75) identified two major perspectives in partnership literature: an institutional and an actor perspective: "The first, the institutional perspective, looks at partnerships as new arrangements in the environmental governance regime. The second, the actor perspective, frames partnerships as possible strategic instruments for the global achievement and problem-solving of individual actors." Or more precisely, the second perspective does "look more *into* partnerships, than *at* partnerships and their functions among other forms of governance" (Van Huijstee et al., 2007, 81). According to this perspective aspects of trust, motives and benefits of actors in as well as success factors of partnerships are analyzed in literature. Whether partnerships can contribute to a traceable and proven progress in reducing negative impacts (externalities), caused by economic activities, is an ongoing research question of sustainable development and also part of the partnership literature. A German research project, funded by the German Environmental Agency identified relevant criteria for impactful NGO–company partnerships (Sperfeld et al., 2018) and analyzed also the partnership of WWF Germany and the retailer EDEKA. The overall common goals of the WWF/EDEKA partnership are a reduction of EDEKA's ecological footprint, especially of their approx. 8.000 private brand products and promotion of sustainable consumption of approx. 12 million German customers per day. The range of conservation goals and activities encompasses the major conservation fields of WWF and includes activities along the whole supply chain and political advocacy in Germany as well as in other regions of procurement. The study showed that this partnership has transformative effects due to its focus on the core business of EDEKA (private brand products), its precise goal setting and annual externally verified progress reporting and its activities above the partnership limits (e.g., collective actions in certain procurement countries).

2.2 Sustainable Supply Chain Management

One research stream in supply chain management is sustainable supply chain management (SSCM). Its relevance has constantly grown, both in academia and in practice. There are several reasons for it: Companies, especially in the food industry like EDEKA as Germany's biggest retailer are more directly vulnerable to changes along their supply chains (e.g., crop failures worldwide can lead to direct delivery and supply bottlenecks), they operate in a highly competitive and dynamic sector with constantly rising consumer awareness and expectations for social and environmental issues, especially in the field of FMCG. Having a secure supply of goods, a positive image, and therefore competitive advantages are motives of FMCG companies to invest in sustainable supply chain management measurements.

Furthermore, more and more initiatives and regulations, induced by NGOs, the EU, and national governments arise and push partly measures on a voluntary, partly binding basis (e.g., SDGs, UN Guiding Principles on Business and Human Rights, the call of the European Parliament for a new law to stop EU-driven deforestation, the EU farm to fork strategy as part of the European Green Deal or the German initiative for a supply chain law or the binding goal of climate neutrality by the year 2050).

One of the most common definitions of SSCM in literature is given by Seuring and Müller (2008). They define SSCM as "the management of material, information and capital flows as well as cooperation among companies along the supply chain while taking goals from all three dimensions of sustainable development, i.e., economic, environmental and social, into account which are derived from customer and stakeholder requirements."

The field of SSCM is constantly evolving in research and in practice. According to Beske et al. (2014) commonly applied measurements in SSCM (and related dynamic capabilities) can be structured in five SSCM superordinate categories (Strategic orientation, Supply chain continuity, Collaboration, Risk management, and Pro-activity). Concrete practices of each category are only outlined in the paper of Beske et al. (2014).

Taking the role of the WWF/EDEKA partnership in context of SSCM into consideration, the partnership itself is an SSCM overall acting instrument for the retailer EDEKA, in particular for the improvement of their private brand assortment and their supplier management. The partnership's individual goals and activities address different aspects which Beske et al. (2014) differentiate, e.g., in the field of supply chain continuity due to the building of resilient long-term relationships, partner development in the sense of capacity building, in the field of collaboration due to joint project development, e.g., the joint citrus project in Spain or in the field of pro-activity due to an intense stakeholder management and constantly coevolving and learning. The partnership allows EDEKA to implement CSR-related requirements and to improve their supplier management, logistics, resource usage, and overall performance.

2.3   Agroecology

Although the production of food is directly and more obviously than other manufacturing industries dependent on functioning ecosystem services, the industrialized agricultural mainstream is producing in a so-called conventional, environmentally and partly socially harmful, non-resilient manner (FAO, 2012, 2020). To change that, many, partly ambiguous reasons for it must be considered if solutions shall develop positive effects in agricultural practices, in markets and consumer shopping behavior. In many Western countries and in particular, in Germany, the public divides agriculture generally into mainly two competing concepts: conventional and organic agriculture. In practice also mixed types exist, and the conceptual boundaries are not always clearly separated.

Two of the currently three agricultural projects within the strategic partnership between WWF Germany and EDEKA are based on conventional agriculture. One is the joint project on conventional citrus production in Spain. The second is a project on conventional banana production in Columbia and Ecuador. Taking into consideration that enhancing organic agriculture is linked to various prerequisites (on and off farm), plus knowing that the main negative externalities result due to conventional agriculture, we aim to improve conventional agriculture in the most ecologically harmful and economically relevant cultures as much as possible. In doing that, we lean on ideas of Agroecology.

According to Gliessmann (1998), Agroecology is defined as "the application of ecological concepts and principles to the design and management of sustainable agroecosystems." The concept of Agroecology is not strictly defined in every detail, which means that you hardly find universally applicable agricultural practices. It is rather an umbrella for various approaches under which certain practices can be assigned. The overall idea—same in our joint projects—is to work with nature and not against it and to (re)build stable agroecosystems that have a high structural and functional similarity to natural ecosystems in its common ecoregion. Bellamy and Ioris (2017) differ between different streams within the Agroecology movement and refer to Francis et al. (2003, 100) that define Agroecology as a cross-disciplinary and integrative approach. Thus, Agroecology is "the integrative study of the ecology of entire food system, encompassing ecological, economic and social dimensions, or more simply the ecology of food systems." This perfectly correlates with the general setup of the strategic partnership of WWF and EDEKA and the joint citrus project in particular.

3   The Role of Citrus Production in Spain

In terms of social and environmental value, the fruit and vegetables sector plays a crucial role in the Spanish agriculture and the Spanish economy in general. The diversity of agricultural products is high and so is the use of applied systems: from conventional to organic, from outdoors to greenhouses, including a constantly increasing use of irrigation systems (Expósito & Berbel, 2017). According to de Castro (2018, 21) "the sector is continuously increasing its economic value, and Spain is the main fruit and vegetables producer of the European Union and the 5th worldwide (in terms of value). Around 50% of production is exported, but in some particular cases even more than 70% are destined to foreign markets. The main products that are exported are vegetables from greenhouses (tomatoes, pepper and cucumber), citrus fruit, peaches and nectarines. From exports, 93% go to the EU market (Germany, France, United Kingdom and the Netherlands)."

From a global perspective the role of Spanish citrus production is not prior, but according to Sanfeliu (2018, 32) "it is the first exporter of oranges and tangerines worldwide. Exporting on average 3 704 000 tonnes in 2017 and dedicating a total of 299 518 ha to cultivate citrus. From those, 50% are used for oranges, 37% to tangerines, and 13% to lemons. Inside Spain, Valencia produces 54% of all Spanish citrus production, followed by Andalusia (28%), Murcia (14%), Cataluña (4%) and the rest in Baleares and other regions. In the case of the European Union, Spain supplies 80% of tangerines and 69% of oranges making it the largest supplier of citrus to the continent. The main destination markets are Germany (28%), France

(23%), The UK (9%) and the Netherlands (6%)." Also, to the latest records, 2019/ 2020, Spain was Europe's largest citrus fruit producer (Statista, 2021a, b) with production mainly located in four regions as Mari and Peris (2005) state: Catalonia, the Comunidad Valenciana (CV), Murcia, and Andalusia.

The joint citrus project is located in the Guadalquivir River Basin and close to the Coto de Doñana, a national park in Andalusia and Spain's most important wetland. With about 4.4 million hectares of agricultural land, Andalusia is one of Europe's food production centers, providing not only neighboring countries with oranges, strawberries or, more recently, also with tropical avocados or mangos. About 22% of the production in this region is organic. With approximately 7%, the organic share in citrus cultivation is significantly lower (WWF progress report, 2019).

Andalusia is an important migration area for migratory birds, with more than six million migratory birds each year in the Guadalquivir estuary and the Coto de Doñana national park.

Furthermore, Andalusia is one of the last retreats for endangered species such as the Iberian lynx or the Spanish imperial eagle. However, conflicts between conventional farming and biological diversity exist for years and dramatically intensify. Land conversion, the intensive use of agrochemicals and groundwater pollution, the loss of soil fertility and biodiversity are related to Andalusian conventional farming.

Furthermore, the excessive and often illegal use of water for irrigation poses challenges for highly valuable ecosystems: 88% of water withdrawals in the Guadalquivir river basin is used in agriculture. This has led to immense ecological stress for the UNESCO World Heritage Site Coto de Doñana that is highly dependent on the availability of water in the river basin.

Expósito and Berbel (2017) state that "The Guadalquivir RB, which is representative of the Mediterranean region, contains 23% of the total irrigated area in Spain. The competitiveness of its agriculture, with a focus on high value crops in a context of water scarcity, explains the remarkable expansion in irrigated areas and modernization of irrigation systems in recent decades, to the point where there are now more than 850,000 ha of irrigated land. The Guadalquivir is the longest river in southern Spain, with a length of about 650 km and a basin extension of over 57,527 km^2."

WWF Spain analyzes and collects data on water and other environmental indicators for several decades in that region. The documented overexploitation of the aquifer, or "water theft," has been a problem for more than 30 years and, although the Guadalquivir River Basin Authority is now sealing illegal wells, it is essential that concrete measurements get implemented (the so-called "Strawberry Plan").

On April 05, 2010, WWF Spain filed a formal complaint with the European Commission denouncing the abusive and unsustainable extraction of water in the Doñana area for the intensive cultivation of soft fruits. This activity was seriously damaging the aquifer as well as the protected habitats and species which were the reason this area was included in the Natura 2000 network. The European Commission initiated infringement proceedings against Spain, which resulted in a reasoned opinion being issued on April 28, 2016, for the obvious infringement of the Water Framework Directive, the Habitats Directive and the Birds Directive, and which, in view of the alarming situation of both the aquifer and the biodiversity, culminated in action against Spain in the EU Court of Justice in 2019 (WWF España, 2020b).

The general relevance of the Guadalquivir river basin for that region, but also for agriculture exists from two perspectives: poor agricultural practices that lead to negative effects as the silting up of streams and the clogging of the marshland that cause eutrophication in various lagoons and water bodies that provide unique habitats for highly threatened species, such as the marbled teal, a critically endangered waterbird (WWF España, 2020a). At the same time, agriculture is highly dependent on stable and existing ecosystem services and resources like water. This is of high relevance not only in the short-term perspective of individual growers, but also in the long-term for the economy of Spain due to the high importance of the agricultural sector in general. In this economic context, the social side of labor-intensive food production is also important to mention. According to Rye and Scott (2018, 928) "it is now clear that the rural realm has been, and is being, transformed by immigration, and that low-wage migrant workers in the food production industry are playing a particularly prominent role in this transformation." The authors furthermore emphasize that there is a clear link between a growing demand for migrants and European food production and that "migrants' integration into rural communities has often been problematic, with the emphasis being on the need for, rather than needs of, low-wage migrant workers."

4 Development of a Joint Project Framework within an Intersectoral Partnership

The joint citrus project is part of the strategic partnership for sustainability, which the NGO WWF Germany and the German retailer EDEKA entered into in June 2012. The two main goals of the partnership are to significantly reduce EDEKA's ecological footprint and to promote and to encourage sustainable consumption on the part of consumers. With respect to the main conservation goals of WWF, detailed and targeted goals are defined and both partners jointly work on achieving them to make a positive contribution toward the main current, global challenges of climate change, freshwater protection, loss of biodiversity, and building a circular resource management. Together and each in their particular sphere of influence. The focus areas with respect to the assortment are fish and seafood, timber/paper/tissue, palm oil, soy/sustainable feedstuff, freshwater, climate mitigation, packaging, and sustainable procurement of critical commodities.

In November 2014, a resolution was passed to continue this partnership in the long-term and EDEKA subsidiary Netto Marken-Discount got included in the partnership. The same happened to Iwan Budnikowsky GmbH & Co. KG, a drugstore as part of the EDEKA Group in June 2020.

A relevant component of the partnership is to transform EDEKA's, Netto Marken-Discount's, and Budnikowsky's own range of private brand products into more sustainable alternatives.

The EDEKA Group is the largest German supermarket corporation. It consists of more than 4.000 cooperatives of established merchants all operating under the umbrella organization (cooperative) EDEKA Group. The national coordination is done by the EDEKA Zentrale Stiftung & Co. KG in Hamburg, responsible for all national purchasing, brand management as well as marketing, and communication in relation to national campaigns and brands. The established merchants are organized in seven regional wholesale houses that coordinate the wholesale and marketing of regional brands. They all own EDEKA as cooperatives.

In literature, global and national retail markets are often classified as key actors with respect to sustainability challenges. Especially food retailers are in focus due to their role as Hermsdorf et al. (2017, 2532) with reference to Midgley (2014) and Cicatiello et al. (2016) state: "Food retailers play a pivotal role as brokers between producers and consumers." Scientists further show that there is a constant trend, globally and on national levels of increasing capital concentration in the food retail sector. Which means that more and more only few actors dominate the retail markets. And that correlates with rising power, also in political regard. Fuchs et al. (2009, 51 f.) describe three main trends with respect to the global food governance and the role of the globally most relevant retailers: Firstly, the trend of oligopoly, secondly, the control of the product chain from farm to shelf and thirdly, the development of competition that is not only based on price but also on quality. Finally, Fuchs et al. (2009, 82) delineate that "retailers have acquired rule-setting power as a result of their material position within the global economy. Due to their control of networks and resources, retailers have gained the capacity to adopt, implement, and enforce privately set rules (standards), which then take on obligatory quality." Furthermore, the authors point to the increasing media presence and marketing activities of retail companies that also increases their "structural and discursive power."

EDEKA with its cooperative structure is only located in the German market, therefore globally trading but not selling or operating in other countries than Germany. Therefore, not all of the in literature described characteristics are applicable. Although, also the German retail market is characterized by a high market concentration and a strong competition between mainly four big retailers in the German market: Edeka, Rewe, Schwarz-Group and Aldi as discount-format capture a market share of around 70% (Statista, 2021a, b).

4.1   Project Background

Food retailers are often in public spotlight when it comes to sustainability. From their perspective, it is very challenging to act and to act with highest possible positive sustainability impact. Why is that?

EDEKA, for example, has around 25.000 private brand suppliers and sources in more than 90 countries worldwide. As every other company also EDEKA and WWF must set strategic and profound priorities, asking themselves: Where are sustainable transformations needed the most? Where is the highest match between: high negative ecological impacts caused by the cultivation, processing, and transportation of a particular fruit or vegetable on the one side and a high assortment relevance on the other side? Aiming that transformative change shall evolve the greatest possible positive effect.

Additionally, taking into consideration whether for a certain commodity, fruit or vegetable already sufficient sustainable approaches, for example, in form of standards/certification schemes exist or not. And finally, concepts shall bring relevantly and the most effective improvements. Therefore, the most harmful negative impacts need to be identified.

In order to generate methodically assured answers to these questions, WWF and EDEKA decided to conduct an impact analysis in the very beginning, using the method of ecological extended input–output analysis (EEIO). Which also included empirical data from EDEKA's purchasing department of 94 fruits and vegetables from more than 60 different countries.

The method EEIO is based on a combined approach with its origin in Ecological Footprinting. Ecological Footprints are merely used to describe the demand of humans on nature (Wiedmann et al., 2006, 30). Therefore, the EEIO combines two approaches, the Ecological Footprinting and economic input-output-analysis which "allows the disaggregation of existing National Footprint estimates by economic sector, final demand category, sub-national area or socio-economic group, whilst ensuring full comparability or results" (Wiedmann et al., 2006, 30). The fundamentals of this approach go back to Leontieff (1966, 1970), among others. Beermann and Wick (2019) delineate that using this combined approach within corporate sustainability management offers several advantages. Especially, if the analysis results are supplemented with external cost rates. This allows to compare and communicate results with the help of a uniform relevance measure (e.g., Euro €) across different impact categories like global warming potential (GWP), water, or land use.

The impact analysis showed that oranges and mandarins that EDEKA sources from Spain have the highest environmental costs in total within their whole fruit and vegetable assortment (conventional bananas from Ecuador and Columbia are in second place). The main reasons are the intensive cultivation of citrus including the use of pesticides and fertilizers both with a high GWP (global warming potential) and the high use of water in a region with high water scarcity. Therefore, also compared to other reference countries, the cultivation of oranges and mandarins in Spain resulted in the highest environmental costs by far. Plus, oranges and mandarins are particularly popular with German customers during winter, and this is reflected in the quantities imported by EDEKA. More than 80% of the fruit comes from Spain. Together with the region of Valencia, the Andalusia region in southern Spain is the most important production area for the retailer.

4.2   Participatory-Driven Project Design and Agroecology Principles

On basis of the results of the EEIO analysis, a profound research of possibly already existing approaches or standard/certification systems for a more sustainable cultivation of conventional citrus was conducted. Besides organic certification, no progressive sustainability certification system for citrus got identified, which supported the decision to start a joint project.

In order to describe and explain better the mindset and conceptional framework of this (and also the other two existing joint agricultural projects within the strategic partnership of WWF and EDEKA) joint project, the author refers to Méndez et al. (2013, 6) and their differentiation between two encompassing approaches within Agroecology. Besides a research and methodology-driven approach, they state a second approach that "developed from firm roots in the sciences of ecology and agronomy, into a framework that seeks to integrate transdisciplinary, participatory, and action-oriented approaches, as well as to critically engage with political-economic issues that affect agro-food systems." See Fig. 1.

The joint citrus project is characterized by many aspects by Méndez et al. (2013) so-called a transdisciplinary, participatory, and action-oriented approach. Table 1 illustrates how and which aspects of participatory action research principles (PAR) on the one side, and which aspects of Agroecology principles on the other side get addressed in the joint project.

The initial step that made the project possible in the first place, came from EDEKA and the first pilot grower—even before the EEIO analysis was conducted. Both had the interest to improve the production system in a more environmentally friendly way. The pilot grower had already independently established initial measures to promote native birds in its production areas. He wanted to do more, but unknowing how exactly and how this could be reflected in the commercial cooperation with EDEKA. Thus, with more far-reaching production changes, there would also be increased production costs without a possibility to market the oranges differently.

Supported by the results of the EEIO-based impact analysis EDEKA and WWF Germany decided to start a joint project to improve the citrus production in Andalusia. With the help of WWF Spain, a group of experts in different fields of good agricultural practices was set up and the pilot grower—open-minded and courageous—was convinced to be part of this multi-stakeholder project. Furthermore, a common understanding of the shared mission had to be developed, considering the different interests, limitations, and risks of each stakeholder. In the beginning, overall conceptional questions dominated. For example, concerning the main goals exactly. Whether the project aims to enhance organic farming of citrus production in Andalusia in order to reduce the ecological threat in a specific region or whether transformations are more effectful in conventional farming systems. With respect to organic farming, the project group asked themselves whether it would be

"Agroecologies"

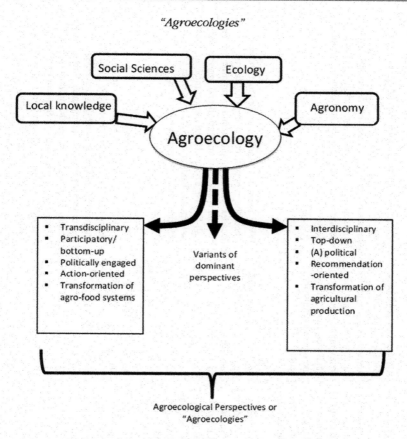

Fig. 1 Schematic representation of the evolution of different types of agroecologies. Source: Méndez et al. (2013, 7/Fig. 1)

possible to change the consumers' consumption patterns (of German customers) in order to create a much higher demand for organic oranges and mandarins in order to close the loop of supply and demand.

In the end, the decision was made to optimize conventional citrus cultivation. The project measures were elaborated and further developed through a participatory multi-stakeholder process. The involved actors decided to harmonize conventional and organic cultivation practices. Therefore, reducing the specific negative environmental impacts of conventional farming as much as possible without a decrease of the amount of harvest. Promoting and enabling resilience as overall goal, differentiated into four main impact drivers: responsible water use, efficient fertilization and soil fertility, more sustainable crop protection and conservation, and promotion of biodiversity.

Table 1 Referring to Méndez et al. (2013, 10/Table 1) and extended by the author

Participatory action research principles (PAR) (Méndez et al., 2013)	Application within the joint WWF/EDEKA citrus project	Agroecology principles (Méndez et al., 2013)	Application within the joint WWF/EDEKA citrus project
PAR foregrounds empowerment as community partners play key roles in defining the research agenda.	The project consists of a transdisciplinary team that has regular project meetings and workshops of several days duration with all project actors. The workshops and the general mindset of the project follow a participatory approach that is characterized by a bottom-up and often iterative decision process.	Agroecologists work with farmers, food consumers, communities, agricultural ministries, food advocates, and others to empower people.	All actors are directly involved. Agricultural ministries only indirectly, same applies to food consumers and the broader local communities in Andalusia.
PAR processes are context dependent as they bring together interdisciplinary teams responding to stakeholder aspirations.	The project consists of a transdisciplinary team that involves people from the following institutions: WWF Germany and Spain/NGO; various citrus growers and their employees in Andalusia as well as their related distributors/agricultural production; technical advisers and conservation experts / science, consultancy; EDEKA and Fruchtkontor/retailer.	Agroecology establishes farming and food systems that adjust to local environments.	WWF Spain is working in the production area where the joint project is mainly located for decades. The development and environmental state of local ecosystems and agricultural impacts are well-known. The goals and project measurements on farm and landscape level are specific and recognize the local circumstances and requirements.
PAR research processes inform action at multiple scales for positive social change.	Iterative and recursive feedback loops are given on a at least weekly basis between involved actors. Partly also actors from the citrus sector, media, and politics—In Germany, as well as in Spain are involved.	Agroecology seeks to manage whole systems.	The prior aim is to reduce the ecological footprint of EDEKAs private brand oranges and mandarins. Doing that many parts of the supply chain get transformed, including Point-of-Sale (POS) and consumer marketing/communication. Furthermore, the project wants to change the production system of oranges and mandarins in Spain in general. Therefore, involved citrus growers and WWF Spain draw attention to the existing challenges and necessary changes within the sector and with local governments.

PAR processes deepen long-term relationships are formed and multiple iterations of the cycle occur.	The joint project began in 2015 and lasts for several more years. The participating suppliers have been in cooperation with EDEKA/Fruchtkontor for years. The strategic partnership of EDEKA and WWF Germany began in 2009 and is going to be extended till 2032.	Agroecology develops strategies to maximize long-term benefits.	The overall goal of the joint project is to foster local resilience of ecosystems and growers to ensure living ecosystems and stable economic conditions. EDEKA/Fruchtkontor is interested in resilient supply chains with a premium and unique product, including a product branding that includes WWF logo.
PAR processes listen to a diversity of voices and knowledge systems to democratize the research and social change processes.	Decision-making takes part on different levels, mainly bottom-up. In critical situations or conflicts, a steering committee consisting of WWF Germany and EDEKA has the final vote.	Agroecology implies processes to diversify biota, landscapes, and social institutions.	The main focus is the farm and river basin level, although due to the constantly increasing number of participating growers also landscape benefits arise. A diversification of social institutions is not part of the project.

4.3   Project Aims and Results So Far

The citrus project was launched in 2015 by EDEKA, WWF national offices in Germany and Spain and one pilot grower. Table 2 contains the most relevant general project information, illustrating its development from 2015 to 2019. The number of farms, the production area as well as the distribution of project oranges and mandarins continuously increased and increases, also in the future.

As a result, in the 2018/2019 season almost every fifth orange/mandarin sold by EDEKA originated from a project farm. In the future, EDEKA and WWF intend to source the entire conventional Spanish orange and mandarin assortment from project farms (Cf. Figure 2).

The main strategies of intervention are based on four fields: (1) More sustainable water use on farms and in the river basin, (2) More sustainable crop protection, (3) Conservation and promotion of biodiversity and ecosystems, and (4) Measures reflecting good agricultural practice (incl. more sustainable use of fertilizers and promoting soil fertility). The objectives of the four areas are described below. The until 2018/19 reached results are then presented. The results are based on a regular

Table 2 General project information. Source: Own illustration

General Project Information	
Project name	"Citrus project": Joint project for better oranges and mandarins
Cultivation region	Andalusia, Spain, mainly in the water catchment area of the river Guadalquivir
Project objectives	• More sustainable water use on farms and in the river basin. • More sustainable crop protection. • Conservation and promotion of biodiversity and ecosystems. • Measures reflecting good agricultural practice (incl. More sustainable use of fertilizers, promoting soil fertility).
Number of project farms	• 2016: One farm • 2017: Eight farms • 2018: Nine farms • 2019: Twelve farms • 2020: Fifteen farms.
Total area under cultivation	• 2016: 167 ha • 2017: 571 ha • 2018: 716 ha • 2019: 939 ha • 2020: 1.132 ha.
Distribution of project oranges and mandarins in Germany	• Oranges: Available throughout Germany in EDEKA stores (since 2017) and Netto stores (since 2018), approximately from October to May. • Mandarins: Available throughout Germany in EDEKA stores and intermittently in Netto stores (since 2018), approximately from December to February.

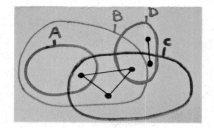

- All A's are B's.
- Some B's are C's.
- No A is D.
- Some D's are not B.

Fig. 2 Share of project-grown oranges in total quantity of oranges sold by EDEKA in the 2018/2019 sales season. Source: EDEKA/WWF Progress Report 2019

progress report, which gets published on an annual basis. The results presented in the progress report are in turn part of annual farm audits and periodic checks on the ground, conducted by WWF Spain. In addition, WWF Spain and external experts carry out continuous farm visits and monitorings several times a year. Results are documented and incorporated into a data system that can be viewed by all project participants and enables overall project management. This includes a performance check at the individual farm level which is based on action plans that are available to each farm, EDEKA and the two WWF offices. The action plans constantly evolve due to the overall project mission of continuous improvement.

1. *More sustainable water use (on farm and in the river basin):* This encompasses legal and efficient water use in irrigation, adaption of irrigation on regional climate change forecast, joint engagement for a more sustainable water use in the Guadalquivir river basin within the Water Stewardship approach and AWS certification (Alliance for Water stewardship) and ensuring a legal use of water.

In this citrus project, EDEKA and the WWF have set themselves precise targets. To adapt the volume of water used in their irrigation systems in relation to the decreasing water availability due to climate change, the project farms are to save at least 8% of their allotted water volume. Eight percent represents the lack of water availability in the future due to climate change caused impacts.

The project is aiming to achieve this saving by optimizing the irrigation systems. In the year 2018, a total of 806 million liters of water were saved in this way (Cf. Figure 3), exceeding the target of 8% by a considerable margin. Aided by above-average rainfall during the assessment period, the growers only used an average of 75.3% of their water allowance for irrigation.

Furthermore, all producers undertake to document water consumption for the irrigation of their plantations on a daily basis and to comply with the water concessions allocated by the authorities. To prevent illegal use of water on the farms, wells without a water concession are shut down. In addition, the foundations for targeted improvements are put in place, for example, by installing soil moisture probes where above-average water consumption occurs. Moreover, the project pilot-

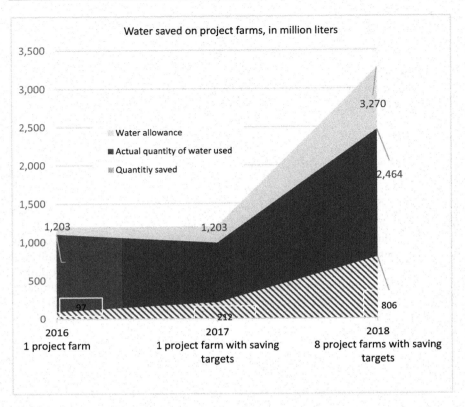

Fig. 3 Water saved in irrigation in relation to the water use rights to the project farms. Source: EDEKA/WWF Progress Report 2019

farm became the first agricultural enterprise in Europe to achieve AWS certification with Gold status, which has only been awarded for the third time worldwide till 2018. The standard of the Alliance for Water Stewardship (AWS) helps producers implement a more sustainable water management system in their local river basin in cooperation with other players. The project pilot-farm has been taking its commitment beyond the boundaries of its operation, by engaging with other users, with relevant authorities and other interested parties, and they succeeded in encouraging other farmers to implement the AWS principles. The certification represents confirmation from independent auditors that the farm uses water economically, avoids pollution, and thus contributes to protect the ecosystem and the region.

2. *Integrated pest management:* Substitution of Highly Hazardous Pesticides (according to HHP list of Pesticide Action Network/PAN), promotion of beneficial insects and natural plague control, application of pesticides only based on plague monitoring and in the area of necessity.

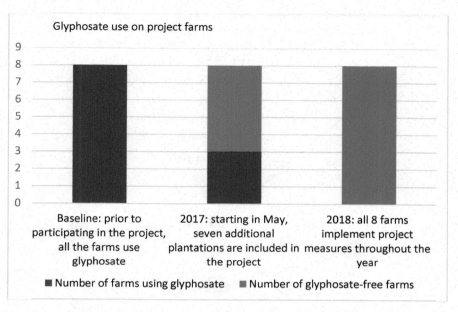

Fig. 4 Glyphosate use on project plantations before and after the project. Source: EDEKA/WWF Progress Report 2019

Specifically for citrus production in the Andalusian region customized pesticide plan was developed and successfully trialled in the project. This pesticide plan was developed with the help of external experts and in feedback from the involved citrus growers. The plan prohibits the use of particularly hazardous crop protection products such as the total herbicide glyphosate (Cf. Figure 4) or the fungicide mancozeb. These two chemicals are still used frequently in conventional orange cultivation in Spain. The plan was again followed by all the growers in 2018. Generally speaking, pesticides are now—due to the joint project requirements— used in a more targeted manner. Whereas before the start of the project eight project farms applied an average of 14 kilograms or liters of active pesticide ingredients per hectare, this amount has since been reduced to one-fifth, or only three kilograms or liters per hectare on average (Cf. Figure 5). Furthermore, the project decided to apply the method of the toxic load indicator (TLI) to calculate the toxic load in the project/ on project farms. The TLI is an indicator that evaluates the toxicity of an active substance to a living organism and thus does not relate solely to the amount of pesticide used. To determine the toxic load, TLI is based on a classification of active substances used in a pesticide. The classification considers the impact on humans, mammals, and the broader environment. This can result in a maximum score of up to 200 for each active ingredient. The higher the score, the more toxic the active ingredient. In the citrus project, crop protection agents with a toxic load of over 100 are no longer used.

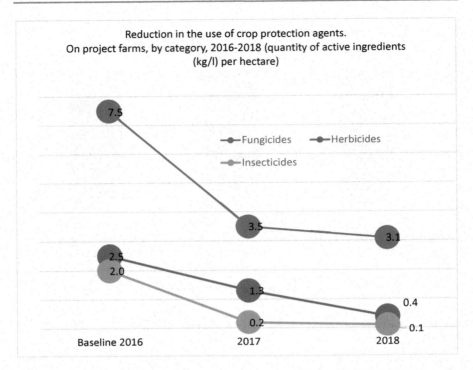

Fig. 5 Reduction in the use of crop protection agents. Source: EDEKA/WWF Progress Report 2019

3. *Efficient fertilization and promotion of soil fertility:* Application of mineral fertilizers adapted on regular water, soil and leaf analysis, application of organic matter, and fertilizers in plantation parts with low soil fertility.

This thematic focus became an official project component in 2018. An illustrated presentation of project successes is not yet possible at the current time. Currently, project partners are working on efficient application of mineral fertilizers based on soil, leaf, and water analyses, as well as soil maps and root studies. Soil fertility is promoted in particular through the massively reduced use of herbicides in the cultivation area and the incorporation of organic matter (e.g., tree pruning residues) into the soil. In cultivation areas with very low organic matter in the soil, additional liquid fertilizer that contains organic matter is applied, in rare cases also compost.

4. *Conservation and promotion of biological diversity:* Strong limitation of herbicide use, establishment of green corridors and further habitats, promotion of the ecological infrastructure (e.g., perches, nesting boxes), protection protocols for species of special interest (e.g., fish otter, ground-nesting birds).

The ground vegetation and the ecological infrastructure are enhanced on all the project farms in order to create refuges for flora and fauna and to promote the presence of beneficial organisms in a targeted manner. Weekly monitoring of the pests and beneficial organisms occurring on the plantations tracks their development and—if necessary—allows for timely intervention. For example, if lice appear on young trees, it is checked whether the "associated" predators such as ladybirds or green lacewing larvae can bring the situation under control. Only if damage thresholds are reached that can no longer be biologically controlled by beneficial organisms and production is threatened will the producers intervene with chemical crop protection agents, based on the pesticide plan.

The ladybird species "Coccinellidae" was selected as a bioindicator for the project because not only are they able to fight pests naturally, but they can also adapt to different habitats. In addition, ladybirds are highly susceptible to microclimatic or physical changes and to chemical contamination. The monitoring of the ladybird populations is therefore particularly well suited to the observation of changes in the plantations over time. In the first year of the project, a total of five different species of ladybirds were found on the pilot farm—2 years later in 2018 already 21 and in 2020 23 different species.

Additionally, a range of different project measures are designed to protect and improve biodiversity and existing ecosystems. The project has set itself the goal of giving nature a place on the plantations once again. Remarkable successes have already been achieved by working with the growers. The protection and expansion of the vegetation cover is one of the most important measures in the project, and it is mandatory for all the growers. The objective here is to reserve as much space as possible for natural vegetation, along the edges of the plantation, along paths and between the rows of orange trees. Because this is where beneficial organisms can find shelter—especially where different plant species are grown (Cf. Figure 6)—and so will help control pests. Before they joined the project, these areas in the plantations were mostly treated with herbicides (Cf. Figure 5). Wherever this is feasible, additional areas on the project plantations are set aside to promote biodiversity. For example, particularly functional plant species such as composites are sown, thus creating additional oases for beneficial organisms. Natural vegetation is allowed to spread largely unhindered between the rows of orange trees on the project plantations. Green corridors are thus created through which animals such as foxes can move. The number of plant species growing in the vegetation cover has increased steadily since the project began. In the 2018 reporting period, some 21 different species were already found there. As a further measure, a total of 35 perches for birds of prey were installed on eight plantations. As a result, the biodiversity expert working for the project was able to observe more than ten species of diurnal and nocturnal birds of prey on the plantations between June 2017 and June 2018. They included osprey, kestrel, peregrine falcon, Bonelli's eagle, black kite, little owl, and barn owl. In 2018, a total of five floating biodiversity islands were created on the four project plantations with their own water retention reservoirs. Floating mats covering a total area of 24 square meters were planted with a total of 950 water plants and then allowed to drift on the surface of the water basins. Little

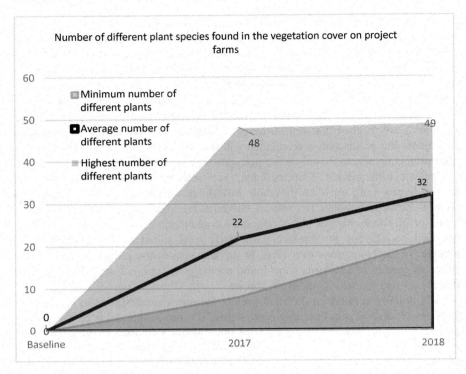

Number of different plant species found in the vegetation cover on project farms

Fig. 6 The number of different plant species growing in the vegetation cover has increased steadily since the project began. In the 2018 reporting year, they consist of at least 21 different plant species on the various project plantations. Source: EDEKA/WWF Progress Report 2019

and great crested grebes and other water birds are now using these islands as nesting sites, and kingfisher, osprey, and the protected Spanish pond turtle have also been spotted on the islands. The project also adopted separate conservation protocols for some animal species. For example, as soon as the nightjar is sighted on a farm, tractors are no longer allowed to enter the sectors where these groundnesting birds' nests are located.

Thus, based on the results of the WWF/EDEKA progress report 2019 overall decisive results are:

- Responsible water use: In 2018, eight project farms saved 806 Million liters of water in relation to their water usage rights due to efficient irrigation, among others based on soil humidity probe sensors. As an overall objective, the project farms adapt their irrigation to regional climate forecasts, which assume an 8–10% decline in water availability for the project region by 2030.
- More sustainable crop protection: Highly Hazardous pesticides are prohibited. The application of pesticides is based on weekly plague and beneficial insect monitoring and is only allowed if beneficial insects cannot control the rise of

plagues. This results in a strong reduction of pesticide use in relation to the year before the project starts (in kg/ha): fungicides (7.5–3.1), herbicides (2.5–0.4), insecticides (2.0–0.1) = Data from 2018.

- Conservation and promotion of biodiversity: Various project measures serving to improve biodiversity and their effectiveness are monitored by a biodiversity expert. Green corridors are created to promote beneficial organisms. Seating poles and nesting boxes are installed. Till 2018, 21 different ladybird species (project's biodiversity indicator) have been spotted on the project farms. Also, the rare fish otters have resettled on one of the farms. The animal is an indicator of the excellent quality of water bodies.

- Proportion of project fruits in EDEKA markets: During autumn and winter 2016/ 17, selected supermarkets in an EDEKA region were supplied with oranges from the pilot farm for the first time. In autumn and winter 2018/19, every fifth orange sold at EDEKA markets was cultivated on one of the project farms. The amount of more sustainable produced oranges at EDEKA has increased significantly within 3 years.

5 Conclusion

Several aspects of this joint project could neither be described and deepened nor scientifically analyzed in this chapter. One of them is the stewardship approach. Here its general idea, its potential of application for other common goods like climate or biodiversity as well for sectoral initiatives. But also, its potentially existing limitations due to antitrust law, for example, or the willingness of stewards/user groups of a certain common good to collaborate. In general, the stewardship approach is characterized by the idea that also companies as well as other user groups shall become good stewards with respect to the resources they rely on and which they use. Taking water as example, studies have shown that implementing water efficiency measurements at farm level can be effective for a while but normally end in a rebound effect. Due to the made savings often more water-intense crops are grown or there is an enlargement of the farm area, which in turn must be irrigated. The stewardship approach strives to convey the general idea of common goods, the effects of rebounds and the necessity to act as a good steward. To protect water resources and to secure water sources in future, it is necessary to broaden the picture from a farm level to the river basin. The stewardship approach, as it is fostered by the Alliance for Stewardship is exactly doing this with the help of analysis, capacity building, and collaboration of different user groups within one river basin. This stewardship mindset has led to collective actions and a common platform in another agricultural joint project of EDEKA and WWF in Columbia (WWF, 2021).

Another not further discussed aspect is the role of marketing and communication of the project oranges and mandarins in EDEKA and Netto stores in Germany. Although the footprint of the project oranges and mandarins got reduced, project goods are still conventional. This leads to the question of marketing, especially regarding POS communication and price-setting. So far, project goods that originate from project farms that verifiable apply all project goals and that have successfully passed the annual audit are branded with the WWF logo. The WWF logo offers the possibility to indicate that these oranges and mandarins are more sustainable, "better" than other conventional oranges and mandarins. Due to a different POS communication and price setting, a confusion regarding organic oranges and mandarins is unlikely. Nevertheless, the overall goal of EDEKA and WWF is in medium term to offer solely conventional Spanish oranges and mandarins from more sustainable sources with as much volume as possible that got produced under the joint project requirements. This leads to the question of rollout and upscaling. Both in the context of the strategic partnership and EDEKAs assortment, but also with respect to the conventional cultivation of citrus goods in general. The joint project has proven since its beginning in 2015 that immense savings effects and sustainable transformation are possible without endangering production and harvest amounts. However, it must also be critically noted that the joint project with all its measurements, participatory project structures, ambitious goals, continuous improvement processes, and additional certification requirements is combined with investments that somehow need to be refinanced.

Additionally, the changes made in agricultural practices on farm level also come along with immense changes to the required skills, knowledge, and capabilities. For clarification, employees once responsible for pest management and therefore have applied pesticides after a clear application plan, now have a very varying daily working routine. Meanwhile, they became experts in identifying and analyzing beneficial insects. On this basis, a possible pesticide application—mostly after feedback from one of the agricultural external experts in the project—might take place.

Furthermore, questions arise about how learnings and best practices can be shared so that an upscaling process is pushed—beyond the borders of EDEKA suppliers. WWF Spain will continue to pursue thematic lobbying at municipal and wider levels in order to improve the Water and Agricultural Policies in Spain. The farmer acts as an ambassador within his agricultural community and offers opportunities for exchange in coordination with the project team. A building was even renovated to facilitate networking and exchange.

For further development and transformation of the Spanish conventional citrus production, it seems to be necessary to accompany the project on an interdisciplinary, scientific basis. Results could help foster to verify, clarify, and spread project insights and best practices.

Considering developments on the EU level concerning decarbonization, climate neutrality as well as the "farm to fork strategy," the current project design and goals might either not be in line with future requirements or it is necessary to analyze the potential of certain project measurements in this respect. It is planned to work on an LCA-based baseline understanding of environmental hotspots across the entire supply chain. So that aspects of waste and GHG reduction are analyzed and optimized above the farm-level.

References

Beermann, M., & Wick, K. (2019). Erweiterte input-output-analyse (EEIO), 251–268. In A. Baumast, J. Pape, S. Weihofen, & S. Wellge (Eds.), *Betriebliche Nachhaltigkeitsleistung messen und steuern*. Verlag Eugen Ulmer Stuttgart.

Bellamy, A. S., & Ioris, A. A. R. (2017). Addressing the knowledge gaps in agroecology and identifying guiding principles for transforming conventional Agri-food systems. *Sustainability, 9/330*.

Beske, P., Land, A., & Seuring, S. (2014). Sustainable supply chain management practices and dynamic capabilities in the food industry: A critical analysis of the literature. *International Journal of Production Economics., 152*(2014), 131–143.

Cicatiello, C., Franco, S., Pancino, B., & Blasi, E. (2016). The value of food waste: An exploratory study on retailing. *Journal of Retailing and Consumer Services, 30*(1), 96–104.

de Castro, L. (2018). *Spanish fresh fruits and vegetables sector*, 21–31. Retrieved from https://www.oecd.org/agriculture/fruit-vegetables/publications/Proceedings-2018-HNIS-meeting.pdf

Expósito, A., & Berbel, J. (2017). Agricultural irrigation water use in a Closed Basin and the impacts on water productivity: The case of the Guadalquivir River basin (southern Spain). *Water, 9*(2), 136.

FAO. (2012). *World agriculture towards 2030/2050*. The 2012 Revision. Retrieved from http://www.fao.org/fileadmin/templates/esa/Global_persepctives/world_ag_2030_50_2012_rev.pdf

FAO. (2020). *Tracking progress on food and agriculture related SDG indicators 2020*. A report on the indicators under FAO custodianship. Retrieved from http://www.fao.org/sdg-progress-report/en/#sdg-12.

Fuchs, D., Kalfalgiani, A., & Arentsen, M. (2009). Retail power, private standards, and sustainability in the global food system, 29–60. In J. Clapp & D. Fuchs (Eds.), *Corporate power in global Agrifood governance*. The MIT Press.

Francis, C., Lieblein, G., Gliessmann, S., Breland, T. A., Creamer, N., Harwood, R., Salomonsson, L., Helenius, J., Rickerl, D., & Salvador, R. (2003). Agroecology: The ecology of food systems. *Journal of Sustainable Agriculture., 22*, 99–118.

Glasbergen, P. (2007). In P. Glasbergen, F. Biermann, & A. Mol (Eds.), *Setting the scene: The partnership paradigm in the making. Partnerships, governance, and sustainable development: Reflections on theory and practice* (pp. 1–28). Edward Elgar.

Gliessmann, S. R. (1998). *Agroecology: Ecological processes in sustainable agriculture*. Ann Arbor Press.

Hermsdorf, D., Rombach, M., & Bitsch, V. (2017). Food waste reduction practices in German food retail. *British Food Journal, 119*(12), 532–2546. Emerald Publishing Limited.

IPBES. (2019). In E. S. Brondizio, J. Settele, S. Díaz, & H. T. Ngo (Eds.), *Global assessment report on biodiversity and ecosystem services of the intergovernmental science-policy platform on biodiversity and ecosystem services*. IPBES secretariat.

IPCC. (2018). In V. Masson-Delmotte, P. Zhai, H.-O. Pörtner, D. Roberts, J. Skea, P. R. Shukla, A. Pirani, W. Moufouma-Okia, C. Péan, R. Pidcock, S. Connors, J. B. R. Matthews, Y. Chen, X. Zhou, M. I. Gomis, E. Lonnoy, T. Maycock, M. Tignor, & T. Waterfield (Eds.), *Global warming of 1.5°C*. An IPCC special report on the impacts of global warming of 1.5°C above pre-industrial levels and related global greenhouse gas emission pathways, in the context of strengthening the global response to the threat of climate change, sustainable development, and efforts to eradicate poverty. In Press.

Leontieff, W. (1966). *Input-output economics*. Oxford University Press.

Leontieff, W. (1970). Environmental repercussions and the economic structure: An input-output approach. *Review of Economics and Statistics, 52*(3), 262–271.

Mari, S., & Peris, E. (2005). A general view of the citrus sector in Spain. *Journal of Food Distribution Research, 36*(1), 249.

Méndez, V. E., Bacon, C. M., & Cohen, R. (2013). Agroecology as a transdisciplinary, participatory, and action-oriented approach. *Agroecology and Sustainable Food Systems, 37*(1), 3–18.

Midgley, J. L. (2014). The logics of surplus food redistribution. *Journal of Environmental Planning and Management, 57*(12), 1872–1892.

Rye, J. F., & Scott, S. (2018). International labour migration and food production in rural Europe: A review of the evidence. *Journal of the European Society for Rural Sociology, 58*(4), 928–952.

Sanfeliu, I. (2018). *Spanish citrus industry–production, challenges and perspectives*, 32–42. Retrieved from https://www.oecd.org/agriculture/fruit-vegetables/publications/Proceedings-2018-HNIS-meeting.pdf

Seuring, S., & Müller, M. (2008). From a literature review to a conceptional framework for sustainable supply chain management. *Journal of Cleaner Production, 16*(15), 1699–1710.

Sperfeld, F., Blanke, M., Mohaupt, F., & Hobelsberger, C. (2018). *Kooperationen von Unternehmen und Umweltorganisationen erfolgreich gestalten*. Texte 14/2018. Retrieved from https://www.umweltbundesamt.de/sites/default/files/medien/1410/publikationen/2018-02-20_texte_14-2018_nro-kooperationen.pdf

Statista Research Department. (2021a). *Volume of citrus fruits produced in Spain in 2019/2020, by type*. Retrieved from https://www.statista.com/statistics/440983/citrus-production-in-spain/

Statista Research Department. (2021b). *Bruttoumsatz der führenden Unternehmen im Lebensmittelhandel in Deutschland im Jahr 2019* (in Milliarden Euro). Retrieved from https://de.statista.com/statistik/daten/studie/153723/umfrage/groesste-unternehmen-im-lebensmitteleinzelhandel-nach-gesamtumsatz-in-deutschland/

Steffen, W., Richardson, K., Rockström, J., Cornell, S. E., Fetzer, I., Bennett, E. M., Biggs, C., de Vries, W., de Wit, C. A., Folke, C., Gerten, D., Heinke, J., Mace, G. M., Persson, L. M., Ramanathan, V., Reyers, B., & Sörlin, S. (2015). Sustainability. Planetary boundaries: Guiding human development on a changing planet. *Science (New York, N.Y.), 347*(6223).

Van Huijstee, M. M., Francken, M., & Leroy, P. (2007). Partnerships for sustainable development: A review of current literature. *Environmental Sciences, 4*(2), 75–89.

Wiedmann, T., Minx, J., Barrett, J., & Wackernagel, M. (2006). Allocating ecological footprints to final consumption categories with input-output analysis. *Ecological Economics, 56*, 28–48.

WWF (2019). *EDEKA and WWF strategic partnership*. Progress report 2019.

WWF. (2020). In R. E. A. Almond, M. Grooten, & T. Petersen (Eds.), *Living planet report 2020– bending the curve of biodiversity loss*. WWF.

WWF (2021). *Collective Action*. "The plataforma de cooperación y custodia del agua" in río frío and río sevilla–a case for water stewardship. Retrieved from https://wwfint.awsassets.panda.org/downloads/wwf_flyer_kolumbien_web_einzelseiten.pdf

WWF España (2020a). *A Truce for Doñana map by WWF SPAIN of environmental threats in Doñana and the Guadalquivir river estuary*. Retrieved from https://wwfes.awsassets.panda.org/downloads/trucedonana_report_wwf_spain_july2020.pdf

WWF España (2020b). *A lost decade*. Report on the state of the Doñana aquifer. Retrieved from https://wwfes.awsassets.panda.org/downloads/wwf_report_donana_aquifer_2020_en__1__1.pdf

Marina Beermann Ph.D., has a degree in ecological economics from the Carl von Ossietzky University in Oldenburg/Germany. There she also worked for 5 years as a lecturer and researcher in the field of climate impact research and completed her dissertation at the Chair of Strategic Management ("Developing Entrepreneurial Resilience in Times of Climate Change") in 2012.

Marina Beermann has been a head of department at WWF Germany since 2016 and also has extensive experience in teaching and supervising students due to various teaching assignments.

Patrick Freund He has a degree in environmental and forest sciences (major) and climatology from the Albert Ludwigs University in Freiburg/Germany. After working in Costa Rica on biodiversity conservation in agriculture and on a solidarity-based farm in Berlin, he is currently a referee for sustainable agricultural supply chains at WWF Germany.

As part of his activities, he strives to make supply chains more ecologically beneficial, at present for citrus fruits and oil palms. He is also involved in the strategic development of WWF's cooperation with the retailer Edeka and in the work of multi-stakeholder initiatives.

Felipe Fuentelsaz He is an agronomist and he has been always linked to environmental aspects. Currently, he is the coordinator for agricultural projects at WWF Spain where he works for projects at field levels with markets and companies, such as ZITRUS project with WWF Germany and EDEKA.

Also, Felipe Fuentelsaz is in different boards and Technical Committee of Agricultural Standards with the proposal to improve environmental criteria on them with the goal to protect natural resources and promote sustainable agriculture.

Challenges in Malaysian's Sustainability Efforts: The Role of Traceability in the Food Industry

Sam Sarpong, Ali Saleh Ahmed Alarussi, and Faizah Shahudin

Abstract

The food industry has made significant strides towards improving its sustainability in recent years. But whilst the industry continues to play a vital role in improving its own sustainability record, it is yet to effectively address the complex array of issues it is still subjected to. Particularly in the case of Malaysia, it is still grappling with a lot of concerns mainly pertaining to issues like traceability in its meat products, the unsustainable nature of its palm oil production and food waste to which the authorities are urgently seeking solutions to. The chapter explores these issues and examines how food security can be ensured in the country.

Keywords

Malaysia · Sustainable issues · Traceability · Food chain · Waste · Halal · Oil palm · Food waste

1 Introduction

The need to increase food production to help feed the soaring numbers of the world's population remains a major challenge for many countries. It is estimated that by 2050 the world's population will reach 9.1 billion (EUFIC, 2015). As a result, EUFIC estimates that food production would need to increase by 70% to feed the larger and most likely more urban population. Key issues that emanate from this projection include the necessity to produce more food using less land. There is

S. Sarpong (✉) · A. S. A. Alarussi · F. Shahudin
School of Economics and Management, Xiamen University Malaysia, Sepang, Malaysia
e-mail: samsarpong@xmu.edu.my

© The Author(s), under exclusive license to Springer Nature Switzerland AG 2022   185
S. O. Idowu, R. Schmidpeter (eds.), *Case Studies on Sustainability in the Food Industry*, Management for Professionals,
https://doi.org/10.1007/978-3-031-07742-5_8

also the added concern regarding the global production methods and how to limit waste throughout the food supply chain. Alongside that, there is apprehension from consumers regarding whether they can have food supplies from sources that they could trust (Sarpong, 2014).

The food industry, to a large extent, has been lagging behind in the world of sustainability. Notwithstanding these disquiets, the last few years have seen an increase in global efforts to improve the sustainability of the food sector, as the threats to its future viability have become ever more apparent. In Malaysia, sustainability issues have picked up quite late. In the last few years, critics have accused Malaysia of running out of natural resources and being engulfed by waste. Critical to this is that, *sustainability* issues have not, in essence, been tackled as much as possible, hence leaving a vast array of less resource inputs and considerable waste in the system.

The aim of this chapter is to highlight the sustainability issues Malaysia is faced with. It mainly focuses on Malaysia's food waste and its food supply chain which has recently come under scrutiny, especially in the wake of allegations of improprieties in the country's halal brand. It also wades into its oil palm business which has been deemed in some quarters to be associated with unsustainable practices. Mindful of the use of traceability as a tremendously impactful tool for advancing sustainability objectives, the chapter explores the key issues in the light of that. We also highlight the policy challenges and suggest actions that can help mitigate the issues regarding sustainable development issues. The chapter hopes such a focus would lead to improvements in the understanding of sustainability initiatives and good practices in Malaysia's food sector.

The chapter, thus, responds to emerging calls for the adoption of a more systemic approach to food security that takes into account sustainability concerns (Lang & Barling, 2012). By focusing on the traceability elements in the sustainability debate, the chapter identifies the challenges to 'sustainable food security'—a concept based on the fundamental assumption that the long-term capacity of the food system to provide an adequate amount of food will depend on its ability to respond to the challenges that threaten its resilience and to minimise its impacts on human and environmental conditions. By joining the security and sustainability lenses, the chapter argues that, it is becoming increasingly necessary to address a wide range of large ecological footprints that are threatening the resilience of the global food system.

The chapter is organised as follows: First, it examines what sustainable food is. It then delves into sustainability issues in the food industry in Malaysia. From there, it provides a review of the situation. Subsequently, it looks at the different levels of food waste, then the palm oil and halal businesses. The next section discusses the issues under consideration. The final part is the conclusion which sums up the outcome of the issues the paper explored.

2   Defining 'Sustainable' Food System

The concept of food system has gained prominence in recent years amongst both scholars and policymakers. There are, however, many different views as to what constitutes a 'sustainable' food system, and what falls within the scope of the term 'sustainability'. The FAO (2018), however, maintains that food systems encompass the entire range of actors and their interlinked value-adding activities involved in the production, aggregation, processing, distribution, consumption and disposal of food products that originate from agriculture, forestry or fisheries, and parts of the broader economic, societal and natural environments in which they are embedded. It further argues that sustainable food system is a food system that delivers food security and nutrition for all in such a way that the economic, social and environmental bases to generate food security and nutrition for future generations are not compromised.

Thus, for food, a sustainable system might be seen as encompassing a range of issues such as security of the supply of food, health, safety, affordability, quality, strong food industry in terms of jobs and growth and, at the same time, environmental sustainability, in terms of issues such as climate change, biodiversity, water and soil quality. This has been given credence by Eco and Beyond (2021) which argues that sustainable food is not only about the food itself but also rather comprises a combination of factors including how the food is produced, how it is distributed, how it is packaged and how it is consumed.

In this context of urgency and uncertainty, the debate on food security and sustainability has been enriched by different considerations that depict a range of plausible futures. It is now recognised that traceability is embedded in the sustainability efforts. In the last decades, researchers have extended the sustainability concept at supply chain level, going beyond the common boundaries of a company dedicated to the goods production (i.e. product sustainability) (Germania et al., 2015). For instance, BSR and UNGC report (2014), describes sustainability through traceability as, 'the ability to identify and trace the history, distribution, location and application of products, parts and materials, to ensure the reliability of sustainability claims, in the areas of human rights, labour (including health and safety), the environment and anti-corruption'. Sustainability should, therefore, be regarded as the ability to generate food at a productivity level that is enough to maintain the human population (Syed, 2021).

Currently, food production has been deemed to be inadequate to feed the projected nine billion people the world could have by the year 2050. As the direct economic consequences of food wastage (excluding fish and seafood) is about $750 billion annually (FAO, 2014), the United Nations estimates that one in nine people in the world would not have access to sufficient food to lead a healthy life. Hence, more people could die from hunger every day if efforts are not made to strengthen food production. Though factually tons of food are produced, they are poorly distributed and wasted. The *contradiction here is that there is a disparity between abundance and scarcity*. In addition, the way food is produced and consumed also tends to have a toll on the environment and natural resource base. According to the FAO (2019), social, demographic and economic factors are contributing to changing lifestyles and eating patterns, and subsequently putting pressure on resources for food production.

The idea of sustainable food is, therefore, built on principles that further the ecological, social and economic values of a community as a whole. Sustainable food production thus provides a means of using processes and systems that are non-polluting, conserve non-renewable energy and natural resources, are economically efficient, are safe for workers, communities and consumers, and do not compromise the needs of future generations (Foresight, 2011). It also implies the use of resources at rates that do not exceed the capacity of the earth to replace them (European Commission, 2019). Thus, the context in which *sustainability is seen in the food industry, calls for* a comprehensive review of the *industry's* current approach to balancing *environmental*, economic and social considerations throughout the supply chain.

3   Importance of Sustainable Food

Critics, including environmentalists and the UN agencies, point out that the current food system has contributed to climate change, deforestation, soil loss and soil pollution, alongside a huge demand on water supply, pollution and exploitation of certain species such as fish, to name a few. Contemporary food systems have exerted undesirable pressure on terrestrial and aquatic ecosystems and they are failing to provide adequate nutrition to billions of people. The way people shop and eat also play a huge role in terms of environmental impact and their choices can have a profound effect on food sustainability (Bushell, 2020). Sustainable food production is, therefore, the key to building a better world for today and the future as it can help the world's growing population, spur economic development, fight climate change and promote stability and peace around the world.

It also necessitates the development of sustainable food value chains in order to offer innovative pathways out of poverty, e.g. by local value addition through local processing, and by linking farmers directly to higher-value export markets (UNEP, n.d.). Within such a context, the UNEP maintains that sustainable, resource-efficient agricultural practices can help farmers adapt to change, sustain their livelihoods and reduce greenhouse gas emissions from farm activities.

4   Drivers of Sustainability

There are some important factors driving the increased attention to sustainability in the food industry. For instance, consumer demand for more sustainable and ethical products has led to commitments and good practices in product marketing in recent times. The commitment to this has been quite significant. Many consumers have therefore signalled that sustainability is a key issue for them and are, therefore, prepared to avoid products that do not measure up to that. Consumers, particularly millennials, increasingly say they want brands that embrace purpose and sustainability (White et al., 2019). Such consumers are oftentimes more than willing or able to pay the premium that certified products command or bring on board. This tendency has also led to product certifications which are highly regarded. The certification of products is a means of providing assurance that the product in question conforms to standards and/or other normative documents.

In many developing countries, a greater part of food is lost during production, handling/storage and processing (FAO, 2014). The list of causes of food insecurity is long and multifaceted: they range from political instability, war and civil strife, macroeconomic imbalances and trade dislocations to environmental degradation, poverty, population growth, gender inequality, inadequate education and poor health (Smith et al., 2000). Some food is also lost before the crops leave the farm. This means that reducing farm-level losses may be a path towards sustainable intensification (Garnett & Godfray, 2012). Generally, food waste sources can be sorted into three groups which are food losses, i.e. food materials lost during preparation, processing and production phases in the food supply chain, unavoidable food waste, i.e. the inedible parts of food materials lost during consumption phase (pineapple peel, fruit core etc.) and avoidable food waste, i.e. the edible food materials that were lost during consumption phase (surplus and wastage) (Thi et al., 2015).

5 Food Sustainability in Malaysia

Food sustainability remains a much talked about issue in Malaysia in view of its severe impact on the country's economy. The Solid Waste and Public Cleansing Management Corporation—Malaysia (SWCorp Malaysia), which is in charge of the management and regulation of solid waste and public cleansing in Malaysia, maintains that food waste has become quite enormous in the country. Common causes of food waste in Malaysia include overbuying, overproduction and spoilage amongst restaurants. The food waste management and policy for food waste treatment are considered less efficient in the country due to limited budget for food waste management (Thi et al., 2015). In addition, there is no coordination or effective means available to enable the Malaysian food sector to evaluate its sustainability performances (Maestrini et al., 2017).

Since the country started off with the Movement Control Order (lockdown measures) in March 2020, which came out as a result of COVID-19, there has been a marked decrease though in the quantity of food waste that has ended up in landfills. The decrease has been attributed to most restaurants catering for takeaway orders, the short operating hours, as well as the cancellation of all wedding functions and other events and gatherings. Throughout this period, most people have stayed at home and either cooked or opted for packed food from outside.

On the contrary, food imports have increased rapidly in Malaysia. From 2013 to 2018, the cost of food imports in Malaysia increased from RM43bil to RM54bil (Market Review on Food Sector in Malaysia, 2019). The imported items included rice, beef, lamb, mango, cabbage, chilli and coconuts. In its Market Review on Food

Sector report, the Malaysia Competition Commission (MyCC), the authority responsible for the report, said the country had been highly dependent on beef imported from Australia (90% of its fresh and frozen beef) and India (80% of its frozen beef). The huge imports, of late, have also opened up a lot of discussions regarding the source of these products, the processes of the manufacturing, treatment and the food supply chain. Such scrutiny has led to the revelation that the Malaysian food processing and production sector has been lagging behind in terms of its ability to work through its systems and processes.

6   Methodology

Emergent literature from the sustainability field, news sources and academic sources were used in this review. We also sought publications from government agencies to explore the issue under consideration. The purpose of this study was to investigate how the government of Malaysia is grappling with sustainability issues. In order to get a holistic perspective of the extent of the situation, we adopted an approach to qualitative research based on document analysis. First, we analysed secondary data from government institutions, policy statements from state agencies, media reports and announcements made in the light of sustainability issues, food security and the general concern regarding the food industry in Malaysia.

7   The Situation with Malaysians

This section deals with issues pertaining to food waste, the palm oil business and the halal business in Malaysia. It begins with the issue surrounding food waste, then the oil palm and finally halal.

7.1   Food Waste

Aside from food waste being a tragic loss of produce, it also takes a lot of precious resources to produce in the first place, such as land and water to grow crops. But managing food waste seems to be wrought with many difficulties as it is extremely time sensitive and also imbued not only with economic and environmental, but also with ethical issues. In addition to these direct effects of wasted or lost food volumes, it is critical to recognise that food wastage has more complex interactions in the food system. These include interactions that affect food prices and availability, production patterns and input use.

It is at this complex level that the connection between food wastage and hunger or reduced livelihoods (e.g. due to reduced access to natural resources) can be looked at. On average, Malaysia generates 42 million metric tonnes of municipal waste a year, 60% of which consists of food and plastic waste, a figure which continues to grow.

According to the Malaysian Housing and Local Government Ministry, in 2018, the waste separation and subsequent recycling rate was 24% and the remaining 76% went to landfill. Meanwhile, it costs the Malaysian government an average RM2.4 billion (435 million pounds) annually to manage waste. The challenges Malaysia faces in waste management and recycling include:

- The huge amount of waste produced.
- Lack of separation at source.
- The public's negative perception of recycled goods.

According to SWCorp, Malaysians generate 16,687.5 tonnes of food waste daily. This generated food waste arguably fills a number of landfills and has also led to increase in managing costs too. Indeed, the need to reduce food waste is gaining more attention today because in Malaysia, consumer behaviours and expectations have now changed quite considerably. Now people have started expressing concerns over food wastage, creating the need for a more realistic attitude towards this development.

But food waste is a complex issue that needs a collaborative solution, with foodstores, producers, government and of course consumers, working together to help resolve the inherent issues. To date, the government's approach still lacks fulfilment, in the eyes of many stakeholders. Much has therefore fallen on NGOs who are doing their best to help out in a bid to avoid food wastage. For instance, the COVID period strengthened some NGOs in Malaysia to undertake a number of humanitarian gestures. Some of these NGOs are therefore re-distributing food before the food's best-before date or by transforming food into biofuel. More NGOs have been encouraged by this to go to places where unsold items are, to collect such food items like unsold vegetables, fruits and raw food that are still in good condition for distribution to the vulnerable in society. This, to a great extent, has also saved such items from the landfill.

Food manufacturers and their trading partners can also help in ensuring that more edible food stays out of the garbage by pre-empting food waste from happening by implementing better traceability processes. Whilst achieving a whole perfect system is admittedly unlikely, food manufacturers and their trading partners can do more to ensure consumer confidence in the safety and freshness of their food without sacrificing product quality. Adopting standards-based traceability procedures or expanding upon the ones already in place could lead to more precise inventory planning and category management.

Ultimately, it might take both businesses and consumers to understand food waste issues by using available data to modify behaviours and help reduce food waste. As the consumers become more empowered, manufacturers and their trading partners ought to realise that there would be a huge demand for greater food industry transparency and accurate product information shared along the supply chain to influence and exceed consumer expectations. With new tastes and ethical considerations being the norm, inventory visibility has become quite appealing to Malaysian consumers, many of whom desire appropriate quality items. By identifying, capturing and sharing information instantaneously and in an interoperable way, transparency in the supply chain would set out some trust in the food people purchase. In all the examples outlined, time and technology seem quite vital.

Though waste has become a contentious term, it has also given rise to a new way of doing business, leading to a more innovative and sustainable dream for the near future in Malaysia. Popularly referred to as the circular economy, this model of production and consumption comprises a process for a regenerative economy that converts waste to wealth and creates enormous possibilities in its wake. It allows the lifecycle of products to be extended through a wide range of practices that involve utilisation of waste, improvement in efficiency and continual use of natural resources. But the obvious need is for Malaysians to create inventions in waste management in order to create wealth.

7.2   Palm Oil

Palm *oil* has become a major, global agricultural commodity, which is used in a host of *foods*. The oil palm industry involves over five million people globally, most of whom are smallholder farmers who depend on this crop for their livelihood. Malaysia is the world's second-largest supplier of palm oil. Its palm oil industry has often been dubbed as having untenable practices with many unexploited areas cut down to make way for oil palm plantations. Through this means, many habitats and primates have been disturbed over the past years. This situation has also given rise to problems for the climate, environment, habitat and indigenous people living in the forest. Indeed, the biggest environmental concern of the palm oil industry is the expansion of cultivation to rainforests and peatland, known not only as a source of biodiversity but also as a major storage point for carbon emissions. As a result, some western governments and NGOs have advocated for a boycott of the Malaysia's palm oil.

In view of above concerns, Malaysia has now placed much importance on sustainability efforts within the industry. In line with that, it has been striving to fall in line with various certification schemes that have been set up over the years to ensure sustainable cultivation of oil palm and production of palm oil. It is seriously engaging with and working under schemes like the Roundtable on Sustainable Palm Oil (RSPO), Rainforest Alliance, International Sustainability & Carbon Certification and the Roundtable on Sustainable Biomaterials. The Malaysian government has, meanwhile, used various stringent policies and guidelines to ensure that the industry upholds the very highest standard of sustainable practices. The aspirations remain high, and today we can also see the Malaysian Sustainable Palm Oil (MSPO) scheme which certifies companies that have complied with global demand for sustainable palm oil production.

But one of the big challenges facing corporations is tracing palm oil through a massive and complicated supply chain. Smallholdings are the ones which most of the time are deemed to be the perpetrators of the violations in the palm oil industry.

They have often been accused of using dangerous pesticides and having workers who work under very trying conditions.

7.3   Halal

The issues on Halal consumption are becoming more complex as consumer expectations are increasing with the growing knowledge of food safety and quality issues. Many Malaysians have been quite livid about food processes and the adulteration of its halal branding. In this Muslim-dominated country, the concept of halal is an absolute key to consumption. First and foremost, halal is able to go beyond just the indication of pure and healthy products and services. Secondly, the concept of halal branding is intended to assist companies to venture into new markets by inculcating a value system to their products and services in the current competitive market.

To this extent, service providers have to ensure that the products offered are based on Islamic values (Ahmad, 2015). Producers, therefore, have to ensure that their products are free from any forbidden ingredients (Zainudin et al., 2019). Furthermore, it is very crucial for these providers to ensure that the resources and the distribution channels are pure and free from illegal components that might jeopardise the purity of the products and services (Shaari & Arifin, 2010). The concept of purity is deemed crucial because it allows consumers to feel assured about the products that they are consuming as it is not questionable.

In December 2020, media reports gained ground in Malaysia that a cartel had smuggled, repacked and re-labelled non-halal frozen meat items as halal certified. Malaysians were therefore shocked by the December 2020 reports of a fake halal meat syndicate that had purportedly been operating for years. The cartel was said to be bribing officers with money and sex to allow non-certified meat including potentially diseased kangaroo and horse meat to be smuggled in from overseas and sold as certified halal beef (Adam, 2021). Rumours were also rife that the illegal activity involved a wide network with local and foreign syndicates. However, investigations conducted by the government revealed that there was no horse, kangaroo or pork meat illegally brought in and that the issue involved only one company. Instead, government revealed that the company involved was registered in 2014 and obtained its licence to import frozen products in 2017.

Whilst the government has debunked the earlier rumours, questions regarding transparency and accountability have been demanded of the key government agencies in charge of regulating food imports and halal certification. These include the Department of Islamic Development Malaysia (Jakim), Veterinary Services Department (DVS), Malaysian Quarantine and Inspection Services Department (Maqis), Customs Department and port police.

The Malaysian case resonates with the horse meat scandal which erupted in Europe in 2013. The incident brought to the fore two particular areas—traceability and supply chain complexity—within the EU. Although there seems to be an ability to trace a number of products that have fallen short of expectation, the complexity of the supply chain sometimes makes such exercises quite problematic (Sarpong, 2014). The issue with the globalised world, offshoring and lean-based efficiency has created a basis that calls into question this whole process in the supply chain. Meanwhile, the pandemic has also led to demand and supply ripples; chaos and resonance effects across global networks (Guan et al., 2020). As Sarpong (2014) explains, the vulnerabilities in the supply chain are considerable, in part, because there is a huge flow of information that can be difficult to analyse, predict, or even measure.

The meat issue has alarmed many Malaysians, with some quarters calling on the country to expand its own cattle industry, to reduce dependency on imports. Some politicians have also voiced concern that the illegality perpetuated on Malaysians calls into question the integrity of Malaysia's halal industry and certification processes. The Islamic Development Department of Malaysia (Jakim) which is in charge of certification has since promised to tighten enforcement and its procedures for halal certification for imported goods.

Jakim's main role is to ensure slaughterhouses, in the case of meat products, meet the criteria that are required in accordance with protocols and procedures in halal certification processes. As the agency responsible for certifying if products are halal, Jakim has found itself at the centre of recent furores fuelled by social media claims that even certain certified restaurants served pork. However, this was found to be false. Generally, whilst Muslims are obligated to make sure the food they consume is halal, cases of fake certificates or unhygienic production processes frequently make the headlines in Malaysia. Malaysia's halal industry contributes 7.8% of the country's gross domestic product.

8   Discussions

The discussion section focuses on the food waste and the positionality of sustainability in the Malaysian food chain. It focuses on the waste issues within the Malaysian food system hierarchy as well as sustainability problems in the oil palm and halal industries.

8.1   Food Waste

It is quite profound that food waste represents a *relentless problem*, which cannot be solved once and for all (Weber & Khademian, 2008). For instance, food waste can be seen as a problem to be addressed, as food or as a resource for further processing. Food waste, therefore, occurs at the intersection of several influences across the food system. These include the myriad ways in which food is, for instance, produced, transported, processed, packaged and stored on the supermarket shelves and at home.

The complexity of this issue is made manifest by the multitude of factors impacting on such waste along the food supply chain, from farm to the eating table. Food waste, meanwhile, can also be discussed in relation to the ethical, moral and value-laden aspects of food waste (Thompson & Haigh, 2017), CSR and power (Devin & Richards, 2018). There is every indication that values, virtues and practices can foster food waste reduction. For instance, ethical issues underlying food waste behaviour in the food chain and how virtues relate to food waste practices have manifested themselves in mature consumer society.

Another focal topic too is the responsibilisation and mobilisation of consumers (Evans et al., 2017). Consumers, NGOs, governments, suppliers and buyers are increasingly demanding more information about the origins of their products and materials and the conditions under which they were produced and transported along the value chain. With the increase in demand for organic, fair trade and environmentally friendly products and materials, well-functioning *traceability systems* and new technologies have been developed to meet stakeholder needs (The ID Factory, 2019).

Traceability can therefore be considered as a driver to set and develop a more sustainable supply chain, engaging all players in improving the sustainability of their sourcing policies and integrating them into their supply chain management systems (The ID Factory, 2019). If the supply chain is complex, this can sometimes mean that parts of the chain are not traceable. Traceability becomes effective if there is reliable information coming from a traceable system as many customers are quite keen to know how and where their products are being made. End-to-end traceability systems should address the core demands of sustainability. Companies, therefore, need to demonstrate how ethical their sourcing policy is and link it to perceivable efforts towards sustainability (EONIC, 2021).

Traceability by its definition is a collaborative effort between companies and their chain of suppliers. In an era where customers want to know more about the products they consume, their origin, the conditions under which they were made, and how were they delivered, are key areas to control well (EONIC, 2021). Gellyncket al. (2004) have argued that consumers occupy a crucial position in the food chain in view of their position at the start and end of the chain. This position provides an inspiration for a consumer-driven or market-oriented chain organisation. The arrangement makes consumer demand for safe and wholesome food in general and meat in particular, the greatest driving force for the introduction of a variety of information systems such as branding, traceability and quality assurance schemes (Gellynck & Verbeke, 2001; Leatet al., 1998). For instance, the ability to follow meat products to its source is important not only from an integrity standpoint but also from a safety standpoint (Frame, 2013).

Whilst manufacturers and their suppliers increasingly have traceability procedures in place for major food components (for food safety reasons even more than for sustainability), it is not always possible to trace back each and every minor ingredient. This is because the length and complexity that the supply system creates offer opportunities for fraudsters to exploit the supply chain (Sarpong, 2014).

Consequently, suppliers that deal only in minor ingredients are unlikely to feel the same degree of pressure to adapt from their customers, yet their sourcing of raw materials can still have major environmental and social impacts.

8.2 Palm Oil

Indeed, over the years, various certification schemes have been set up to ensure sustainable cultivation of oil palm and the production of palm oil. The Malaysian government has, meanwhile, instituted various stringent policies and guidelines to ensure that the industry upholds the very highest standard of sustainable practices. The certification asserts that a company has complied with global demand for sustainable palm oil production. But one of the big challenges facing corporations is tracing palm oil through a massive and complicated supply chain.

Generally, very few companies pursue the highest level of sustainability certification because of the complexity and length of the supply chain. Moreover, the vast majority of palm oil is traced only as far as the mill where it is processed, not to the field where it is produced. There is now a growing effort to deploy technology to trace each bunch of fruit to a field and farmer, which would finally ensure new deforestation does not occur in the production of palm oil. Certification schemes are good for the industry since such schemes will continue to create greater transparency in the value chain, including imposing a variety of standards with the intention of reducing rainforest clearance, amongst others.

Part of the solution would therefore be to help small-scale farmers adopt more responsible practices, which will make it possible for them to obtain global certification standards and gain improved access to international markets.

There is also the need to create a multi-stakeholder platform, where buyers and consumers could be assured that the palm oil in the products they use and consume has indeed been sourced sustainably. The aspirations remain high, and today we see the Malaysian Sustainable Palm Oil (MSPO) scheme taking up this task.

8.3 Halal

In Malaysia, the halal certification is done by the Malaysia's Department of Islamic Development (JAKIM) which is under the purview of the Ministry in the Prime Minister's Department. But the damning expose of the domestic halal meat industry in the last few months has brought to light a systemic and institutional corruption of some people comprising foreign exporters and with the alleged connivance of government agencies.

Apart from the reactive need to fortify the Malaysian Anti-Corruption Commission (MACC) with more resources, and the Ministry of Agriculture and Food Industry (MAFI) to increase its governance on the issuance of approval permit for meat importers into Malaysia, there is a pressing and proactive need to be ahead of what can be taken advantage of in order to prevent the recent incident from recurring again.

One of the main challenges faced by Malaysia is in the area of enforcement. The lack of enforcement by the Department of Islamic Development's (JAKIM) personnel in monitoring the use of the certified halal logo has caused the public to question the validity of some of the products or services claiming to be halal. Indeed, there seems to be a lack of assistance given to JAKIM by the Ministry of Domestic Trade and Consumer Affairs with the latter focusing on its own problems, than those that do not necessarily relate to halal products.

In addition, the lack of collaboration amongst the world's halal certification authorities has created 'doubts' amongst Muslim consumers on the authenticity of the halal certification process. The speed of issuing a halal logo is another challenge facing JAKIM. Currently, JAKIM does not have a full-fledged research and development (or technical) unit which is able to process each halal application promptly. Halal integrity in the supply chain is now held with a bit of misgiving in some quarters because halal food travels a great distance before reaching the end user, thus, allowing the possibility of cross-contamination along the way.

8.4   Blockchain Technology to the Rescue?

Some consumers have touted the adherence and use of blockchain technology to help in the face of the difficulties being encountered with the halal certification and food safety in general. Blockchain technology may represent a possible answer to this consumers' request because it is a trust-proof system that can enhance food safety assessment through a higher traceability means (Creydt & Fischer, 2019). Also known as distributed ledger technology (DLT), blockchain technology is deemed to have the capacity to disrupt any intended devious and cunning strategies of actors to achieve their fraudulent objective.

In the last years, several case studies have assessed that a blockchain traceability may have several advantages for the agrifood supply chain (Mao et al., 2018). One of the most significant real-world applications of blockchain is data provenance—the documentation of where a piece of data comes from and the processes and methodology by which it is produced. Indeed, the ability to instantaneously trace the entire lifecycle of food products from origin through every point of contact on its journey to the consumer clearly can help bolster credibility, efficiency and safety (Sarpong, 2014).

This process, for instance, can help trace all the locations of the halal supply chains that began with the original halal farm or livestock farm, including the identity of the farmers or livestock breeders and also ensure a halal standard, verify halal compliance and enforce the performance of halal supply chains to guarantee halal integrity. This is necessary because the food chain worldwide is highly multi-actor based and distributed, with numerous different actors involved, such as farmers, shipping companies, wholesalers and retailers, distributors and groceries. In effect, the blockchain technology is quite promising towards a transparent supply chain of food.

At present, however, some barriers and challenges still exist, which hinder the adoption and implementation of blockchain technology amongst farmers and food supply systems. In particular, it seems that inadequate knowledge of the differences in terms of cost and complexity of the available blockchain technology is making farmers and food supply systems not confident in adopting them. On the contrary, the technology is secure, reliable and transparent tool to ensure food safety and integrity.

8.5 Food Security and Sustainability

From the foregoing, it is obvious that Malaysia faces tremendous challenges in ensuring its sustainable development ideals. Solid waste, which relates to waste generated from products purchased by the public for household use, such as garbage, refuse, sludge and other discarded solid materials, is one of the major environmental problems in the country. Many Malaysians now know that their purchasing behaviour has a direct impact on ecological problems, yet, the public often deflects charges of their complicity in this, instead, they accuse the government and businesses of not doing much in that regard.

In fairness, government agencies and municipal authorities have not helped either in terms of making provisions that could enable the public to pursue a high level of environmental awareness. At the same time some businesses have not shown much keeness to promote and ensure a high level of environmental awareness and adherence in the country. A key challenge, therefore, is for stakeholders to promote aggressively the understanding and awareness towards environmental protection. It is worth noting that one of the obstacles to recycling in Malaysia is the lack of awareness about the importance of separating household solid waste and a lack of interest in the local community to participate and contribute to the recycling programmes. A shift in the level of awareness and attitudes has to exist first in order to implement a shift in behaviour.

Although various strategies have been carried out by the government over the years, in an effort to educate and encourage environmental awareness amongst the public, there exists little or perhaps not much knowledge about environmental issues within the local communities. What one finds is that consideration for the environment comes only from a small group of well-informed and environmentally conscious citizens who are fully committed to their rights to a quality environment. More active and leading roles would be needed in order to encourage greater environmental responsibility amongst communities to bring about effective changes in the approach and attitude of the public towards environmental concerns.

Meanwhile, the lax in enforcement of environmental infractions also constitutes a major blow to environmental advocates who believe the government must do more in spreading awareness and also in taking decisive action against those who degrade the environment. Furthermore, Malaysia needs to have a collective social imperative requiring government agencies, NGOs and the food industry to engage with consumers to raise awareness and elicit change in environmental behaviour. Whilst the exact mechanisms and conditions to elicit such change may not be that easy to accomplish, it may be anticipated that current consumer awareness is highly acute and receptive, therefore, there could be a window of opportunity in engaging people to appreciate the health benefits and improvement of sustainability in the food system. The call for sustainable development has been championed over the years to drum home the point that mankind needs to save the world from its extravagance and ill-pursuit of resources. This is all in a bid to ensure that the current and future generations would be in a position to enjoy well-preserved, fully utilised and well-managed resources.

It is worth noting that food security and self-sufficiency have become increasingly urgent amid the COVID-19 crisis during which the movement of goods and people has either been halted or restricted. During this time as well, there has been lax in the physical halal auditing and certification processes, potentially leading to the risk of shortages of halal certified ingredients, foods and beverages, as well as an ability to have authentic halal supplies. This has added to the concerns about the integrity and validity of the halal certification of products of animal origin.

Notably, one of the most critical drivers for transparency in supply chains is an increased consumer demand to know more about the products they are buying: what is in them, where they came from, the conditions under which they were made, how they got to the consumer, and how they will be disposed of. As already argued elsewhere in the paper, traceability can verify certain sustainability claims about commodities and products, helping to ensure good practices and respect for people and the environment in supply chains. Innovations and technologies supporting traceability are also on the rise, and can be used to strengthen traceability and transparency in the system. Advances in this area would allow traceability to support the concerns of consumers and also ensure the sustainability of products.

9   Conclusion

The chapter discussed the issues and challenges pertaining to food security and sustainability challenges in Malaysia. It provided an appraisal of the situation and also identified problems inherent in them. The chapter noted that in order to accomplish systemic change towards sustainability, traceability issues should be quite defining and articulated in any systemic change. It also stressed that traceability has become a viable and appealing way for businesses to meet both sustainability requirements and expectations of their customers. That said, a better approach to the issue of traceability in the system would be to ably support it with technology and the active involvement of the various actors in the field. In this way, sustainability efforts would be strengthened for the betterment of Malaysians and the world at large.

References

Adam, A (2021, January 4). 'How Malaysia's 'meat cartel' scandal unfolded: A timeline,' *Malay Mail*. Accessed Mar 19, 2021, from https://www.malaymail.com/news/malaysia/2021/01/04/how-malaysias-meat-cartel-scandal-unfolded-a-timeline/1937007

Ahmad, M. F. (2015). Antecedents of halal brand personality. *Journal of Islamic Marketing, 6*(2), 209–223.

BSR and UNGC. (2014). *A guide to traceability–a practical approach to advance sustainability in global supply chains*. United Nations Global Compact Office.

Bushell, S. (2020, February 3). *'What is sustainability'*. Accessed Feb 12, 2021, from https://www.childrensnutrition.co.uk/full-blog/food-sustainability#:~:text=This%20could%20reduce%20greenhouse%20gas,for%20human%20health%20(3)

Creydt, M., & Fischer, M. (2019). Blockchain and more-algorithm driven food traceability. *Food Control, 105*, 45–51.

Devin, B., & Richards, C. (2018). Food waste, power, and corporate social responsibility in the Australian food supply chain. *Journal of Business Ethics, 150*(1), 199–210.

Eco & Beyond. (2021). Accessed Mar 11, 2021, from https://www.ecoandbeyond.co/articles/food-sustainability/

EONIC. (2021). *'Traceability in supply chain management: Why it matters'*. Accessed Mar 13, 2021, from http://eonic.com.hk/traceability-supply-chain-management/

EUFIC. (2015). *Food production: A sustainable food supply*. Accessed June 2, from https://www.eufic.org/en/foodproduction/article/food-production-3-3-a-sustainable-food-supply

European Commission. (2019). *'Sustainable food'*. Accessed Feb 8, 2021, from https://ec.europa.eu/environment/archives/eussd/food.htm

Evans, D., Welch, D., & Swaffield, J. (2017). Constructing and mobilizing 'the consumer': Responsibility, consumption and the politics of sustainability. *Environment and Planning A: Economy and Space, 49*(6), 1396–1412.

FAO. (2014). *'Food wastage footprint–full cost accounting final report'*. Accessed Mar 12, 2021, from http://www.fao.org/3/i3991e/i3991e.pdf

FAO. (2018). *'Sustainable food systems concept and framework'*. Accessed Mar 15, 2021, from http://www.fao.org/3/ca2079en/CA2079EN.pdf

FAO and WHO. (2019). *'Sustainable healthy diets guiding principles'*. Accessed Feb 12, 2021, from http://www.fao.org/3/ca6640en/ca6640en.pdf

Foresight. (2011). *The Future of Food and Farming: Challenges and Choices for Global Sustainability. Final Project Report*. Government Office for Science.

Frame, A. (2013). Improving meat traceability in the US: A look at tracing meat through the supply chain. *Food Safety News*. Accessed May 30, from http://www.foodsafetynews.com/2013/05/meat-traceability-preventingfraud-in-the-u-s/

Garnett, T., & Godfray, C. (2012). Sustainable intensification: Defining terms. *Food Ethics, 7*(2), 8–9.

Gellynck, X., & Verbeke, W. (2001). Consumer perception of traceability in the meat chain. *Agrarwirtschaft, 50*, 368–374.

Germania, M., Mandolinia, M., Marconia, M., Marilungoa, E., & Papettia, A. (2015). The 22nd CIRP conference on life cycle engineering–a system to increase the sustainability and traceability of supply chains. *Procedia CIRP, 29*, 227–232.

Guan, D., Wang, D., Hallegatte, S., Davis, S. J., Huo, J., Li, S., Bai, Y., Lei, T., Xue, Q., Coffman, D. M., & Cheng, D. (2020). Global supply-chain effects of COVID-19 control measures. *Nature Human Behaviour, 4*, 577–587.

Lang, T., & Barling, D. (2012). Food security and food sustainability: Reformulating the debate. *Geographical Journal, 178*, 313–326.

Maestrini, V., Luzzini, D., Maccarrone, P., & Caniato, F. (2017). Supply chain performance measurement systems: A systematic review and research agenda. *International Journal of Production Economics, 183*, 299–315.

Mao, D., Wang, F., Hao, Z., & Li, H. (2018). Credit evaluation system based on blockchain for multiple stakeholders in the food supply chain. *International Journal of Environmental Research and Public Health, 15*(8), 1627.

Market Review on Food Sector in Malaysia. (2019). *Malaysia competition Commission (MyCC).* MyCC., Malaysia.

Sarpong, S. (2014). Traceability and supply chain complexity: Confronting the issues and concerns. *European Business Review, 26*(3), 271–284.

Shaari, J. A. N., & Arifin, N. S. B. M. (2010). Dimension of halal purchase intention: A preliminary study. *International Review of Business Research Papers, 6*(4), 444–456.

Smith, L. C., El Obeid, A. E., & Jensen, H. H. (2000). The geography and causes of food insecurity in developing countries. *Agricultural Economics, 22,* 199–215.

Syed, Z. (2021, February 7). 'What is sustainable food?' *The daily meal.* Accessed Feb 12, 2021, from https://www.thedailymeal.com/healthy-eating/what-sustainable-food-and-why-and-how-should-you-cook-it-slideshow

The ID Factory. (2019). *'Does traceability matter'.* Accessed Mar 12, 2021, from https://www.theidfactory.com/blog/does-traceability-matter/

Thi, N. B. D., Kumar, G., & Lin, C. Y. (2015). An overview of food waste management in developing countries: Current status and future perspective. *Journal of Environmental Management, 157,* 220–229.

Thompson, K., & Haigh, L. (2017). Representations of food waste in reality food television: An exploratory analysis of Ramsay's kitchen nightmares. *Sustainability, 9*(7), 1139.

UNEP. (n.d.). *'Sustainable food production'.* Accessed Mar 18, 2021, from https://www.unenvironment.org/regions/asia-and-pacific/regional-initiatives/supporting-resource-efficiency/sustainable-food

Weber, E. P., & Khademian, A. M. (2008). Wicked problems, knowledge challenges and collaborative capacity builders in network settings. *Public Administration Review, 68*(2), 334–349.

White, K., Hardisty, D. J., & Habib, R. (2019). 'The elusive green consumer'. *Harvard Business Review.* Accessed Mar 14, 2021, from https://hbr.org/2019/07/the-elusive-green-consumer

Zainudin, M. I., Hasan, F., & Othman, A. K. (2019). Halal brand personality and brand loyalty among millennial modest fashion consumers in Malaysia. *Journal of Islamic Marketing, 11*(6), 1277–1293.

Dr. Sam Sarpong is an Associate Professor of Management and Marketing at the School of Economics and Management, Xiamen University, Malaysia. He obtained his PhD from Cardiff University and his MBA from the University of South Wales (UK). He has since taught at Cardiff University, University of London (Birkbeck College) and Swansea Met University, respectively. His research interests lie in the relationship between society, economy, institutions, and markets.

Dr. Ali Saleh Ahmed Alarussi is an Assistant Professor at the School of Economics and Management, Xiamen University. He has experience in accounting scholarship. He had his MSc and PhD in International Accounting from University Utara Malaysia (UUM), Malaysia.

Dr. Faizah Shahudin graduated from the Universiti Putra Malaysia, Malaysia in 2017 in Agricultural Economics and is currently an Assistant Professor at the School of Economics and Management, Xiamen University Malaysia. Her research interest spans economic impact of agricultural and tourism products, environment science and economic analysis.

Food Waste in Romania from an Individual and a National Perspective

Silvia Puiu and Nicoleta Mihaela Florea

Abstract

Food waste is an important problem that should be carefully addressed in order to contribute to the achievement of sustainable development goals (SDG) of the 2030 Agenda for Sustainable Development, mainly the 2nd—Zero Hunger and the 12th—Responsible Consumption and Production. The main objectives of the paper are: establishing the food waste context in Romania comparing to other European countries, identifying the main causes of food waste and the measures that could be implemented to reduce the waste and promote a more responsible consumption. The methodology consists in conducting a triple analysis, one focused on individuals, another on nongovernmental organizations (NGOs), public authorities, and private sector, and the third on the initiatives of the most important retailers in Romania in terms of turnover and net profit. The results offer a clearer image on the food waste in Romania presenting many perspectives in order to better understand the phenomenon and find solutions to reduce food waste.

Keywords

Food waste · Responsible consumption · Responsible production · Sustainable development goals · Waste · Food retailer · Responsible behavior

S. Puiu (✉)
Department of Management, Marketing and Business Administration, Faculty of Economics and Business Administration, University of Craiova, Craiova, Romania

N. M. Florea
Department of Finance, Banking and Economic Analysis, Faculty of Economics and Business Administration, University of Craiova, Craiova, Romania

© The Author(s), under exclusive license to Springer Nature Switzerland AG 2022
S. O. Idowu, R. Schmidpeter (eds.), *Case Studies on Sustainability in the Food Industry*, Management for Professionals,
https://doi.org/10.1007/978-3-031-07742-5_9

1   Introduction

The second and the 12th goals of the 2030 Agenda could be achieved if people, governments, and businesses would adopt a more responsible behavior. There is a specific target—12.3 Global Food Loss and Waste—according to which food waste should be reduced by 50% until 2030. The definition of the term food waste is important because it influences the statistics within various countries, the comparability and the policies that could be implemented to achieve the 12.3 targets of the 12th goal. Thus, Food Use for Social Innovation by Optimising Waste Prevention Strategies (FUSIONS, n.d.) offers a broader meaning of the term: "any food, and inedible parts of food, removed from the food supply chain (FSC) to be recovered or disposed." For a clearer understanding of the term, Food and Agriculture Organization of the United Nations (FAO) distinguishes between food loss and food waste. Food waste implies a "decrease in the quantity or quality of food resulting from decisions and actions by retailer, food service providers and consumers" (FAO, 2019) and "food loss is the decrease in the quantity or quality of food resulting from decisions and actions by food suppliers in the chain" (FAO, 2019). Food wastage comprises both food waste and food loss.

Food waste is defined by the Waste Framework Directive (European Parliament and Council, 2008) as a part of the biowaste produced by households, retailers, and restaurants and is responsible for greenhouse gas (GHG) emissions generated by biowaste disposed in landfills. The revised version of the Directive (European Parliament and Council, 2018) was introduced on March 31, 2019, as a deadline for the Commission to have a proper legislation on food waste and a common methodology for all member states that will report the food waste levels. Starting with 2020, member states should monitor the levels of food waste according to this methodology ensuring comparability. At the EU level, a platform dedicated to food loss and food waste was created in 2016 (European Commission, 2019) with the aim to help member states better understand the causes of food wastage and implement adequate policies to reduce it.

According to Caldeira et al. (2017), there is an important variation regarding the share of food waste generated by each sector in EU, especially if the reporting institutions took into account the edible food or also the inedible part. Their research reveals that the households' share varies between 34% and 65%. Even if there are differences between the studies, the impact of food waste on the environment is important and should be addressed accordingly. The estimation made by European Commission (2018) is that 20% of the food produced in Europe is wasted. According to FAO (2015a), food wastage is responsible for 8% of the total GHG emissions. The report mentions the carbon footprint of various categories of food (cereals more than 30%, vegetables and meat more than 20% in the total carbon footprint of food wastage) and for each stage in the supply chain (consumption is responsible for 37% of the total carbon footprint of food wastage even if its share is only 22%).

The Waste Framework Directive (WFD) recommends member states to apply the hierarchy of waste: prevent, reuse, recycle, recovery, and disposal by incineration or landfill, from the most preferred to the least preferred. Besides the environmental impact, food waste should be seen also in relation to the second SDG. According to United Nations (2018), limiting food waste is beneficial for reaching the Zero Hunger goal. More than 820 million people in the world are hungry and this figure increases (World Health Organization, 2019). "The food currently wasted in Europe could feed 200 million people" according to FAO (2015b, p. 173). In Europe, the European Food Banks Federation (n.d.) was founded for reducing hunger by redistributing the food where it is needed and avoiding its waste. As Corrado and Sala (2018) highlight, food waste reflects a "considerable inefficiency of the global food supply" affecting the environment and depleting the limited resources.

2   Food Waste in European Union

According to the European Parliament (2017), the food loss estimates at the level of 2012 show that there are 173 kg of food per person wasted annually in the EU and 170 million tonnes of carbon dioxide emitted from this waste. The situation estimated in 2010 for each EU member state in kg of food waste per person each year reflects that the first five countries, Netherlands (541 kg), Belgium (345 kg), Cyprus (327 kg), Estonia (265 kg), and Poland (247 kg) have the highest food waste. The countries with the least food waste are Romania and Malta (76 kg each), and Slovenia (72 kg). European Parliament (2016) mentions the causes for food waste at consumer level: a developed country registers higher levels than a developing one, the urbanization, the diet preferences, globalization, changes in the behavior of modern consumers who buy good-looking fruits and vegetables, have higher portions than needed. Urbanization and consumer behaviors are also mentioned as increasing factors for food waste by Secondi et al. (2015) who appreciate that there is a need to address these particularities using targeted policies. A study by the European Commission (2015) revealed that 49% of Europeans appreciate that there is a need for more education regarding food safety, food quality, and date marking that can help them to reduce the food waste.

European Commission (2017) presented statistics for a few European countries revealing that the share of inedible food in the food waste is more than 60% (64% in Slovenia, 63% in Iceland, and 80% in Austria). This explains the differences between the levels of food waste reported by various countries and strengthens the need to implement a common methodology for reporting food waste in the EU.

Regarding food waste hierarchy and the possibility to redistribute food by donation, the EU Platform on Food Losses and Food Waste (2019) presents the situation in the member states. Among the 27 countries that were analyzed, 22 have a national strategy including food donations as a way to reduce food waste, and two other countries (Croatia and Estonia) have it in progress. According to HOTREC (2017), even if households represent the first source of food waste, all actors in a circular economy should implement proper measures to help reach the target of 50% reduction in food waste until 2030 according to the 12th SDG. HOTREC proposed a few guidelines: reconsider menus, make better choices when buying food, implement a better food management to ensure consumption of food before expiring, prioritize quality not quantity in the menus, redistribute and recycle food whenever possible, have partnerships with food banks and other NGOs that can help in the process. In the public opinion, even if households are the primary source of food waste, 62% of Europeans perceive retailers as well as hospitality sector as playing an important role in preventing and reducing food waste (European Commission, 2015).

In Romania, the law that targets food waste is Law no. 217/2016 (Romanian Parliament, 2016) that suffered many amendments and became operational in February 2019 with Law no. 200/2018 (Romanian Parliament, 2018). According to it, retailers can donate food close to its expiry date to NGOs and Food Banks (FB). The National Plan for Waste Management (Romanian Government, 2018) also refers to food waste management as a part of other waste streams. A problem of the Romanian legislation is that there is no clear distinction between the date marking of food (best before and use by) and donors still have to pay value-added tax for the donated food. EU Platform on Food Losses and Food Waste (2019) mentions that countries like Romania, Bulgaria, Poland, and Hungary do not allow donation of food that is past the best before date.

According to the National Institute for Public Health (2019), 2.55 million tonnes of food is wasted annually in Romania and 4.5 million people have difficulties in procuring their food. The report also mentions as a barrier the limited presence of FB in Romania and recommends good practices from Italy, Belgium, and France meant to encourage retailers to donate food to FB and NGOs.

3   Methodology

The objectives of the chapter are: establishing the context for food waste in Romania, identifying the causes for food waste in order to propose adequate solutions, highlighting the initiatives of public authorities, NGOs, and other organizations in combating food waste. There were established three research questions (RQ):

RQ1: What are the main factors responsible for food waste in Romania?
RQ2: What consumption and buying patterns in Romanian households lead to or
     prevent food waste?
RQ3: What initiatives have been taken to reduce food waste?

The methodology consisted in conducting a triple analysis: an online survey addressed to Romanian households to identify consumer patterns and the appropriate solutions to combat food waste; a qualitative analysis of NGOs, public authorities, and private sector in Romania that have initiatives regarding the fight against the phenomenon; and a third analysis focused on the initiatives of the main retailers in Romania starting from the data they made public on their websites and for the press because they did not accept the invitation for an interview.

4   Analysis of Individuals' Perspective on Food Waste in Romania

To answer RQ1 and RQ2, we distributed an online questionnaire on various Facebook groups (10,000 members) in July 2020 for reflecting the individuals' perspective. The response rate was 3.35%, equivalent of 335 respondents. Thus, the results cannot be extrapolated toward the entire population but can be used to better understand the consumption behavior and the motives for food waste in Romania. All results are seen in the context provided by national and international studies regarding food waste in Romania and the EU.

The profile of the respondent is: female (89%) between 26 and 35 years old (35.2%) living in urban areas (80.3%), having a university degree (86.9%), being employed (65.1%) with a net income per family between EUR 400 and 800 (26.3%) and a 2-member family (34%). Regarding respondents' buying preferences, most of them do their shopping mostly from hypermarkets and supermarkets (94.3%) followed by farm markets (50.7%) and in terms of frequency they go shopping 2–3 times per week (53.1%) and 26% of them only once per week. When asked about their food waste behavior, 61.2% of the respondents said they rarely and very rarely throw food, followed by those who throw it often and very often (33.7%) and those who never waste food (5.1%). The share of those who waste food is important and more educational campaigns are needed. There could not be identified any correlation between the retail format used for shopping and their food waste behavior but there was a direct correlation between shopping frequency and wasting food (the p-value of chi-square test is 0.001). A direct correlation was also noticed between the net income and the wasting behavior (the p-value is 0.006), half of those with the highest income (more than EUR 1650 per family) said they throw food often.

Asked what they do with leftovers, most of them (53.4%) said they throw them, closely followed by those who mentioned feeding animals (51.5%). Only 29% of them reuse leftovers for future meals and 11% compost them. The share of those throwing away leftovers is important meaning that if better informed, they can reuse them when planning their meals. Most respondents live in urban areas but the share of people who use leftovers for feeding animals is important because many of them have family in rural areas and have animals. Redirecting food to hungry people or animals is one of the solutions recommended by the WFD. Most respondents (48.4%) cook 1–3 days per week, 26.3% daily, and 24.2% cook 4–6 days per week. No correlation was found with the wasting behavior.

Among the reasons for throwing food, most respondents (66.2%) mentioned that the food was altered because they did not eat it in time, followed by 44.6% who mentioned the food altered faster than they thought. Altered food is a sign of buying more food than needed, not storing it properly and not checking the fridge or pantry before going shopping. More education is needed in this respect. A good behavior was noticed when respondents mentioned in proportion of 72.2% that they bought the so-called ugly products or products close to their expiry date. Seventy percent of them mentioned they did that to save money and 45.5% because they wanted to reduce food waste. Asked about the percentage of the food waste in their household, most of them (45.1%) mentioned 1–10% followed by 27.2% of them who mentioned they waste less than 1%. So, 72.3% of the respondents throw less than 10% of the food which is an encouraging behavior. 93.4% of the respondents are aware of the term food waste and 89.3% of them are acting toward reducing food waste. On a scale from 1 (not interested in food waste) to 100 (interested and making efforts to reduce food waste), the average value is 72.5. All data reveal a responsible consumer and are in accordance with data provided by the European Parliament (2017) that states that Romania occupies the second-best position in terms of kg of food waste/person/year.

Among the actions taken to reduce food waste, 74.6% use a shopping list, 72.8% check the fridge/pantry before going shopping, 54.3% check the expiry date of the products they buy, 53.7% buy only what they need, 50.1% pay attention to how they store products, 40.3% freeze fruits, and vegetables to prevent alteration. Only 17.3% of them make a meal plan and 31.9% use leftovers. Regarding the reasons for taking measures to reduce food waste, the respondents mentioned saving money (58.8%), being the right thing to do (54.9%), protect the environment, and reduce the carbon footprint (46.6%), and also the thought of people dying of starvation (46.1%).

At the last open question, some of the respondents (12.54%) shared their thoughts regarding food waste in Romania. Some expressed the wish to have more places to donate food and also a composting solution shared by more apartments; some appreciated that retailers, hotels, and restaurants waste the most. A part of them mentioned the need for more educational campaigns and brought to attention that the consumer behavior in Romania is also influenced by the deprivation in the communist era.

5   Qualitative Analysis on the Perspective of Public Authorities, NGOs, and Private Sector

The qualitative analysis consisted of interviews with public authorities, private sector, and NGOs having a role in combating food waste. The interview was sent to 40 entities and 12 answered positively: Crucea Rosie Romana (Romanian Red Cross—NGO1); O Masa Calda (NGO2); TERRA Mileniul III (NGO3); Act for Tomorrow (NGO4); Food Waste Combat Cluj and Banca pentru Alimente Cluj (NGO5); CeRe: Centrul de Resurse pentru participare publica (NGO6); Banca Regionala pentru Alimente Brasov (NGO7), Banca pentru Alimente Bucuresti (NGO8) and Ateliere fara Frontiere (NGO9) among the NGOs; Romalimenta from private sector; Ministry of Agriculture and Rural Development (MARD); and National Institute for Public Health (NIPH) as public authorities. The interview was also sent to restaurants, hypermarkets, local authorities, and other entities from the private sector. Two supermarkets answered and declined the invitation for the interview and local authorities did not answer in any way.

The answers to the six questions addressed are summarized in Table 1 for the NGOs.

Table 1 NGOs perspective on food waste in Romania

NGOs	Q1	Q2	Q3	Q4	Q5	Q6
NGO1	Retailers are reluctant to donate food, lack of knowledge among population, not enough fiscal incentives for companies to donate food.	More food banks, educational campaigns on food waste.	N/A	France for educating people on the terms use by and best before; Germany for the fiscal incentives granted to companies donating food.	Alerting consumers when entering a store that some products are close to the expiry date and that they can reduce food waste if they buy them.	Programs for a healthy life that also address this subject.
NGO2	Lack of understanding the phenomenon among population.	Education of population and a better legislation.	Definitely a progress because the subject is more debated by civil society and academic community and there is also a law of food waste even if not perfect.	Stores selling food close to expiry date at a lower price or donating it, NGOs organizing educational campaigns on this issue.	Online educational materials for raising awareness; applications connecting food donors and receivers; apps for helping stores to manage the food supplies.	An application for connecting food donors and beneficiaries; partnerships with other entities developing programs for reducing food waste.
NGO3	There is food waste along the entire supply chain but the most part is wasted by households due to ignorance. The higher the income, the higher the waste. In urban areas, the food waste is higher than in rural areas where people grow their vegetables and use leftovers to feed animals.	Educational campaigns to raise awareness, a legislation improved to stimulate retailers to donate food close to expiry date.	A progress at retailers' level which are more aware of the problem but less on households' level.	Mega image, Auchan who reduce prices for food close to expiry date or use them for salads/juices, share Food app of NGO2, food banks, community fridges.	Apps that can be used among the entire FSC to help people and businesses reduce food waste.	Educational and awareness programs; developing a guide to help households prevent food waste.

(continued)

Table 1 (continued)

NGOs	Q1	Q2	Q3	Q4	Q5	Q6
NGO4	Lack of knowledge among population, a better legislation to prevent and reduce food waste.	A legislation addressed to the other actors involved in the process because the present one targets especially the retailers; more educational campaigns in community.	N/A	Several apps in Sweden where people with surplus are connected with those who need food; in Belgium, the platform happy hours market; French legislation forbids retailers to throw food forcing them to donate it before expiring.	Apps similar with olio (a British start-up) that connect donors with receivers.	A Facebook campaign during the lockdown generated by the pandemic to help people reduce food waste.
NGO5	Lack of education and respect toward food; lack of a national strategy for reducing food waste and lack of a proper legislation.	Monitor food waste, its effects, support food banks with specific legislation, the obligation to donate food close to expiry date; education on food waste starting from an early age.	Not sufficient data but at a first look, consumerism increased and probably the food waste too.	Food banks that contribute to reducing food waste and poverty; Sweden Food tech supports start-ups activating in this area, stop wasting Food movement in Denmark, feedback campaign in the UK, WRAP program in the UK for waste management.	A better management of the FSC.	Educational campaigns started since 2013. As obstacles, we mention the lack of financial resources and support from central authorities such as MARD and the Ministry of Education.

NGO6	Excess production and buying, reluctancy to redistribute food close to expiry date.	An improved legislation and penalties for throwing food. Donating food is optional for retailers not mandatory.	Not enough data.	Social stores Somaro, NGOs like NGO2, NGO9 in Romania; the strict law in France; donating food beyond the best before date in many countries.	Apps connecting the food donors and receivers.	Advocacy campaign to improve legislation on food waste in Romania.
NGO7	Lack of education and a proper legislation.	Education from early age; the government should approach more seriously the issue.	The first FB in Romania was founded in Bucharest in 2016 and now there are 5 FBs in the country, this being an important progress.	Legislation in other EU countries can serve as a model for Romanian legislation to encourage retailers to donate food; European FBs.	A platform for connecting people with food surplus, FBs, and other NGOs; online educational campaigns.	Educational campaigns. As obstacles, we mention lack of support from authorities, a legislation with gaps. The recent changes offer more fiscal incentives for companies donating food close to expiry date.
NGO8	Lack of information regarding alternatives (food redistribution, composting).	Educational campaigns, the new improved legislation stimulating food donation.	A progress: There are 5 FBs in the country, this contributing to a reduction in food waste and poverty.	The new improved legislation helpful in reducing food waste; retailers publishing recipes using leftovers.	Technologies are helpful among the FSC to reduce food waste, optimizing the entire process and in educating consumers more easily.	We collect the surplus but not at the level in France or Hungary. The obstacles we encounter are procedural and the lack of information and trust among food donors.

(continued)

Table 1 (continued)

NGOs	Q1	Q2	Q3	Q4	Q5	Q6
NGO9	Lack of education, donation of food close to expiry date is optional for businesses, lack of a system connecting all actors.	Improved legislation, more studies on food waste, composting, and school programs.	No changes.	Food waste Combat in Romania, Mai Mult Verde—NGO.	To better identify the food close to expiry date among the FSC.	We have a bio-farm too and we developed a pilot program in which we used food waste from retailers and restaurants to make compost. As obstacles: Lack of financial support and lack of legislation regarding recycling this type of waste.

Source: Authors' analysis based on the interviews

Q1: What are the causes for food waste in Romania?
Q2: What solutions can be implemented to reduce food waste in Romania?
Q3: Do you think there is a progress or a regress regarding food waste in Romania?
Q4: Can you mention a few good practices on food waste combat?
Q5: How can technology help in food waste combat?
Q6: What type of activities regarding food waste combat did you develop or are
developing now?

The NGOs were selected based on two documents: one published by MARD
(2020) with the organizations registered for receiving the food donated by retailers
and another one with the NGOs having an active role in combating food waste in
Romania that is recognized as good practices by the NGO Mai Mult Verde (2017).
The invitation for this NGO was not answered in any way but the role played by
them is well recognized in the community and their name was mentioned several
times by the other respondents.

From private sector, only Romalimenta, a federation of food industry businesses,
responded to the interview. According to them, the main causes for food waste are:
lack of education, the increased consumerism, outdated technologies among the FSC
which do not prevent food waste. As solutions, they mentioned educational
campaigns, investments among the FSC, promotion of good practices, a better
measurement of food waste, a more flexible legislation (addressing date marking).
In their opinion, legislation improved and the awareness on this subject increased.

MARD, one of the two public authorities that responded, highlighted the main
causes for food waste among households: lack of knowledge, poor meal planning,
and excess buying. As solutions, they identified awareness campaigns, adoption of
good practices, improvement of legislation, encouraging research on the topic, and
developing projects aimed to reduce food waste. Progress has been made with Law
no. 217/2016 and the educational campaigns among the population. MARD is
responsible for publishing the entities that can receive and manage food donated
by retailers and other businesses. The ministry presents a few initiatives regarding
education on food waste: an agreement with FAO to use educational materials
addressed to youngsters, a partnership with InfoCons to translate the materials and
a protocol with the Ministry of National Education to educate children and teachers
on food waste.

NIPH is the second public authority at national level that responded positively to
the interview. Thus, they mentioned the following causes for food waste in Romania:
lack of awareness regarding the importance of this subject, lack of clear legislation,
and strategies to target food waste reduction among households. NIPH also
highlighted the connection with the Romanians' culture and lifestyle, which are
not in accordance with a behavior focused on combating food waste: big portions
served in restaurants, the habit to buy food in excess for the holidays, and cook a
large quantity of food for the extended family reunited during Christmas and Easter,
important consumption of bread and meat. According to NIPH, the most important
categories of food waste in households are represented by cooked meals (24%),
fruits (22%), vegetables (21%), bread (20%), and dairy (20%). Even if NIPH did not

mention the cause for this cultural behavior, we notice a similarity with the answer provided by one of the individuals who mentioned deprivation in the communist era, the changes in consumption and buying preferences after its end and the market liberalization that followed. As solutions for reducing food waste, NIPH mentioned a clearer legislation, better-defined strategies to prevent and reduce food waste, educational and awareness campaigns on this subject, more food banks in the country (only 5 now), more programs to redirect food close to expiry date for animals and people who need it. The institute considers that there is some progress registered in combating food waste but not so noticeable. Since 2016, Romania has a National Day dedicated to the fight against food waste—October 16. Also, the effects produced by the legislation are not so important because it targets mainly the retailers and their share (7%) in the food waste generated in Romania is not so high compared with that of the households (49%). The good practices mentioned by NIPH are: the legislation in Italy and Belgium that encourages companies to donate food, fiscal incentives in France granted to retailers donating food, the application of the food waste hierarchy in France and projects in Romania developed by Food Waste Combat Cluj, the regional centers for public health in Timisoara and Targu Mures, the NGO Facem Romania Bine, all with educational and awareness campaigns on the issue. NIPH developed a few programs focused on educational programs through 22 of its regional divisions.

6   Analysis of Food Retailers' Initiatives on Food Waste Combat

The previous qualitative analysis was meant to present also the opinion of the retailers and other companies in the food industry, but some of them declined the invitation and others did not answer at all. Thus, we continued the research with an analysis of the initiatives or other corporate social responsibility (CSR) projects developed by retailers. We chose the most important international retailers present in the Romanian market and analyzed the section dedicated to CSR projects on their websites but also the press releases and media articles focused on their activity in order to have a better image on their activity. Even if retailers are responsible for a small share in the food waste in Romania, the role played by them is important because they can change buying and consumption patterns through their strategies related to this subject. Households are numerous in Romania (7 million) according to the last report of Euromonitor (2020) and this is the reason the share is so high for them. But each individual buys his or her food mainly from hypermarkets and supermarkets as the analysis on Romanian individuals revealed. Buying decisions are shaped by marketing strategies in many cases so there is a connection between the marketing tools used by retailers and food waste. There is a potential to change the phenomenon using marketing strategies. So, even if the share of the food waste generated by retailers is not so important, they might influence indirectly the food waste generated at the households' level. Aschemann-Witzel, Hooge and Norman (2016, p. 271) consider that "food marketing and retailing contribute to

Table 2 Top 10 food retailers in Romania in terms of turnover in 2019

Retailer	Turnover in 2019 (million lei)
Kaufland	11877.24
Lidl	9765.13
Carrefour	8160.45
Profi	7264.76
Mega image	6650.44
Metro	5824.05
Auchan	5480.38
Penny (REWE Romania)	3961.09
Selgros	3831.43
Cora (Romania Hypermarche)	1723.50

Source: Data gathered from the Ministry of Finance

consumer-related food waste." That does not mean that food marketing always leads to food waste. It can also prevent or reduce food waste. Reducing prices for ugly products or for those close to expiry date can reduce the phenomenon by influencing the buying decisions of consumers. Calvo-Porral, Medin and Losada-Lopez (2016, p. 42) present a few marketing actions that can be implemented by retailers in order to help with food waste combat: giving up on promotions that encourage individuals to buy in excess, developing awareness campaigns on this subject or "clarifying date labelling." It is important for consumers to understand the difference between *best before* and *use by*. Some products can be used beyond the *best before* date meanwhile others cannot be used beyond the *use by* date. Young et al. (2017) present the results of an experiment conducted by a retailer in the United Kingdom meant to show if and which marketing platforms are helpful in changing the consumers' behavior toward reducing their food waste. The results showed that social media and online newsletters contributed to a change in this regard, this proving the important role played by retailers' food marketing. Evans (2015) considers that food waste generated by individuals should be seen in a broader context related to the way they procure their food. Still, the responsibility for food waste belongs to and should be shared by "all stakeholders of the food supply chain" (Göbel et al., 2015, p. 1429).

Even if in Romania there are also initiatives of some local retailers, the most important ones are those of international food retailers present on the Romanian market having bigger budgets to spend for educational and awareness campaigns or other CSR projects related to the food waste subject. According to the data provided by the Ministry of Finance, the top 10 retailers in Romania in terms of turnover and net profit in 2019 are illustrated in Tables 2 and 3.

We sent invitations for an interview to all of these retailers, but no answers were received. Kaufland and Selgros declined the invitation to answer the interview mentioning that the information they want to be public is available freely on their website (Kaufland) and that their policy does not allow to participate in external research (Selgros). The other retailers did not answer in any way to the e-mails we sent. The decision to analyze the strategies of these retailers regarding food waste in

Table 3 Top 10 food retailers in Romania in terms of net profit in 2019

Retailer	Net profit in 2019 (million lei)
Kaufland	847.66
Lidl	491.39
Mega image	272.60
Carrefour	196.81
Metro	93.27
Selgros	82.40
Penny (REWE Romania)	55.14
Auchan	28.16
Cora	−1.14
Profi	−35.65

Source: Data gathered from the Ministry of Finance

Romania is based on their financial power revealed by the financial data in Tables 2 and 3.

6.1 Lidl Romania

Following the model of European Food Banks (FEBA), there was launched a Romanian Food Bank in 2016 in Bucharest. At present, there are only five in the entire country (Bucharest, Cluj, Brasov, Oradea, and Roman). Three of them are presented in Table 1 (NGO5, NGO7, and NGO8). We mention here these NGOs because the role played by the food banks in reducing food waste is very important and in Romania, these NGOs were possible because of the founder of the first Romanian Food Bank which is Lidl Romania who also supported the development of the other FBs in Romania. FBs are a valuable link between retailers, NGOs, and consumers, helping with redistribution of food in the community. Bazerghi, McKay and Dunn (2016, p. 732) appreciate that FBs "improve food security" and Riches (2002, p.648) consider they are a "matter of urgent public debate" in countries where there is not a well-developed network of FBs. The logistics costs needed for creating FBs are important and the support provided by retailers like Lidl Romania (ranked the second in terms of both the turnover and the net profit in 2019) represents a great initiative helping to reduce food waste in the country and also contributing to a reduction in poverty by redistributing food to those who need it.

According to Bucharest Food Bank (2020), all their results (824 tonnes of food collected and redistributed, 22,000 direct beneficiaries) were possible due to retailers like Lidl (founder and donor), Metro (sponsor and donor), Kaufland (sponsor), and Penny (sponsor and donor). Searching the website of Lidl for data and news regarding food waste and FBs, we found the following results presented in Table 4.

We started the presentation of the most important retailers in Romania (in terms of their financial results in 2019) with Lidl because of its role in the foundation of a FB network in Romania that functions in accordance with the principles of FEBA. Other retailers that support the network are Metro, Kaufland, and Penny.

Table 4 Data and news on food waste and FBs on Lidl website

Posts on Lidl website	Content
The second FB in Romania was founded in Cluj (October 2018).	References to the previous 7 months in which there was a pilot program in Cluj and 7 tonnes of food were saved. The FB was founded by junior chamber international with support from Lidl.
The third FB in Romania was founded in Roman (2019).	The article presents the results registered in the first 6 months of activity: 46 tonnes of food collected and distributed toward 4000 beneficiaries in six counties in the northeast part of Romania.
Conference on food waste in Romania and the role played by FBs.	On October 24, 2019, the conference reunited FBs in Romania, retailers such as Lidl, Metro, and penny, the chamber of commerce and industry in Romania, representatives of FEBA, Romalimenta, and other NGOs interested in the topic.
Sustainability report for 2017 and 2018 (published in 2019).	There are 10 references to food waste and nine to FBs. In 2017 and 2018, Lidl donated more than 300 tonnes of goods to FBs in Cluj and Bucharest and to other NGOs. The partnership with the NGO Mai Mult Verde for a national caravan promoting food waste combat is also mentioned. Regarding the responsibility for preventing and reducing food waste, Lidl better organized its inventory and the entire logistics in its deposits. The cooperation with FBs is another way to diminish the food waste by redistributing food to those who need it.
A Christmas campaign in partnership with FBs for donating food to vulnerable people (2019).	Poverty and food waste are a reality, Lidl trying to contribute to a reduction of these phenomena in Romania. This CSR campaign is not directly linked to food waste combat but is a way to make people realize that others are not so fortunate. It is an indirect way to make people become more responsible toward the way food is used in the household and respect it more.
The results of the Christmas campaign from 2019.	In the campaign, people donated 16.7 tonnes of unperishable food distributed to 7000 beneficiaries from FBs network. There were only 4 cities and 12 stores but the combined effort of people made it possible.
Food donation during the pandemic in Romania.	Lidl donated in march and April 4 tonnes of food for vulnerable families distributed with the help of FBs and also 2500 packages of food and hygiene products distributed for Easter.

(continued)

Table 4 (continued)

Posts on Lidl website	Content
A dedicated section on Lidl website for Health and nutrition—Stop food waste.	There are presented many tips and tricks useful for households in order to reduce the food they waste at home, they provide recipes, ways to store and freeze products to keep them fresh and safe to eat, or what to do with leftovers. These tips are also presented on the Facebook page of Lidl Romania.
Responsibility section on the website includes food waste among others (environment, social problems in the community, and supporting local suppliers).	These issues are presented in a concise way, they being detailed in the sustainability reports.

Source: Data from https://corporate.lidl.ro/pentru-media/cautare-avansata, https://surprize.lidl.ro/responsabilitate, https://www.lidl.ro/sanatate-si-nutritie/stop-risipa

6.2 Kaufland Romania

The first position in terms of the financial results in 2019 among the supermarkets in Romania is occupied by Kaufland. This retailer declined the invitation for a more elaborate interview regarding food waste and mentioned to check their website for the available data. The retailer is mentioned on the website of the Romanian FB in Bucharest as a sponsor. The data we found on the website are not so numerous when searching after terms like *food waste*, *food donated*, and *food bank* even if Kaufland is well known in the community for allocating important budgets for CSR. Thus, they have many programs through which they finance projects initiated by NGOs helping them to come up with innovative ideas in various areas (social, education, environment, health, and culture). The results of our research are presented in Table 5.

There were not found any results on Kaufland website on the FBs in Romania and the support offered to them by the company. Thus, we continued the research using the terms *food bank* and *Kaufland* on the Google search engine. There were generated a few articles and documents mentioning the contribution of Kaufland in 2019–2020 (Kipper, 2020; Sandulescu, 2020; Department for Sustainable Development, 2020). Kipper (2020) and Sandulescu (2020) refer to the same subject: Kaufland donated 11 tonnes of food to the Romanian FB, goods that were distributed around Easter to 8123 vulnerable beneficiaries taking into account the pandemic and the state of emergency in Romania at that time. The Department for Sustainable Development organized in 2019 a conference dedicated to food waste combat as part of the National Strategy for Sustainable Development. This event reunited representatives of Bucharest and Cluj FBs, Kaufland, Romalimenta (interviewed by us), FEBA, FAO, Lidl, and NGOs preoccupied with this issue. The document with the conference's conclusions presents the results mentioned by Kaufland at the conference: projects in partnership with NGOs focused on social activities in the community, the intention to open a social canteen for vulnerable people in

Table 5 Data and news on food waste and FBs on Kaufland website

Posts on Kaufland website	Content
A shopping guide for the pandemic period focused on responsible and safe buying and consumption.	Among the guidelines, there is one referring to the need to avoid food waste and buy only the goods necessary for a family's consumption. The article also mentions the limited availability of products. This might be linked to the start of the epidemic in Romania when people bought important amounts of food worrying that there would be a shortage in the near future.
An article about responsibility focused on education.	The post mentions the project Akademia Kinderland in partnership with an NGO in which a caravan goes to five cities in the country promoting healthy nutrition and ways to avoid or reduce food waste. The program is addressed to children between 4 and 14 years old and to their parents.
A short mention within the responsibility section about food waste combat.	The company pays attention to its inventory in order to make sure that the food waste is avoided, the products on the shelf are always fresh and safe, and that there is no shortage of products in the supermarkets.
Sustainability report for 2016.	There are 13 results for food waste. The report mentions two objectives linked to food waste combat (the second and the 12th goals of the Agenda 2030), objectives they support among others. Other mentions in the report refer to: educational campaigns helping consumers to be more informed and responsible, price cuts for products close to expiry date, food donations, an automatic process for their orders, and stocks for 1–2 days in order to avoid food waste. They highlighted that reduction of food waste is an important commitment and they will continue their efforts to reduce this phenomenon in 2017.
Sustainability report for 2017.	The search for food waste generated 17 results. The same objectives are reiterated, there is also a mention of food donation and campaigns for helping consumers be more responsible in their buying decisions. There is an entire section focused on food waste combat that is seen as an important part of being a responsible retailer. Thus, the report mentions they focus on campaigns on this topic addressed to their stakeholders and also in continuously investing in their systems in order to reduce food waste within their supply chain from suppliers to the final customers (IT systems and a supply Chain management team). Another reference is made to the financial support offered for eat smart conferences that started in 2016.

Source: Data from https://www.kaufland.ro/ and https://despre.kaufland.ro/

Bucharest, a project active during Christmas dedicated to people socially or geo-graphically isolated. This document also presents the results of Lidl mentioned by the CSR manager of the company: price cuts of 30% for food close to expiry date, donation of food with deteriorated packages for animal shelters, educational campaigns for preventing food waste, 300 tonnes of food donated in 2017 and 2018 to food banks, animal shelters, and farms.

6.3   Carrefour Romania

We combined the search on the company's website and social media page with the one using the Google search engine for terms like *food waste* and *food bank*. We found a few posts on Carrefour website which we detail in Table 6 and a few articles in the media covering the subject of food waste combat led by Carrefour, which occupies the third position in terms of turnover and fourth in terms of net profits.

Using Google search, we found a few press articles from 2019 and 2020 men-tioning a few innovations and projects related to food waste combat within the Carrefour Group. Since 2019, the group implemented a smart technology to better organize its food supply chain and thus reduce food waste in their stores (Neagu, 2019). This is available for the entire group, not being specific to Romania. Other articles mention the partnership with NGO1 for the food bank project and the donation of goods with a longer shelf life (Scarlet, 2020).

Table 6 Data and news on food waste and FBs on Carrefour website and social media

Posts on Carrefour website	Content
A post in the responsibility section focused on environmental protection.	In partnership with an NGO, Carrefour developed the project of a bio-farm that used compost from vegetables and fruits from Carrefour (800 tonnes between 2014 and 2017), project financed by Carrefour with 1.5 million lei. The project received a prize at the Romanian CSR awards in 2018.
A post on solidarity.	The company mentions the support offered to the Romanian red cross (NGO1) for its food bank program. There are also mentioned food donations to other NGOs supporting people in need.
A post dedicated to Food banking.	Partner of NGO1 since 2009, the company launches a campaign in the pandemic context to make people donate food for vulnerable families. Their food donated has to have a long shelf life.
A Facebook campaign meant to raise the awareness on food waste and climate change.	Various posts offering people tips and tricks that they can use at home to reduce food waste and become more responsible.

Source: Data from https://carrefour.ro/

Table 7 Data and news on food waste and FBs on Mega Image website

Posts on Mega Image website	Content
A post from 2020 mentions the partnership of 10 years with NGO1.	For the last 10 years, the company donated 45 tonnes of food with the help of its customers through the food bank project of NGO1.
A post from 2019 about an educational program for healthy nutrition.	The program is addressed to children and tries to promote healthy nutrition, teaching them to eat healthily but also how to avoid food waste.
A post related to the mega image for community fund.	The fund financially supports local initiatives in various areas. The post on the website mentions a project focused on educational campaigns on healthy food, food donation, and helping vulnerable communities to adopt a healthier life.
The sustainability report for 2016 (the only one found on the website).	The report mentions 4 directions for their company (employees, community, environment, and developing its own brands). Food waste reduction is mentioned in the environment category. In 2016, the company donated 78 tonnes of food for vulnerable people, contributing to a reduction in poverty and hunger, and also 225 tonnes of food for animal shelters.

Source: Data from https://www.mega-image.ro/

Table 8 Data and news on food waste and FBs on Metro website

Posts on Metro website	Content
A general post on how to reduce costs by reducing food waste.	A useful post as the company describes it regarding the numerous costs related to food waste and about the many benefits companies have if they reduce their food waste.
A post on the new buying habits of the customer post-COVID-19.	The post highlights that the customer is more aware of the environmental issues, including those related to food waste combat and that they will prefer local companies to support them during the pandemic crisis.

Source: Data from https://www.metro.ro/

Mega Image is on the third position for its net profit and fifth for its turnover in 2019. Its contribution to food waste reduction is presented in Table 7 based on the results found searching their website.

Metro Romania is in the fifth position for its net profit and the sixth for the turnover in 2019. We used the same research system (website and Google search) using terms like food waste, food bank, and sustainability report in relation to the company. The results generated from the website search are presented in Table 8.

There were only a few posts on the website regarding this topic so we extended the search on Google and got a few more news on Metro activity on this topic. Thus, Asociatia Marilor Retele Comerciale din Romania (2015) mentions the contribution of Metro to the food donation program of the Romanian Red Cross in which Metro supports vulnerable people from the country with slow perishable food donated by

Table 9 Data and news on food waste and FBs on Profi website

Posts on Profi website	Content
The sustainability report for 2019.	The report mentions that one of the objectives is to have a supply chain that is efficient and thus reduces food waste and the carbon footprint. Food waste is considered by Profi stakeholders as an important issue that needs to be addressed by the retailer among others. Food waste has a dedicated section in the report. The food waste in 2019 reached the level of 3176 tonnes and reported to the number of stores, the waste decreased by 17%. The initiatives of the company to reduce food waste include: Optimizing the food displayed on the shelf, training the employees on how to prevent food waste in the stores, price cuts of 25% for food close to expiry date, and food donations to animal shelters and animals at the zoo.

Source: Data from https://www.profi.ro/

Table 10 Data and news on food waste and FBs on Auchan website

Posts on Auchan website	Content
A post from 2020 on the project zero Risipa (zero waste).	The project Zero Waste was launched in August 2020 and encourages its customers to choose the products close to expiry date that have considerable price cuts in order to contribute themselves to the reduction of the food waste. The company mentions an IT system meant to identify these products in the store and label them accordingly to be easily noticed by customers. The project is at the same time one of awareness for both the retailer and the customers.

Source: Data from https://romania.auchan.ro/

its clients since 2010. Metro is also mentioned on the website of the Romanian Food Bank from Bucharest as a sponsor and food donor.

Profi Romania is in the fifth position for its turnover but in the tenth for its net profit. The results generated on its website searching for news on food waste, food bank, and sustainability reports that might mention these aspects are presented in Table 9.

Only a sustainability report was found on the website, the search for the terms not generating any other posts neither on Profi website nor on a Google search where we used the same combinations of terms but in relation to the name of the retailer.

Auchan Romania is in the seventh position for its turnover and in the eighth for its net profit in 2019. The search results revealed only a post detailed in Table 10. The Google search generated two sustainability reports for 2012 and 2013 and many recent news on the latest project on Zero Waste initiated by Auchan Romania. No recent sustainability report was found on Auchan website or on a simple Google search.

Penny (REWE Romania) is in the eighth position for its turnover and in the seventh for its net profit. The search on its website for terms related to food waste, food bank, and sustainability revealed three results detailed in Table 11. Penny is an important sponsor and donor of food for the Romanian Food Bank developing many social projects to support people in need.

Table 11 Data and news on food waste and FBs on Penny website

Posts on Penny website	Content
A general post on food waste in the sustainability section.	The post is very general, not presenting anything specific, any numbers but only a commitment to reduce food waste along the entire food supply chain. The post mentions as solutions price cuts for products close to expiry date.
A post during the pandemic on the food donations made by penny to support vulnerable people.	The post mentions its partners to whom they donate food for people in need: The Romanian food Bank, Romanian red cross, world vision, save the children Romania, etc. the donations during the first months of pandemic in Romania reached 100,000 euros.
Another post related to the pandemic and the support to the community.	The article is similar to the one above, mentioning the partners to whom the company donated food.
The latest report on sustainability for 2018.	The report addresses the issue of food waste. The company mentions an improvement of the entire food supply chain in order to have an optimal inventory in their stores and thus reduce food waste. This measure is also combined with the possibility that consumers have to buy products close to expiry date with an important discount.

Source: Data from https://www.penny.ro/ and https://sustenabilitate.penny.ro/

Table 12 Data and news on food waste and FBs on Selgros website and Facebook page

Posts on Selgros website and Facebook page	Content
A post on Food revolution day by Jamie Oliver.	The company presents an event from 2017 dedicated to food waste combat and to educating young generations on healthy nutrition.
A section on social responsibility.	The post mentions a few projects developed and supported by Selgros: One dedicated to educating children about how to eat healthier, one for food donations in partnership with the Romanian church.
A Facebook post on masterclasses by Selgros Romania.	An event where chefs show people how to have a successful buffet and also reduce food waste.

Source: Data from https://www.selgros.ro/

Selgros Romania is in the ninth position for its turnover but in the sixth for its net profit in 2019. We searched its website for terms related to food waste, food bank, and sustainability to have a clear image on the contribution played by the retailer. The results are presented in Table 12. No sustainability report was found on the website or on a Google search. We also searched for food waste and food bank in relation to Selgros and found a few more data: the retailer is a partner for Somaro—a network of social stores in Romania and during the pandemic crisis in Romania, the

Table 13 Data and news on food waste and FBs on Cora website

Posts on Cora website	Content
A post on its foundation Lumea Cora launched in 2014.	The NGO is focused on various projects, one of them dedicated to food waste combat—Cugeta, Mananca, and Salveaza (think, eat, and save).
A post on the blog from 2017 presenting the project mentioned above.	The project is meant to make Cora consumers more aware of the food waste and adopt a responsible buying and consumption behavior.
Three posts on the blog in the lifestyle section offer advice for consumers who want to reduce food waste.	One is for reducing food waste at home, one for living a more sustainable life that includes a reduction in food waste and one for better organizing the fridge to keep food fresh and reduce alteration.

Source: Data from https://www.cora.ro/ and https://www.blog.cora.ro/

retailer supported the initiatives of volunteers who bought food for people in need (Ariton, 2020).

Cora (Romania Hypermarche) is in the last position for its turnover and the penultimate position for its net profit in 2019. The results on Cora website and blog related to food waste are presented in Table 13. No sustainability report was found. Using Google search, we found the company mentioned as a partner for Somaro (social stores in Romania) and the Romanian Red Cross for food donations (Crucea Rosie, 2018).

Only four of the 10 food retailers (Lidl, Kaufland, Mega Image, and Profi) have sustainability reports on their websites addressing the food waste issue among others. The search results related to food waste were the most numerous on Lidl website, followed by Kaufland, these two retailers being in the first two positions in terms of their financial results.

7   Conclusion

The factors responsible for food waste (RQ1) revealed by both studies are: the lack of knowledge on food waste, an increased consumerism, not making meal plans, not using leftovers, food alteration as a consequence of buying too much, an improper legislation, and improper date marking. The consumption patterns leading to food waste (RQ2) are going shopping too often, buying more than needed, cooking more food than eaten, and not using leftovers for future meals or donations. The consumption patterns preventing food waste (RQ2) are: using a shopping list, checking the fridge and pantry to avoid buying unnecessary products, checking the expiry date, adequately store or freeze items to keep them edible longer. Answers in Table 1 offer a clear picture on RQ3. Five of the eight NGOs mentioned they organized educational campaigns, three of them collect the food surplus and redirect it to people who need it, one created an application connecting donors and receivers and another one is focused on advocacy campaigns to generate an improvement of the legislation.

MARD was active during the last years, organizing educational campaigns and supporting the initiatives of the NGOs fighting against food waste. The other public authority (NIPH) developed only a few awareness campaigns, one in partnership with a private company.

Taking into account that households represent the most important source for food waste, we can notice from the individuals' perspective that the general perception of the population is that restaurants and retailers are the sectors where food is wasted the most. As NGO2 also highlighted, this distorted perception is caused by the lack of understanding among population regarding the way small quantities of food wasted daily in a household multiply with all the households in the country resulting in important quantities of food wasted.

The analysis on the top 10 food retailers in Romania revealed an important presence and many initiatives regarding food waste combat. Lidl had an important role being one of the founders of the first Romanian food bank in 2016. Besides price cuts and awareness campaigns, retailers also donate food close to expiry date to food banks and animal shelters encouraged by the recent legislation on food waste. The partnership with the Romanian Food Bank or the food bank of the Romanian Red Cross is not limited only to food close to expiry date but also to other goods needed by vulnerable people. Regarding RQ1, retailers try to better optimize the entire FSC in order to reduce food waste because factors responsible for this phenomenon are present from production to the final consumer. Related to RQ2 and RQ3, retailers organize awareness campaigns and guidelines for consumers and offer important discounts to orient the consumer in buying products close to expiry, donating the food that is not sold and risks to expire to FBs for helping vulnerable communities.

The triple analysis on individuals, NGOs, FBs, public authorities, representatives from the private sector and food retailers reveal the important progress made in Romania in the last years (the legislation, a national day dedicated to food waste combat, more awareness campaigns, price cuts, and donations for the food close to expiry). There is a lot of space for improvement as the interviews with the NGOs revealed: a better and clearer legislation, more fiscal incentives for retailers who donate food, developing more FBs covering all important regions in the country, and more educational campaigns starting from early age.

References

Ariton, A. (2020). Kaufland si Selgros se alatura initiative civice *Cumparaturi la usa* ta care livreaza gratuit alimente la domiciliu pentru persoanele vulnerabile. *Ambasada Sustenabilitatii.* Retrieved from http://ambasadasustenabilitatii.ro/kkaufland-si-selgros-se-alatura-initiativei-civice-cumparaturi-la-usa-ta/

Aschemann-Witzel, J., Hooge, I., & Normann, A. (2016). Consumer-related food waste: Role of food marketing and retailers and potential for action. *Journal of International Food & Agribusiness Marketing, 28*(3), 271–285. https://doi.org/10.1080/08974438.2015.1110549

Asociatia Marilor Retele Comerciale din Romania (2015). *Metro CSR.* Retrieved from www.amrcr. ro/wp/wp-content/uploads/2015/07/METRO-CSR.pdf

Bazerghi, C., McKay, F. H., & Dunn, M. (2016). The role of food banks in addressing food insecurity: A systematic review. *Journal of Community Health, 41*, 732–740.

Bucharest Food Bank. (2020). *About section.* Retrieved from http://bancapentrualimente.ro/despre/

Caldeira C., Corrado S., & Sala S. (2017). *Food waste accounting–Methodologies, challenges and opportunities.* EUR 28988 EN. Luxembourg, Publications Office of the European Union, JRC109202.

Calvo-Porral, C., Medin, A. F., & Losada-Lopez, C. (2016). Can marketing help in tackling food waste?: Proposals in developed countries. *Journal of Food Products Marketing, 23*(1), 42–60. https://doi.org/10.1080/10454446.2017.1244792

Corrado, S., & Sala, S. (2018). Food waste accounting along global and European food supply chains: State of the art and outlook. *Waste Management, 79*, 120–131.

Crucea Roșie. (2018). *Caravana de Paste in sprijinul persoanelor nevoiase din Bacau.* Retrieved from https://crucearosie.ro/stiri-si-comunicate/caravana-de-paste-in-sprijinul-persoanelor-nevoiase-din-bacau/

Department for Sustainable Development (2020). *Impreuna combatem risipa alimentara.* International conference. Retrieved from http://dezvoltaredurabila.gov.ro/web/wp-content/uploads/2020/01/Livrabil_compressed.pdf

Euromonitor. (2020). *Romania country factfile.* Retrieved from www.euromonitor.com/romania/country-factfile

European Commission. (2015). *Flash Eurobarometer 425, Food waste and date marking.*

European Commission. (2017). *Food waste statistics.* EU Platform on food losses and food waste: Sub-group on food waste measurement. Retrieved from https://ec.europa.eu/food/sites/food/files/safety/docs/fw_eu-platform_20170331_statistics.pdf

European Commission. (2018). *Communication from the commission to the European parliament, the council, the European economic and social committee and the committee of the regions on a monitoring framework for the circular economy.* COM(2018)29.

European Commission. (2019). *EU platform on food losses and food waste.* Retrieved from https://ec.europa.eu/food/sites/food/files/safety/docs/fw_eu-actions_flw-platform_tor.pdf

European Food Banks Federation. (n.d.). *Our mission.* Retrieved from www.eurofoodbank.org/en/mission-vision-values

European Parliament. (2016). *Tackling food waste: The EU's contribution to a global issue.* Retrieved from www.europarl.europa.eu/RegData/etudes/BRIE/2016/593563/EPRS_BRI(2016)593563_EN.pdf

European Parliament (2017). *Food waste: The problem in the EU in numbers.* Retrieved from www.europarl.europa.eu/news/en/headlines/society/20170505STO73528/food-waste-the-problem-in-the-eu-in-numbers-infographic

European Parliament and Council. (2008). *Directive 2008/98/EC of the European parliament and of the council on waste and repealing certain Directives.*

European Parliament and Council. (2018). *Directive 2018/851 of the European parliament and of the council amending directive 2008/98/EC.*

EU Platform on Food Losses and Food Waste. (2019): *Redistribution of surplus food: Examples of practices in the member states.* Retrieved from https://ec.europa.eu/food/sites/food/files/safety/docs/fw_eu-actions_food-donation_ms-practices-food-redis.pdf

Evans, D. (2015). Blaming the consumer–once again: The social and material contexts of everyday food waste practices in some English households. *Critical Public Health, 21*(4), 429–440.

FAO. (2015a). *Food wastage footprint and climate change.* UN FAO. Retrieved from www.fao.org/3/a-bb144e.pdf

FAO. (2015b). *70 Years of FAO.* Retrieved from www.fao.org/3/i5142e/I5142E.pdf

FAO. (2019). *The state of food and agriculture–moving forward on food loss and waste reduction.* Retrieved from www.fao.org/3/ca6287en/ca6287en.pdf

FUSIONS. (n.d.). *Food waste definition.* Retrieved from www.eu-fusions.org/index.php/about-food-waste/280-food-waste-definition

Göbel, C., Langen, N., Blumenthal, A., Teitscheid, P., & Ritter, G. (2015). Cutting food waste through cooperation along the food supply chain. *Sustainability, 7*(2), 1429–1445. MDPI AG.

HOTREC. (2017). *European hospitality industry guidelines to reduce food waste and recommendations to manage food donations.* Retrieved from https://ec.europa.eu/food/sites/food/files/safety/docs/fw_lib_2017_hotrec-guidelines_en.pdf

Kipper, L. (2020). *Peste 8000 de personae au fost hranite cu alimente donate de Kaufland Romania.* PRwave. Retrieved from www.prwave.ro/peste-8000-de-persoane-au-fost-hranite-cu-alimente-donate-de-kaufland-romania/

Mai Mult Verde. (2017). *Romania fara risipa.* Retrieved from http://foodwaste.ro/wp-content/uploads/2018/10/FoodWasteRO-Anexa21-CatalogONG.pdf

MARD. (2020). *Lista operatorilor receptori.* Retrieved from madr.ro/docs/ind-alimentara/risipa_alimentara/2020/Lista-operatori-receptori-13.05.2020.pdf

National Institute for Public Health. (2019). *Analiza de situatie.* Retrieved from https://insp.gov.ro/sites/cnepss/wp-content/uploads/2019/10/Analiza-de-situatie-2019.pdf

Neagu, L. (2019). *Carrefour foloseste Inteligenta Artificiala pentru optimizarea distributiei si reducerea risipei alimentare.* Green Report. Retrieved from www.green-report.ro/carrefour-foloseste-inteligenta-artificiala-pentru-optimizarea-distributiei-si-reducerea-risipei-alimentare/

Riches, G. (2002). Food banks and food security: Welfare reform, human rights, and social policy. Lessons from Canada? *Social Policy & Administration, 36*(6), 648–663.

Romanian Government. (2018). *Decision no. 942/2017 regarding the national plan for waste management.*

Romanian Parliament. (2016). *Law no. 217/2016 regarding food waste management.*

Romanian Parliament. (2018). *Law no. 200/2018 amending Law no. 217/2016.*

Sandulescu, L. (2020). *18 tone de produse donate prin Banca pentru Alimente.* Revista Biz. Retrieved from www.revistabiz.ro/18-tone-de-produse-donate-prin-banca-pentru-alimente/

Scarlet, A. (2020). *Banca de Alimente, 14500 kilograme de hrana colectate intr-o luna. Revista Biz.* Retrieved from www.revistabiz.ro/banca-de-alimente-14-500-kilograme-de-hrana-colectate-intr-o-luna/

Secondi, L., Principato, L., & Laureti, T. (2015). Household food waste behaviour in EU-27 countries: A multilevel analysis. *Food Policy, 56*, 25–40.

United Nations. (2018). *Zero hunger: Why it matters.* Retrieved from www.un.org/sustainabledevelopment/wp-content/uploads/2018/09/Goal-2.pdf

World Health Organization. (2019). *World hunger is still not going down after three years and obesity is still growing.* Retrieved from www.who.int/news-room/detail/15-07-2019-world-hunger-is-still-not-going-down-after-three-years-and-obesity-is-still-growing-un-report

Young, W., Russell, S.V., Robinson, C.A., & Barkemeyer, R. (2017). *Can social media be a tool for reducing consumers' food waste? A behaviour change experiment by a UK retailer.* Resources, Conservation and Recycling, 117, Part B, 195–203.

Websites

https://carrefour.ro/
https://corporate.lidl.ro/pentru-media/cautare-avansata
https://despre.kaufland.ro/
https://romania.auchan.ro/
https://surprize.lidl.ro/responsabilitate
https://sustenabilitate.penny.ro/
https://www.blog.cora.ro/
https://www.cora.ro/
https://www.kaufland.ro/
https://www.lidl.ro/sanatate-si-nutritie/stop-risipa
https://www.mega-image.ro/

https://www.metro.ro/
https://www.penny.ro/
https://www.profi.ro/
https://www.selgros.ro/

Silvia Puiu is a PhD Lecturer in the Department of Management, Marketing and Business Administration, Faculty of Economics and Business Administration, University of Craiova, Romania. She has a PhD in Management and teaches Management, Ethics Management in Business, Public Marketing, Creative Writing in Marketing, Marketing in NGOs, and Marketing of SMEs. Silvia Puiu followed postdoctoral studies on *Ethics Management in the Public Sector of Romania*. During the scholarship, she conducted comparative research between Romania and Italy regarding the Corruption Perception Index, ethics management in the public sector and regulations on whistleblowing at the University of Milano-Bicocca. Silvia Puiu published more than 50 articles in national and international journals. Her research covers topics from corporate social responsibility, strategic management, ethics management, public marketing, and management. She is a reviewer for journals indexed in Web of Science (Sustainability and Social Sciences) and a member of the Eurasia Business and Economics Society. Some relevant works: *Strategies on Education about Standardization in Romania*, chapter in the book *Sustainable Development*, Springer, 2019; *NEETs—a Human Resource with a High Potential for the Sustainable Development of the European Union*, chapter in the book *Future of the UN Sustainable Development Goals*, Springer, 2019; *Entrepreneurial Initiatives of Immigrants in European Union*, *The Young Economists Journal*, no. 31, 2018.

Nicoleta Mihaela Florea is a PhD Lecturer in the Department of Finance, Banking and Economic Analysis, Faculty of Economics and Business Administration, University of Craiova, Romania. She has a PhD in Finance since 2011 when she presented her thesis "Fiscal and fiscal harmonization in the context of Romania's integration into the financial structures of the European Union." Specialization areas are: Financial Management of the Enterprise, Financial Management, and Taxation. She is conducting a postdoctoral research internship in 2015 at the University of Piraeus, Greece, and an internship at the Universidad de la Rioja, Logrono, Spain.

Scientific representative papers: *Achieving sustainable development goals (SDG): Implementation of the 2030 Agenda in Romania*—Sustainability 2019; *Comparison between Romania and Sweden Based on Three Dimensions: Environmental Performance, Green Taxation and Economic Growth*—Sustainability, 2020; *Taxation and Tax Harmonization*, Universitaria Publishing House, 2013. She is a member of the editorial board of the journal "Finance Challenges of the Future."

Sustainable Food Production in Serbia, an Exploration of Discourse/Practice in Early 2020s

Milan Todorovic

Abstract

The matter of sustainable food production is a convoluted one; it is inexorably linked with ethics, ideology, economic and political arguments, corporate power contrasted with sole enterprises, traditional ways of life and so on. This includes widely publicised issues, such as those of vegetarian and vegan ethics, discourses and realities of animal cruelty, environmental protection, access to clean water and unpolluted land—and by extension, mining, mass exploitation, industrialisation and land ownership by powerful groups and individuals. Among the latter, weighty stakeholder groups, including state actors, justifications of post/industrial, quasi/scientific experimentation and the spectre of genetic modification are used as arguments for progress, eradication of hunger, reach/supply, distribution, cost-effectiveness and access to sustenance among the poor.

At the heart of the ideological discourses underpinning all this are the notions of freedom, utility and choice, as axiomatically inherent concepts of neoliberal thought so deeply embedded in the past 40 years of state and supra-state policies that the neoliberal brand of policymaking, business strategy and their very *logos as thought*, seems discerned only by those initiated in the breach between multiple parties involved, and thus seldom questioned.

Food, land, water and air as the most fundamental needs—and rights—of human and non-human life are the subjects that arouse passion in many—including advocates, activists, farmers and producers—and the detached, utilitarian elaborations of those currently holding dominant positions that cross geographical, organisational and economic strata.

M. Todorovic (✉)
London Metropolitan University, London, UK
e-mail: m.todorovic@londonmet.ac.uk

© The Author(s), under exclusive license to Springer Nature Switzerland AG 2022
S. O. Idowu, R. Schmidpeter (eds.), *Case Studies on Sustainability in the Food Industry*, Management for Professionals,
https://doi.org/10.1007/978-3-031-07742-5_10

This text explores contemporary practices and positions concerning traditional, sustainable and health-promoting food production in Serbia against the backdrop of international, national and local contexts of socio-political, environmental, economic and industrial significance. Fieldwork, conducted online during the pandemic finds that food production in Serbia largely fits within the forceful global patterns, with health food producers displaying a traditional/reflective slant but are by no means backward in their ethically driven stance.

Indeed, a critical rethink of agriculture and health priorities appears to permeate Serbian discourse across ideological and socioeconomic spectrums. Young professional families appear to be embracing health food lifestyles and village life. Arguably though, this relates to the wealthier minority able to make such choices. Scarred by the turbulent century behind it and civil strife awarded by unremitting structural transition, Serbians redefine their sense of Self by either embracing or angrily rejecting tradition and family values.

Such patterns are not unfamiliar to the affluent West. Yet, here they tend to define everyday life in a different way, impacting on health, community and civic duty. Discursive drives range from mining to re-industrialisation, enduring impact of the tumultuous Nineties, 'modernising' regulatory frameworks, imported anticompetitive practices and from resistance to the extinction of inherited egalitarian notions to the capillary attempts to reclaim communal prerogatives. Organisational, regulatory and policy challenges surface, and this essay contextualises those recounted by practitioners, the public and key scholars in such complex circumstances.

Keywords

Ethics · Discourse · Discourse and practice · Discourses of now · Health food discourse · Traditionalist discourse · Agriculture · Ideology · Enterprise · Zadruga · [Agricultural] cooperative/s · Householder/s · Community · GMO · Water · Waterways · Mining · Activist/s · Direct activism · Farmers · Producers · Utilitarianism · Industrial food production · Village life · Regulation · Regulatory frameworks · Anticompetitive practices · Imports · Exports · Certification · Policy · Policy frameworks · Gallant peasantry · Belgradisation · Global factors · Political economy · sociocultural · Rural society · Rural community · European · EU · Empire · Postcolonial · Nonaligned movement · NAM · Rural communities · Industrialisation · Economic tensions · Peasant society · Auto-chauvinism · Reflexive negation · Identity-laden · Self-revulsion · Tradition · Traditional belief systems · First World War · The Great War · Organic food produce · Unadulterated food produce · Agrarian reform · 10 hectare limit · Agricultural land · The new landowners · Landlords · Post-communist 'transition' · 7% growth · [Former- · socialist-] Yugoslavia · 'Yugoslavian industrial miracle' · International debt · The 1965 reform · [Yugoslav workers'] self-management model · Corruption · New elites · Federal state · Civil society · Brain drain · Depopulated villages · Privatisation · Neoliberalism · Ageing society ·

Post-transitional period · Mismanagement · Cultural shift · Cultural change · Critical terminology · Infrastructure · Overpopulation · 'Public intellectuals' · Internal migrants · Displaced refugees · Fieldwork · Respondents [as] practitioners · Thematic contextualisation · Health food production · Marketing · Cultural conditions · Reflexion · Transparent declaration of interest · Horticulture · Biodynamic agriculture · Biodynamic farming practices · Livestock · Traceability · Traceability of origin · Supply chains · Srpska Magaza · Goodwill · Strategic impact · Health · Nature and tradition · Implementation · Eurosceptic · Visegrad Group · Traditional right · Independent producers [and] farmers · Peasant farmer · Banjska Monastery · Craft beer · IFOAM · Organically certified food production · What constitutes food · Critical taxonomy · Nanoelements · Inorganic origin · Mycoprotein and algae · Long-term health effects · 3D-printed 'meats' · Unconventional sources of protein · Traditional stance on health · The origin of food · Inorganic · Experimental imports · Interpretive gap · Village community · Organic status · 'Certification houses' · Certification organisations · [Land] lease agreement · Cannibalism · Soylent Green · Deontological vs utilitarian · Pesticides · GMO soya · Business culture · International retail chains · In-house brands · 'Post socialist tycoons' · Quality control [of food imports] · VAT exempt · "Kobasicijada" · "Kulenijada" · "Slaninijada" · "Kupusijada" · "Projada" · Hospitality · Village tourism · Authenticity · National heritage · Native-produced foods (USA) · Ajvar · Educate through flavour · Donkey stock · Donkey cheese · 'Ecocide' · herbicides · Agricultural pharmacy products · Trust · Subsidising [organic] · Trade bloc · Globalisation · Incentives · State actors · Cell-cultivation · Food [from] synthesised nanomaterials · Modification [of foods] · Direct [local] representation · Exploitation of lithium · Rio Tinto · Kolhoz · Heisenbergian uncertainty model · Orthodox Church · Natural veganism · Spiritual · Mindfulness · Structural forces · Strategic shifts · Postmodern · The new glocal · Privilege

Structure and content:
 From gallant peasantry to "Belgradised" society
 Thematic explorations and purposive sampling
 Regulation vs implementation
 Market farces, global factors
 Business, culture, and business culture
 GMOs, industry, ethics, ideology
 Political economy and the weight of history
 Is future cyclical: fact and metaphor

1   From Gallant Peasantry to "Belgradised" Society

This section is an introduction of sorts, necessary for the grounding of the discursive and the societal structures that impact the subject-matter at hand. This analysis, being critical is not without controversy, dictated by the somewhat scornful[1] commentary found in adjacent fields of sociocultural leaning. As such, it has to set the background for investigation that goes deeper than the specific matters reported in relation to sustainable food themes.

At the start of the previous century, Serbia was a predominantly rural society (Bataković, 2013: 316, 325; Baričević, 2018: 12; Isić, 2008; Lyon, 1997: 1; Pijanović, 2012). As can be found in historical/archival documents as well as more recent and contemporary treatises, it was the rural community that was the backbone of the then freshly independent state whose unifying and liberating ambitions had made it a focus of imperial objectives of the day, both hostile and friendly. Why would these statements have any bearing on a text about health food production in early twenty-first century Serbia? Because the past shaped the present and continues to shape the future and this is the case with the state in question and its society, perhaps even more so than with many more developed European countries which have arguably entered a post-industrial economic age long ago and a postmodern societal structure more recently. *History repeats itself first as a tragedy then a farce* is a much reviled yet well-attributed statement and as such it may apply to the society referenced here in more ways than one. Modernity does not seem to have been very kind to Serbia. Having decolonised herself from two empires straddling central, southern Europe, the Middle East and parts of North Africa, Serbia entered a union to be later known as Yugoslavia, which now, 104 years on seems to have left all its inheritors and now former constituent states dissatisfied, with arguably very contrasting explanations for such discontent. Opinions are divided as to the *why*, but all seem to concur of the *what*. That process included two world wars and one protracted regional conflict which involved her for a good ten years, a conservative estimate and during those few decades, Serbia emerged as a much changing society[2] (e.g. Lazić & Cvejić, 2007; Tomanović & Ignjatović, 2006; Šljukić, 2004; Vukov, 2013).

First, the very traditional and devoutly religious rural communities have changed quite rapidly within the first few years of a fast-modernising supra-state that subsumed it albeit at own choice, arguably (e.g. Ejdus, 2017). The common linguistic and economic sphere had allowed for cross-pollination of cultural identities divided by such linguistic connexions; civil servants and defence personnel were posted

---

[1] It is sometimes startling to see that at in the age of professed equality, authors can engage derisive contextualisation of language, trivialising of entire societies; this being a particular feature of some sociocultural analyses of the nation in question (e.g. the term 'idiocy' as attributed to Marx, Ramet, 1996).

[2] Authors express quite a variance of opinion as to the efficiency, effectiveness and extent of *transition* in Serbia post-2000, even the desirability of its pace, magnitude and direction; which in itself implies change.

across the newly formed kingdom. Industrialisation was to take hold and a new ideology, one of a modern European state that embraced Western ideals of progress and futurism[3] seemed to affect the communities across the still deeply old-fashioned rural populace. However, social contradictions, economic tensions and global totalitarian processes rooted in very different notions of 'freedom from' had made an impact on a newly formed state. Those would play significant roles in decades to come: in the centrifugal tendencies that led to two violent break-ups of the union (e.g. Pavkovic, 2000; Pavlowitch, 2008), and in the imported revolutionary drives that took hold on both extremes of the ideology spectrum, where left and right would have met in their most destructive forms (Minehan, 2006). As this is an analysis that straddles the internal rather than external aspects of what had happened to Serbian society, the brutal contradictions that spelled ill omens for the state of the southern Slavs shall be left behind. Nonetheless, relatedly and internally, the Serb peasant society no longer was rooted in its traditions, either religious, communal or productive (Šljukić, 2006).

Why *gallant*? If one observes the somewhat romanticised views of benevolent writers of the late 19th and early 20th centuries (e.g. Christitch, 1922: 424; Jenkins, 1883; Rastović, 2004), many of whom were rooted in the friendly empires[4] both to the West and to the East of the small landlocked country (incl. Askew et al., 1916; Eyer, 2020; Boeckh, 2016: 109, 127–9; 131–2; Farnam, 1918; Popović-Filipović, 2018; Hirshfield, 1986; Hutton, 1928; Jelavich, 2004; Končar, 2020; Mills, 2011; Sanborn, 2014), this very notion of the simple family-oriented society that knew hardship and love of freedom in equal measure is starkly contrasted by the current view of many a historian with a somewhat revisionist taste (e.g. Gordy, 2013; MacKenzie, 1994; Wedgwood, 2002). However, just as the *discourses of now* are justified in the now of today[5] as they will have been rooted in the present of the past, we need to give due attention to all the worldviews to understand the perception, and self-perception of a society once steeped in tradition.

Almost uniquely in the modern world, and often incomprehensible to many observers, the term *auto-chauvinism* was *coined into* the early twenty-first century Serbian discourses (Bazdulj, 2015; Gligorov, 2018; Glišin, 2018; Kišjuhas, 2020[6];

---

[3]Not to be mistaken for the art movement or its ideological presuppositions but there are numerous examples of such wholesale embracing of progress in a then seemingly optimistic and novel society.

[4]Indeed, *Gallant Serbia*, a 1916 military march written by the Czech-British composer and bass player Adolf Lotter, re-recorded by the Royal Logistic Corps Band in 2001 is a first-rate example of such sentiment of yore.

[5]Fittingly elaborated by Savić, 2001, vis-a-vis the 5 October events in 2000, changing the course of Serbia's contemporary history in ways in which he explored the accounts of those immersed in ongoing happenings.

[6]While Kišjuhas (2020) considers the term a despicable hallmark of *unrepentant Serbian nationalism*, attributing it to a sole individual source of the journalist Zoran Ćirjaković (2021), its full complexity is not lost on the popular Bosnian author Muharem Bazdulj (2015), whose columns in the Belgrade daily, Politika, are well respected by Serbia's public.

Првуловић, 2019) to depict the impact of extreme guilt felt predominantly among her liberal cosmopolitan elites, or those who construe themselves thus, a reflexive negation of any self-worth brought about and justified by the three decades of failure on the national, regional (including post-Yugoslav) and global stage. This, whether in economic, cultural, political or other subtler identity-laden terms, including religion, societal structure or perceived lack of contribution to the world and its civilisation, the predominantly young and middle-aged educated Serbs of all backgrounds but mainly in cities, and primarily Belgrade's geo-cultural heart tend to espouse such sentiments with passion and much emotion. For them, the nation's own self is such a failed project that they disown it in public discourse, web forums, arts and even policies [as argued by authors within, and beyond Serbian contexts[7]].

Such self-revulsion would have been as incomprehensible to the romanticised 'gallant peasant' of a century ago as it is to an external observer from societies unaffected by the destructive experiences that have affected the so-called *Yugosphere*. (A term used by some public intellectuals across the region roughly coinciding with the disappeared federation.) (e.g. Alexander, 2017; Farkas, 2014; Mandić, 2014; Mazzucchelli, 2012; Jagiełło-Szostak, 2013; Judah, 2009; Polajžer, 2013).

Second: As the ideologized education of socialist Yugoslav schools would have it,[8] the peasant community of much of the 1910s Serbia would engage in almost mediaeval means of food production (Angelkova, 2010: 7, 16, 18, 57, 61, 89; Bojičić, 2018; Isić, 2008; Kostić, 1970: 144–6, 148; Kržišnik-Bukić, 1989; Njegovan, 2017; Rakić, 2018: 35, 109). It is not without irony that such critique seems now misplaced in the light of what constitutes unadulterated and therefore often most sought-after food produce. The communist ideology that found fertile ground from the 1920s onward was not entirely unjustified: just as the peasant society was to collapse under the weight of new centralisation and industrial forces, not too immune to the global ripples of what was a world depression following the Great War and then the 1929 crash, so did the proto-proletariat give way to the many discontented and uprooted labourers whose hopes had been dashed and who had to make way into precarious work in the cities (e.g. Davidović, 1990; Kostić, 1970: 144–6, 148).

Moreover, the agrarian reform designed by the Soviet-trained commissaries and highly disciplined party structures had undermined land ownership in Serbian (and overall Yugoslavian) villages. A model less insensitive and sweeping was installed from 1945 and a single household would not own more than 10 hectares of agricultural land until the break-up of the socialist system and the state that supported it. By which time, the party elites had taken over the industries and much like the oligarchs further east, had taken over swathes of the economy. They were to become the new

---

[7] See Alispahić, 2020, Bulić, 2017, Првуловић, 2019.

[8] Some of which persists to date.

landowners, the landlords of a post-communist 'transition' economy[9] (Албуновић, 2017; b92.net & balkaniyum.tv, 2008; Boarov, 2005; Gulan, 2011; Gulan, 2016; Kostić, 2007; Krivokapić, 2018; Palić, 2011; PSD, 2012; Rakić-Vodinelić, 2007; Sudžum, 2021).

The industrialisation of post-war Yugoslavia took colossal steps to do away with the previous economic, political, ideological and spiritual models. Manufacturing in the cities became heavily industrialised; the new working classes and the political class of the single party system became even more mobile, civil service, defence, engineering included. By the end of the 1960s a very confident player on the world's stage was a state that was the European co-leader and co-founder of the Non-aligned Movement[10] (NAM) consisting mainly of former colonies of imperial and pseudo-imperial interest (e.g. Krstić, 2011; Jakovina, 2011, 2017; Monro, 2017). It is often neglected that, despite being based in Europe, most of the states of what was once the South Slavic union or federation were indeed colonies themselves into the second decade of the twentieth century. This would have given the credibility to such egalitarian global alliances. In 1967, the most liberal of all the real-socialist states enjoyed the annual economic growth of 7 percent [Though, in a critical paper by Zec & Radonjić from 2012 it becomes evident that the 7% growth in 1966/7 was the result of *actual decline* in previous years; arguably, much of the 'Yugoslavian industrial miracle' was externally funded through international debt and there was the controversy of the 1965 reform (e.g. Kežić, 2017; Novaković, 1993; Novakovic, 1994; Todorović, 2004; Tomić, 2020; Vuković, 2011) yet the Yugoslav *workers' self-management model*, in place from 1952 on, had evident merit which attracted attention (Chittle, 1975) and the country's recorded growth of industrial output was real (Bajt, 1986; Calic, 2011; Sapir, 1980; Lucas, 1993; O'Rourke & Williamson, 2017)].

Third: For reasons that cannot be addressed in a text of this nature, and likely caused by growing corruption among the new elites of the ruling communist party, the federal state was getting deeper and deeper into debt, just as were its constituent parts, the republics and provinces (notably, Brown[11], 1993: 149–155; 158; also, Babić & Primorac, 1986; Bukowski, 1987; Chossudovsky, 2015; Cvikl & Mrak, 1996; Dyker, 2013, 2014; Lampe, Prickett, et al., 1990; Štiblar, 2019; Woodward, 2003; Yarashevich & Karneyeva, 2013). The violent implosion of a once prosperous state that took almost 10 years to complete had many troublesome consequences,

---

[9]Restitution of capital taken over by the communists since 1944 remains an issue fraught with much difficulty, especially regarding agricultural land (b92.net & TANJUG, 2013; Majdin, 2006; Kišgeci, 2015; Kljajić, 2015; Gnjatović, 2012; Gulan, 2020a, b). To complicate matters further, there is an added international dynamic [including the EU accession], some of tracing back all the way to WWII (Албуновић, 2015; Dimitrijević, 2011; ekspres.net, 2016; Džubur, 2017; Gulan, 2011, 2016).

[10]With Cyprus as the only other European member of the NAM upon liberation in 1960 (Hatzivassiliou, 2005).

[11]Indeed, among the most astute, analytical and unideological by far is Brown's (1993) account. ['The war in Yugoslavia and the debt burden: a comment'. *Capital & Class, 17*(2), 147–160.]

one of which was the impact on Serbia's civil society. Notwithstanding the exceptionally significant impact on its international standing, the destruction of most of its industrial potential (Dinkić & Antić, 1999) and the brain drain that continues three decades on (e.g. Bobić, 2009), Serbia saw its villages depopulated, its industries wound up and privatised, its manufacturing all but extinguished and its society ageing at an alarming rate (Jakovljević, 2017; Jakovljevic, Jovanovic, Mihajlović, 2013; Milovanovic & Radevic, 2019; Penev, 1998, 2014; Stojilković & Devedžić, 2010; Zdravković, Domazet & Nikitović, 2012). Despite much promise in the early 2000s—and a 2001 text of an established Serbian social scientist springs to mind (Savić, 2001), concerning the judging the character of contemporary events—the divisions in Serbia's society grew larger and more pronounced, the economy has been in constant turmoil despite phases of consolidation and calm, and the assets appear diminished, especially at the 'periphery', which appears to have encompassed all but the capital city of Belgrade.

The metaphorical *gravitational pull* of its most populous city, its administrative and economic centre, the seat of central government, the site of entertainment and travel sectors, with the sole fully functional civilian[12] airport in the land, has led to an internal migration of much of Serbia towards Belgrade. Not only did most of the other cities and towns lose much of their prestige, economy and power, owing to the protracted post-transitional period entering its third decade. The villages were already bereft of much self-reliance following *Titoist* agrarian reforms[13] (Baričević, 2018; Hadžić, 2008; Holjevac Tuković (2003); Juriša, 1983; Kršev, 2012; Miljković, 2004); then further affected by industrialisation and the destruction of those industries by mismanagement, external force and corruption alike; losing their attractiveness in the light of a cultural shift towards the cities[14] and appeal of urban

---

[12] The major city of Niš is one notable exception, with its successful Constantine The Great airport which became a regional hub for Ryanair and other low-cost airlines in Serbia. However, this was not without controversy as at the time of Belgrade's airport being sold to the French firm Vincy, there was a reported takeover of Niš's municipally controlled airport by Serbia's central government which allegedly affected such deals which best suited the southern city and its broad catchment area (Brdar & Spasić, 2010; Nikolić, 2019; Miladinović, 2019). Other commercial civilian airports may be in the making by conversion from the existing sports- and decommissioned military airfields—in the long run, but that does remain to be seen.

[13] This of course was not only the case in Serbia. Baričević (2018), Hadžić (2008) and Holjevac Tuković (2003) have indeed identified similar mechanisms in Bosnia and Croatia during Titoist rule, and especially in the context of agrarian reform harshly imposed on Yugoslav farmers by the communist party's apparatchiks.

[14] The post-Yugoslav and Yugoslav treatises, including but not limited to (Baričević, 2018 Hadžić, 2008; Holjevac Tuković 2003; Juriša, 1983; Kršev, 2012; Miljković, 2004) tend to overturn the notion proposed by Ramet (1996) which appears to portray, in a positive light the Titoist drive towards urban-proletarisation of the small landowners. Quite the contrary, the miscarried collecitivistion of farms in the late 1940s and early 1950s had all the methodological hallmarks of Stalinism, ironically at a time of the famous split with the USSR. It had created an urban underclass out of the impoverished peasantry, some of whom were to later become further uprooted through external migration as *Gastarbeiters* and were to fall prey to populist discourses across the internal and external borders and divides of (post) Yugoslavia's nationalistic identities that persist to

life (Dyker, 2013; Kržišnik-Bukić, 1989; Zec & Ranonjić, 2012); mis-grounded through decades of cultural erosion of traditional identity and belief systems[15]; also, sites of much recruitment for emigration towards the West since the 1960s (e.g. Avramović, 2012; Bauer, 1970; Jurić, 2021; Marković, 2005; Todorović, 2004). A cultural change reported by respondents in line with widespread public notions whereby young villagers spend their formative years aiming to move away to the cities, or abroad (Bogdanov, Đorđević-Milošević & Clark, 2007). Inopportunely, the plurality of the term 'cities' is largely misleading.[16] All roads lead to the capital.

Belgradisation of Serbia (e.g. Anđelković, 2018; Mijailović, 2021; Nova Video, 2021) is a term that entered Serbia's public discourse in the late 2010s. While it is hard to exactly identify its first use, it appears in the media and everyday speech around the same time as the controversial *auto-chauvinism*. Since this is not a dictionary exercise, tracing it to exact first mention and source will be left for other treatises, but the fact remains that it is a term now established in critical terminology, sometimes used more loosely but also present in mainstream media and intellectual speech.[17] Bursting at its seams, the city's infrastructure struggles at the weight of overpopulation, most of which is underreported. Attempts to revive and rebuild the infrastructure have not yielded desired effects and many urban planners, 'public intellectuals' and others' swaying opinions remain deeply sceptical

---

this day. Therefore, the forcible uprooting of the peasantry across the federation contributed to intolerant populism.

[15] What looks like an almost romanticised perception of Tito's regime is present in public discourses within and beyond the physical and cultural borders of former Yugoslavia, and this even permeates intellectual treatises. Possibly as contrasted with the brutal reality of its break-up and the unpromising aftermath for most successor states, such idealisation deserves the scrutiny of skilled reflection. Tadić (1993) soberly asserts that Tito's reign was facilitated by well-planned Cesarist coups d'état within the tailor-made system designed for consensual diktat. Though Tadić himself is scrutinised by critics in the Yugosphere and West alike, a governance model designed for and/or by Tito had proven an inept tool for the survival of the federation, or of the continuation of its positive contributions such as its role in the NAM or the Yugoslav model of *worker's self-management*.

[16] References to *Belgradisation* are very much about that notion—as the cities and towns are attempting a pushback against this trend, leading to regional tensions of 'primacy'. For example, Novi Sad, placed firmly on the world's cultural map thanks to the famed Exit festival, rather controversially held at what would elsewhere normally be a protected national monument of the baroque fortress of Petrovaradin. Novi Sad also hosted the European Capital of Culture in 2022. Niš is growing in prominence, or regaining it rather: Once recognised as an advanced industrial hub with major electronics goods production before the collapse of Yugoslavia. Other municipalities, such as Šabac, Jagodina, Užice, Čačak etc., are also regaining a sense of assertiveness.

[17] The term is both contested and heavily politically charged. Belgrade's 'natives' read it as a negative impact of internal inbound migration; municipal leaders in other towns and cities bemoan the investment drive towards the capital, leaving the countryside and former industrial hubs such as Niš behind in financial and infrastructural terms, a notable example being the debates around Belgrade's metro (Knežević, 2021). Opportunistic populists 'of the regions' decry 'stop to Belgradisation!' (e.g. the Nova Video report from 2021 featuring the opportunistic statement made by a politician of the 'nouveau right'), while sociologists and other opinion leaders try to make sense of it in an analytical and proactive manner (Anđelković, 2018).

if not openly opposed to the construction projects in the capital. The official figures stand at well below two million,[18] but many a taxi driver will loosely speak of at least three million inhabitants. It is impossible to ascertain exact figures as many internal migrants, including the still displaced refugees of previous conflicts—as well as a growing external migrant community—remain underreported and largely unobserved. The city boasts large malls, supermarket chains mainly on the periphery, an economy that overshadows much of the rest of the country (Beograd.rs, 2022; Blic, 2017; Caranović, 2021; Mondo/BETA, 2015; MPC Properties, 2019; Vujović, 2014). In addition to entertainment and media, other non-productive economic activities thrive in the city, with some notable exceptions such as the IT, construction and pharmaceutical sectors that seem to be adding real tangible value (IT Serbia, 2022; PC Press, 2020; Републички завод за статистику/РЗС, 2022a, b, c, d).

As will be seen below, it is precise that overburdening of the country's capital that gives rise to renewable potentials, as reported by the respondents consulted for this work. The disillusioned and somewhat affluent young professional families seem to drive this trend.

2 Thematic Explorations and Purposive Sampling

The analysis displayed in the fieldwork aspect of this paper is set in qualitative terms and based on thematic work and emerging contexts (Alvesson & Skoldberg, 2017; Howlett, 2021; Kozinets, 2006, 2015; Leipold et al., 2019; Suddaby, 2006; Tayaben, 2018). There are concurrent themes and contextual tensions that emerge from the data which is both rich and arguably skewed. Because of the latter, it was essential to employ structuralist political economy tools (e.g. Justman & Teubal, 1991) to recount any preconception and reflexive rooting to counter any of their impacts. It is to be stated very clearly that the respondents are practitioners, highly proficient in their fields, educated and experienced in the practice of health food production, its economy, marketing and communications, adept in their appreciation and criticism of cultural conditions and social forces that work for and against the subject-matter and way of life they devoted themselves to. Both highly articulate and fully conversant with the practical aspects of policy and to an extent, politics of the matter; the latter being an aspect of the calling of a very prominent respondent who is an activist, former member of parliament [independent] and whose representative career spans decades in municipal terms (Istinomer, 2020). As such, his self-declaration of status, worldview and position is of help here as it counters any partisan standpoint by the very openness in his transparent declaration of interest (Alvesson & Skoldberg, 2017).

---

[18]Beograd.rs (2022), the official government site cites the number of Belgraders at 1.659.440 according to the 2011 census. However, in the 2022 elections there are 1.597.000 adults registered to vote in the capital (BETA, 2022), which, despite the ageing society indicates that the internal migration would have increased the city's population quite considerable on the 2011 census.

In approaching the field aspect of the matter, the author sought to identify a potential network of respondents. Numerous potential informants were approached but the burdensome nature of the subject meant that very few were able to partake in the interviews for this fieldwork. A total of over 6 hours of conversations were recorded, together with an exploration of networking aspects of the case. Two respondents provided most of the data and extensive thematic contextualisation was employed to assert its dependability. The two chief respondents were Dragan Sajić, and Prof. Miladin Ševarlić. Mr. Sajić is an active entrepreneur in the field of horticulture and the co-founder of the biodynamic-oriented, health food project cooperative *Srpska Magaza*, loosely translated as *Serbian Store*. Prof. Ševarlić is a retired university professor, an agroeconomics specialist, who holds numerous accolades in the trade and industry nationally and internationally and, by own admission a rather uncompromising activist of a complex political persuasion. Other respondents who were either unable or unwilling to go on record, likely due to the thorny subject-matter are among the key members of *The Serbian Householders' Association* (Друштво српских домаћина, 2022), entrepreneurs who run successful farming businesses in the southwest and east of the country, one cattle holder who worked on international projects and food specialists. Some informal data is used cautiously for context but the excess of 30,000 words in transcripts from the two chief respondents provided a synthesis of thematic contexts explored here.

While the two respondents' practical assessments and experiences largely cohere and concur in matters of significance for practice and day-to-day processes defining the field of sustainable health food production in Serbia, the ideological and discursive nuances are pronounced well enough and clearly so thus enabling the analysis of a considerable breadth of positions which they respectively occupy. This is an important matter, as the discourse of what constitutes health foods, what varied definitions of sustainability may apply in practice and what are the intellectual, social and management tenets of these intricate subjects—does affect practice despite its very tangible nature. Management rationality of substantive character effects a strategic impact (Bakir & Todorovic, 2010) and though the practical nature of the matter leaves not much room for non-instrumental thought, the confines of combined demarcations of *health, nature and tradition* tend to group likely answers within a band of discourse narrower than the full breadth of sustainability spectrums. This will be seen to be justified within the context of Serbian experiences in health food production partly owing to the concrete sociocultural, regulatory economic, discursive and historical circumstances defining the field in this particular instance.

3   Regulation Versus Implementation

Prof. Ševarlić is very outspoken on the matter, as an activist, academic and practising farmer himself. As a politician whose career spans decades from municipal—as the youngest elected member of Belgrade assembly—to national, as a recent independent MP in coalition with a party of traditionalist right (Istinomer, 2020), his statements appear pragmatically Eurosceptic though in line with a recognition of the values of effective regulation embodied by the trade bloc. In one very pronounced account, actually quite well known in recent press reports, he is most critical of the double standards in food product quality exported by Europe's *Big 15* (as states, in his emphasis) industrial giants, whereby the Visegrad Group member states openly complained (Gotev, 2017; Reuters, 2017a, b, c; Slobodna Bosna, 2017) about the quality of chocolate and other produce. While, on the other hand being approving of the internal quality control, regulations, and production within the bloc. While his Eurosceptic stance is clearly in tune with the traditionalist discourse, Prof. Ševarlić is not uncritical of the concessions given to the Church in Serbia upon restitution of the land confiscated upon the communist coup of 1946 (Majdin, 2006), '*as they do not pay VAT and are therefore acting as disloyal competition to the small-scale organic agricultural producers'*. Such a critical stance of the Church's position within the agricultural economy clearly stands independently within the discourse of traditional rights and shows a pragmatic approach adopted that seems to favour independent producers, farmers and such. Indeed, as stated in an informal exchange with a marginal respondent for this research, it is well known that monasteries of the Serbian Orthodox Church have returned to the time-honoured organic food and beverage tradition well known in the West, sometimes in a rather new guise. This, not least as the monastic community frequently recruits the urban young whose worldly experiences are utilised.[19] (The Banjska Monastery for example has its own craft beer produced since 2017 (Đurić, 2007; Manastiri u Srbiji, 2022; Novaković, 1892; untappd.com, 2022; Sinkević, 2022; Задужбине Немањића, 2022), a beverage better known to German Franciscans than Serbian monks.) Prof. Ševarlić certainly recognises the pragmatic value of such entrepreneurial approaches in the West, citing the fact that monastic settlements of western churches regularly join IFOAM membership for organically certified food production (ifoam.bio, 2022a, b). (More will be said below on the respondents' positions on the discursive relevance of spirituality in biodynamic, health food and sustainability contexts.)

Regardless of the specific context of these little snippets from Prof. Ševarlić's commentary (*Ševarlić* hereinafter), in a broader thematic sense they reveal what is a running theme in this aspect of the respondents' expressions: a discord, mismatch between the proclaimed aims of regulations and policies, and their respective implementation or lack thereof. For want of a better word, both are guarded against what the analysis identifies as *manoeuvres* in various implementation contexts. While the more project-oriented respondent whose hands-on approach allows him to steer clear of challenging themes does not state so explicitly, the activist informant does not shy away from this. Indeed, Ševarlić remains outspoken on matters of controversy and tension throughout the long conversation: he is particularly critical of what he states is the stealthy import of grey-area regulation allowing for a looser interpretation of *what constitutes food*. In a lengthy elaboration of food classification in Serbian regulation, effectively a critical taxonomy on the subject, Ševarlić raises

---

[19] Including actors, pop musicians, artists and other public figures, a new feature of national culture.

the alert to matters of foods that consist of components *with nanoelements ... or of inorganic origin ... or originating from mycoprotein and algae;* in which he dissects the notion of food, health impact of those and lack of experimental understanding of what long term effects these may have on health of the population.

Note: It is hereby important to note that we may reach a fork in the road of what constitutes sustainability and health in food production and supply. The two are not automatically synonymous or inherently aligned. Hence some of the more inspired reports in media if to a less affected extent, in readily accessible peer-reviewed publications about the sustainable futures of food production such as 3D printed 'meats', those derived from insects and other unconventional sources of protein (Aditya & Kim, 2022; Beall, n.d.; Caporgno & Mathys, 2018; Jandyal, Malav & Chatli, 2021; Handral et al., 2022; Ko et al. 2021; Percival, 2019; Rosmino, 2021; Teng, Zhang & Mujumdar, 2021; Varvara, Szabo & Vodnar, 2021; Wilson et al., 2021). This may well be where discourse meets or rather clashes with, intersects with regulatory policy and implementation: while it is impossible to generalise due to the sampling approach, there is some credence in a likely claim that a more urbane, cosmopolitan discourse existing in a *'Belgradised elite'* would be more attuned to such culinary experimentation under the heralds of greater good. Whether or not there would be a disconnect between discourse and practice is pure speculation. However, what does remain clear is that both respondents, in their set notions of what they consider good practice in food, possess a firmly traditional stance on health. Not being Luddite and dismissing of technological innovation—indeed, both flexibly display an interest in cleaner and greener technologies—the position as to the origin of food, biologically and chemically is where they draw the line. Whether or not this is anything to do with a discursive undercurrent of a more traditional slant, to what extent this may relate to how Serbian society is construed at home and beyond is truly outside the scope of this text. However, the modern experiment that constituted Serbia's twentieth-century experience did not seem to enamour the majority populace to the post/modernist thought experiment and this can be reflected in many aspects of purchaser and citizen behaviour outside of the narrow confines of the cosmopolitan circles.[20] The former, partly due to a diminished purchase power, tracing back decades; the latter, where even the organisations with progressive or democratic suffixes do need to look after the traditional corroboration of their programmes. It will be seen from both accounts that a new appreciation of countryside emerges from cities, itself more postmodern and post-industrial than any pronounced rethink of what it is to be environmentally aware.

While Ševarlić openly critiques the legislation that concerns him as a backdoor to GMO and inorganic, experimental imports of what he sees as a threat to health of the populace, Dragan Sajić looks at matters more closely to implementation in everyday

---

[20]The synopsis of a TV show produced for the national RTS network (Svetličić, 2022) neatly contextualises the „Krug dvojke', the signifier of Belgrade's trend-setting cosmopolitan centre (*The Circle Line of Tram No. 2*, epitomising the heart of the capital city, much maligned in inter-Serbian debates akin to 'Belgradisation').

practice. What is a particularly important contribution he provides to this research is centred on the following disconnect: The one between the entrepreneurial opportunities *that may in theory empower the peasant farmer—the term 'peasant' does not have pejorative connotations, quite the contrary and the consumer, the regulatory framework designed to assist and not hinder such enterprise and the legal acts and statutes that leave an interpretive gap.* Such interpretive gaps, for instance, what constitutes legal frameworks for food production and sales and how such food is best sold by a minor production entity, he views as quite problematic in terms of implementation. In short, the small landowner, farmer and peasant from a village community are disadvantaged against the bigger producers who seem too able to feign organic status while Sajić clarifies that large-scale production of such nature remains unlikely organic or biodynamic by definition. Furthermore, he addresses the matter of consistency and the constant changing of who are the trusted 'certification houses' i.e. certification organisations working with the ministry of agriculture and what impact that has on how the certification is implemented, what best practices are adopted, who are the people to turn to when working with small organic or biodynamic farmers.

For the sake of definition, Sajić looks at biodynamic farming practices from a practical perspective as it is not principally dissimilar from organic ones: the one where the full cycle is covered by a single producer and farm and where the manure from animals is used to fertilise the land. Where the vegetable and fruit, crop and herb aspects are intrinsically connected to the animal aspect of farming. In that, he remains clearly within the realms of traditional farming practices whereby the national decimation of livestock Sajić decries forms part of the problem of organic or biodynamic[21] farming of small independent producers (Ćosin, 2021; Ersado, 2006; Pejanović, 2011). [Indeed, in expertly conducted in-depth analysis of the matter, Pejanović finds that, by 2009, Serbia's livestock was at 50% of its 1980s strength, a problem much exacerbated by the slashing of agricultural budgets from 5.6% in 2004 to 2.92% in 2009 and a mere 2.5% in 2010 of the national budget during the controversial transition and zealous neoliberal economic remodelling of Serbia's economy[22]]. Similarly, and far more vocally than Sajić, Ševarlić states that the meat available in Serbia's shops has no clear certification of traceability of origin and could therefore be suspect on a number of levels. Sajić's main concern with *traceability*—(e.g. Aung & Chang, 2014; Ćoćkalo et al., 2019; Glintic, 2012; Kneafsey et al., 2013; Kovačević & Lazić, 2012; Posavec, Tomiša & Šarkanj, 2022; Sič, 2013; Тодоровић, 2021; Turudija Živanović, 2015)—again involves small-scale farming he works with regularly: How does this leave an impact on what may or may not be sold in Serbia's markets and green markets when produced,

---

[21] Moreover, as Sajić goes on to explain, there is no true biodynamic farming without survival of livestock as the manure and waste are reused in the organic cycles of food production, occurring through natural means.

[22] Such fervent neoliberalism often critcised as both irrational and brutal across Serbia's public spectrum.

distributed and sold independently within the same chain or network of cooperatives[23] such as the project he once cofounded.

> We 'set up camp' at the ministry offices while trying to explain to them that this is on the regulatory level, not even the level of government whereby the government would bring forth the law but on the level of ministerial statute. Meaning, that one decree, statutory decision by the minister, you make a decision that the traceability of the produce such as potato if that is considered, having the location, manure and the rest, you will say that the peasant had produced, say 1000kg. And those 1000kg can be offered in a supermarket—any supermarket—as it has the traceability, and it is known that it is produced by him, and where. Well, you cannot do that now, as it is not legally an option, saying that 'no, if he wishes he may sell it on his doorstep but if he wants to supply a supermarket, he needs to set up an independent retail store and/or a ltd co., they cannot offer it as a farm but another store may buy from him.

With the problematic notion of regulatory implementation consistency, Ševarlić seems to have a similar standpoint. He cites that only a handful of producers in Serbia actually have a full Organic certificate, especially highlighting Biofest from Subotica (visitsubotica.rs, 2021) under the auspices of Subotica's Terras (Subotica. com, 2022; Terras, 2022), the oldest organic certified business in Serbia situated in the north of the country. Those five organisations contrasted with many small farm producers who cannot attain it due to the reasons outlined by Sajić (below). However, Ševarlić remains pragmatical once more stating that 'the biodynamic food need not be certified—but it is a symbiosis of plant and animal production'. Drawing the roots of distinction between organic and biodynamic and tracing biodynamic agriculture to a Swiss model developed by the influential *Anthroposopher* Rudolf Steiner, who postulated on biodynamics (Kovačević & Lazić, 2012; Paull, 2020; Turinek et al., 2009; Zubović & Domazet, 2008).

Once more, discourse meets practice and suitable interpretations of regulatory frameworks allowing the small independent producers of healthy foods to survive the squeeze between regulation and competition from big players who can afford the economies of scale, importers who work with established global organic brands and those who successfully cut corners and manipulate as indignantly observed by Sajić whose altruistic Srpska Magaza project had failed to deliver its promise chiefly because of unfavourable regulatory environments and mounting costs: an excerpt below—

---

[23] In this treatise, the word 'collective' is intentionally not being used for what is a village/ agricultural cooperative—so as to remove any ambiguity. The term 'collective' is thus a clear reference to the organisational and economic model imposed on the villagers and farmers in socialist Yugoslavia and the USSR, one where obligation and coercion rather than volition was involved: While Kostić (1970), Kolin (2010) and Ševarlić & Nikolić (2010) use the word (s) 'zadruga' which *may* be translated as 'collective' in terms that not subject to ethical judgment, Lazić & Cvejić (2007), Nedeljković et al. (2018) and especially Stojanović (2020) clearly connect the 'collective' connotations of 'zadruga' to the mentality and/or 'collectivisation' to corruptive and coercive, even totalitarian aspects of society and community at stake; a slant settled at another time.

We had a problem, we can buy—we could buy from the peasant farmer, we were a limited company, but we had no supply link to show to allow us to have a link with that farmer, so that we may sell it. But we could, ourselves, enter into a rental, lease agreement for the land, the fields, together with them [farmers] then they do the work, but the ownership of the business and the lease is registered as ours. But that is a wholly different concept. . . .so the marketing can be very effective and with an opening potential that is great but a great deal of manipulation . . .at some point the trust is eroded among people, because whoever finds themselves well-placed to offer that? Those who knows of the need, who has no real connexion with the villages and who successfully manipulated through some marketing means—to manipulate what the decent farmers did work on, and then offer that to a marketplace.

And to show that the definition of food is not infinitely stretchable, that there is a line that is drawn even in the most hard-nosed of concepts: with an expression of horror, Ševarlić cites the case of the Swedish scientist Magnus Soderlund who presented a case for human cannibalism as a remedy absolving us from climate change (Bendix, 2019; Gaynor, 2019; Houser & Tangerman, 2019; MacGuill, 2019; Ratner, 2019; Timpf, 2019). The disturbing concept reminiscent of Science Noir *Soylent Green* (Knipfel, 2018; IMDB, n.d.) is where Ševarlić raises the alert, being very deontological despite otherwise evident practical thinking when it comes to effective resource management. This includes the notions of GMOs, and uncontrolled use of pesticides—on the other end of the debatable experimentation spectrum—in equal measure, citing studies of the health impact of airborne pesticide distribution also in other countries, as well as his own sombre statistics on the likely illegal growing of GMO soya in the nation's farmland. The drive of ethics will be duly explored below as it is central to the subject-matter; however, the notion of marketing, its exploits and matters coming from beyond the borders are examined first.

4   Market Farces, Global Factors

The above wordplay is intentional. While market forces and the power of marketing may remain a force for good, both respondents identified matters where this is not the case, so Sajić refers to big businesses feigning organic credentials and Ševarlić exposes what he believes are anticompetitive practices and abuse of market power undermining the national standing and identification of source products. Equally well, both Ševarlić and Sajić cite positive examples.

The matter of global factors of course becomes more complex as it does involve market positioning, coercion of big players but also the as yet undetected opportunities to place small producers within international markets and also to learn from good practices in well-established economies.

Firstly, in relation to the regulatory disconnect but also the matter of business culture to be addressed below, Sajić relates to the inherent matters at hand:

And . . . again we have a new certification organisation, they switched again, whatever, of no consequence who, and now we have some producers who are *gigantic* in terms of 'health food' production—and that is totally impossible . . . it's not doable on that scale . . .*and then people [customers] who care they can't understand, 'come on, that's healthy food, it's written there, see'.*

As for the limitations imposed by international retail chains that have a decisive foothold in the Serbian market, having acquired the national chains that arose through privatisation of state assets since 2000 (Arsić, 2015; Begović, Živković & Mijatović, 2000; BIZLife, 2015; Dondur, Radojević & Veljković, 2007; Drašković, 2010; Đorđević, 2009; Kostić, 2010; Mićić & Zeremski, 2011; Nikolić, 2014; Plavšin, 2013; Stošić, 2014), Ševarlić looks at what he interprets as a trickle-down effect of market dominance on part of big international players from the EU. Stating that this is based on such prerequisites as *'legislation adapting to foreign influence and to the 'post socialist tycoons', and that random checks lead to 97% of imports from EU into Serbia lacking quality control'* ... Relating this further to the marketing of imported agricultural produce, the privatisation of high-street food retail and seeing this as a problem with national branding of own produce built into those foreign retailers' supply chains: EU shopping chains insisted on the use of the Latin alphabet (as opposed to Serbian own, Cyrillic). Chains' own brands without any reference to Serbian producers but for the small print:

> you'd need a microscope to see the name of the manufacturer, and that they are from Serbia; that it is produced and packaged by a manufacturer from Serbia. But the trademark is ... [the chain's own] whereby the retail chains introduce their own inhouse brands... and that is then 10–15% cheaper a product. I wonder, how is it then possible, are they exempt from VAT ... why does [a specified retail brand] ask of you to bring over a document that you are an [enterprise] ... which is verified and signed off by the secretary [of your employer] and then you get a store card that lets you in to the store, and then you pay with a bill that has an invoice relating to [your employer] and then they [the chain] get VAT repayments so [the employer] goes into debt and the tax office does not intervene; that way billions of uncompetitively generated profits are exempt from tax.

The 'value' offers, he argues, the food produce at reduced prices generate incentives for the promotion of those chains' in-house brands. This, he claims undermines the local producer and local production on fundamental levels. Then he muses, do these chains pay VAT, speculating on the costs of tax being retrieved and *exported into the chains' native land*.

This may seem a contradiction to such fierce observations, but the Eurosceptic statement is not out of context; while showing the appreciation for the trade bloc's own self-protection, this respondent remains lucidly critical of the perceived double standards of self-interest. On the other hand, the same author of numerous campaigns, initiatives and papers explores good marketing practice at home and abroad: Ševarlić refers to the *Belgrade ethno-fair of foods and drinks* (RTV, 2021; etnohip.rs, 2021).

> 'Not just a showcase but also for sales purposes' .... 'Where I am the organisational board member' ... espousing 'open market orientation ... for those oriented toward sales not just self-sustained or for friends and family' ...

As will be seen, in equivalence with a separate observation by Sajić, such independent producers reach out directly to customers:

"selling online by special postal delivery for example" . . . "then there are those village fairs and local events with an '-iade' suffix" [themed around a type of produce]— "kobasicijada",[24] "kulenijada[25]" etc etc" . . . "which are exhibitions-showcases, competitions and sales events all at once" . . . ""slaninijada[26]", for instance" . . . "imagine when a thousand exhibitors come, not just from Serbia but from surrounding nations". This involves feats of engineering "earthen pots for cabbage" . . . "we managed a pot containing 600 portions" . . . "technology solutions, how to cook it" . . . solving a "design problem". Then the "kupusijada[27]" and "projada[28]" were both done on the same day".[29]

As the respondent clearly implies, the significance of such food-themed events is one of combining hospitality, culinary expertise, panel competitions, showcases and indeed, direct-to-customer marketing, product placement and even elements of diplomacy. Of value is also that such festivals predated the current market-economy orientation of the country and are therefore authentically connected to the national heritage though of more recent date.

Sajić also speaks of such authenticity and direct relations between farmers and customers; his take is also organisational but transnational beyond that. As someone who keenly cares about the food experience as one of culture and life-enhancing value, it transpires that the Srpska Magaza was a project particularly aimed inwards, at the inhabitants of Serbia herself.

We at some point. . . we worked the concept. To this day I see that, the limitations and the advantages, we did packages under the Magaza brand, such as lard and the bacon and ajvar [herbal/vegan relish], the package we called "the village package"—we would sell such assembled produce to, say, banks, were buying those for their staff, that would be a package which would enlighten them, when they see it—firstly, those flavours they have not encountered, we must first educate them through the flavours, what they actually mean. . . that is never a problem as you can do all sorts of things with it, all points to a great marketing potential.

---

[24] Sausage fair or festival (Vukić, Kuzmanović, & Gligorijević, 2014; КОБАСИЦИЈАДА ТУРИЈА, 2022). All these gastronomic events form an important part of village tourism in Serbia, as covered in academic treatises.

[25] Kulen, a renowned type of traditional spicy sausage ring produced in Serbia as well as Croatia, which some may compare to Spanish chorizo. Kulen, however, is autochthonous produce with its own Appellation of Controlled Origin in both countries (Завод за интелектуалну својину, 2022; poslovni.hr. & HINA, 2013).

[26] Bacon fest (Pivac, 2011; SLANINIJADA, 2022). See also 'Pršutijada' – The Cured Ham Fest (Пејовић, 2019).

[27] Cabbage and sauerkraut fest (izletijada.rs, 2022; Milićević & Petrović, 2017; Štetić & Pavlović, 2014).

[28] An annual festival event devoted to traditional Serbian corn bread (Ć. G./Novosti, 2018; Vujičić & Ristić, 2016).

[29] The Norwegian ambassador was a guest of honour on the day and was served a 'proja' corn bread shaped as the map of Serbia. Apparently, this involved a little diplomatic pun staged by the hosts.

Sajić elaborates on a case of international sales:

a friend of mine who does a mushroom, a woodland one—fully organic and natural, he did not even have a partner here, someone who knows how to appreciate that over here, but he went straight to Germany and now he does dried mushrooms, it is placed in Germany, Belgium and Holland, I have seen it as a friend had shown me ... he used to store them in brine, large 200l barrels, now it's tiny little packages, not 250g even but 30g each ... so he just clearly skipped [it all here], he has only one, just that mushroom is what he produces, and the Germans are buying it all out and there is no problem to export it.

This is to show how this concept not only crosses the borders of state and taste but how the frustration expressed elsewhere by this respondent is not one of commercial nature but that the project mission of *Srpska Magaza* was to educate and enhance the quality of life through food within the country's borders, as it transpires that exporting such goods may face fewer boundaries than internal retail does as reported in the next section.

To conclude this segment, two short examples are quite illustrative of good practice best adopted, the value of acculturation to food experiences learnt from distant communities. Prof Ševarlić reminisces on his experience of direct sales of Native organic foods in the USA. He warmly recalls visiting Penn State University where colleagues took him to see the stalls set up by Native farmers near local schools for sales to children. Edifying children to appreciate organic Native-produced foods, and how the longitudinal effect of such educational and experiential value when the youngsters once progressing through the educational system had grown fond of the produce they first learnt about as children and became lifelong customers of the same farms and their organic produce.

Sajić refers to the entrepreneurs in the region of:

Stara Planina [in eastern Serbia], there are several people there who we worked with, I went there many times, and a man who is a DSc in veterinary science, I can give you his contact, as he still produces the donkey stock, trying to sell the milk with medicinal properties and so on.

Donkey cheese appears to be the world's most expensive product in its range, at £1000/kg, as Ševarlić referred to another well-known farmer Slobodan Simić of Zasavica in Mačva in the north-west of the country (Business Insider, 2021; Gubić et al., 2014; Kalenjuk et al., 2016; Stojanović et al., 2021; Stojanović & Milošević, 2017; Šarić et al., 2016).

Apparently, the world's unique product developed by this entrepreneurial farmer, where one of the greatest challenges is coagulant fermentation from a very rare milk and a protected species of donkey bred only in a handful of places in the Balkans (Agro TV Srbija, 2019; Stančić & Stančić, 2013; Stojanović et al., 2021; Stojanović & Milošević, 2017). The exclusivity of this product is such that it is impossible to export to the EU as the milk must remain unpasteurised to preserve its medicinal properties. Moreover, the sole technology that ever worked with donkey milk is Simić's trade secret.

Of similar levels of exclusivity is Serbian mead brandy distilled and fermented rather than macerated with grain or grape alcohol. In the words of a local beekeeper/producer, *the trick is to make honey ferment, and honey is inert by very definition.*

5   Business, Culture and Business Culture

Ševarlić paints a bleak picture of honey production with a simple and short remark, referring to the unregulated use of airborne pesticides in the country as 'ecocide'. An activist who is most staunchly opposed to GMOs and invasive mining and a firm promoter of green and firmly traditional approaches to what agriculture is at its essence, he speaks of imports and global regulatory influence as *budgetary euthanasia of Serbian agriculture,* then smilingly remarking 'I have already been told off for using medical terminology in my presentations'.

He also complains about the ineffective implementation of pharmacological regulation in Serbia for the destruction and recycling of herbicides and other agricultural pharmacy products which in turn poisons the waterways and blames an underdeveloped business culture for it, an unawareness of the impact of such poor practice.

Sajić is similarly unflattering yet a little more diplomatic (as will be seen), about another example of unethical business culture among some farmers. He is embarrassed to quote an example of the trickery performed on the unassuming city folk, as:

> ...see, I recently heard—it may sound banal but just to explain, I heard recently how a young married couple from Belgrade went to Avala [mountain near Belgrade] to go and buy from them organic eggs, they regularly bought them from these farmers. And at some point, they, people just told me so it's still funny, once they passed by from the way of [mount] Kosmaj they came over, not having announced themselves but just stopped by those people. And now, the parents weren't in so this kid who knows them as he'd seen them often as they came ...they asked, 'where are mum and dad?' and he says 'there they are behind the house at the farm, *crapping up the eggs*'—what an ugly phrase, *to crap up the eggs*, what that means is the eggs they bought at the farm they soiled so that they're sold to the Belgrade folks as organic." "So when I hear something like that, I don't know, what really, so that is what it is..." "so the awareness of those down there who do such things who soil those eggs, they don't think or ask themselves what and why, well those people need it as they need it, they need [the health foods] because they do. No certification organisation can iron that out.

Previously explaining his mission statement and the purpose of the project of Srpska Magaza that cost him dearly but which he does not regret, this respondent elaborates at length. The below quote is much insightful concerning the ethics, the raising of awareness, and the attempts to formulate a new business culture from the old foundations in traditional heritage progressively integrated with most notable contemporary international practices:

Our intention with the Magaza was to come and collect the produce from those small farms and offer it to the big market and make that margin work as we know how to position it/place it and how to package it and how to present it *and to explain it as a story* and to offer it to people in urban environments and who think differently and whom at the same time are aware that *they do need* health foods—to offer it to them as they cannot come to the countryside and buy it off someone's doorstep—nor do they know whom to buy it from *nor, . . . nor do they speak the same language*—when that village man wants something and he is surprised that the city folk would come to him and why, he doesn't get it—he does not understand what is this guy doing there at his place if he's got a Maxi (supermarket) to buy whatever he may need, like chicken for RSD190.[30] That is totally. . . and we wanted to abridge that, but the legal regulative environment was not at that moment, nor is it now, . . .to allow for that to happen.

As a near-textbook example of organisational culture, policy frameworks and marketing or consumer behaviour cycles of mutual influence, this statement is also a testament to how hard it is to enact change, to reshape strategy even with support of major players, let alone in the context of such entrepreneurial projects motivated by altruism and self-financed at a loss.

in the end it is it is down to the consciousness that I see, those people who have awareness built of it, they do arrive at, reach the people who do know what that actually is, to find that synergy in the means by which—I now have some fine examples of people who actually do produce such health foods here, some of the produce from the range, they have and they offer that service, and they offer it at a higher prices as it is otherwise impossible for them to survive, simply—if something is RSD190 in a shopping mall or a [mainstream] state-owned[31] butcher's—they really have to raise the price as realistically it costs them more. And these people [buyers] who are aware know and care, they created this they made, the first thing they did was create a positive awareness on the level of socialising, of friendship, they *respect each other*—and on the basis of that respect and understanding, they produce for them, the buyers understand what the producers are doing to achieve that, they befriend and socialise, *and that part of friendship and respect and of trust, creates a cycle that suits the ones and the others alike.* And that is the model that I see as the only one possible, and I have come across that many times and it can be seen in the manifesto of their business mission statement '*if we don't have that trust and respect, no economy of that kind would function'*, matters not if it's a firm or anything else. If that is missing. If there is no synergy of that kind, letting them interconnect, connect and interact.

Sajić is indeed not only willing to learn from the examples of the noteworthy successes in related fields of organic and biodynamic farming. In a manner similar to his more outspoken political activist, academic-and-practitioner counterpart, Sajić sampled the best in good practice from abroad as well as within the country's traditional and contemporary ways. He speaks of informal biodynamic solutions based on relations of trust and mutual respect between an interconnected customer/producer circuit and a set of unwritten rules that work elsewhere:

---

[30]Approximately £1.50.

[31]Very little remains state-owned of the retail sector in Serbia; this is more of a figure of speech.

...we followed how that was done, we had the opportunity to do so, for example in Belgium and the Netherlands, Austria and Germany when we were developing the project of Srpska Magaza/ Serbian Store, we had the opportunity to travel and see their ...so in Holland at some point things went so far that they stopped acknowledging those certification houses [*organizations*] BUT GOOD AGRICULTURAL PRACTICE INSTEAD. Therefore, as the certification houses [*organizations*] are changing and the politics, policy should I say as a figure of speech changes, they said '*who has the people's trust and esteem, they will sell*'... even if they have no certificate and that is how it seems to function over here these days. I know a number of households that work with us that still work to this day as these are small numbers and they do the production but the placement they do by ... as the production is small scale it cannot be big, to the people they know they address them and sell to them directly as those people know they can expect of such farms those products.
    ... "*without any paperwork, without any certification*" ...

Coming from a different background but with a similar pragmatic viewpoint, Ševarlić remarks:

...currently in Serbia there are 24,000 ha certified for organic plant production... that is relatively modest compared to the 3.5m ha which is the total in Serbia and relatively minimal as compared to Austria which in some regions has over 35% of land dedicated to organic production... certified. Italy streaks ahead by organic means of agriculture as they have subsidies for—the biggest problem of entering European agriculture is the conversion period from conventional to organic agricultural production, which takes from 3–5 years... during that time there is lower yield per ha as there is a depletion of fertiliser, increase in pest activity and so on, and there are no subsidies for it. And the Italians have started subsidising everything including the conversion period, of 3–5 years. Currently Italians have over 1m ha in the process of conversion underway ... well the Italians are the grandmasters of food production. That includes drinks of course.

Before exploring in detail some examples of best organic growers in Serbia, such as *Ecoagri Serbia d.o.o*[32] from Bela Crkva. The discrepancy grew clear between Serbia and its developed neighbours whose agricultural policies enjoyed the trace bloc's support. Regardless of any matters of material nature, including financial limitations and economic imperatives, international position and globalisation pressures, remained on the internal level of awareness, goodwill and education, relationships within and trade organisation, namely of organisational cultures and institutional learning on all levels of practice.
    The most astute summary of the problems and solutions, the change for the better in the face of perennial challenges came from Sajić:

On the other hand there is a big problem, with the incentives as you say, they want something, well there could have been incentives for small ones—that is possible, and I know how that is done within the ministry, that is a very hard job to do when you announce a credit boost for a large group of small people, who aren't educated who need extra effort, that is suffering for you, it's hard to implement [for the state actors]. It's hard to implement the means to arrive at the addresses of the right people who need them most.

---

[32] [d.o.o. ↔ ltd]

. . .

that is why they say, they only cover the big ones, the big producers come to them and collect the funds and then that is it, that is how it seems to function.

. . .

part of the projects we did, we would gather the small producers, then we announce what they need as we knew what matters to them, say a small dairy, stables are needed and so on—and then we go to the ministry and say 'these people need such and such, we get the means and distribute them *and then we iron it out* between the ones and the others [govt and producers] . . . so then some of them say 'now you need an electrical connection' and the reply is 'no, where did you hear that, we don't know' . . . so we literally were doing that, ironing out those communications and differences, and prior to those years there were no problems, *it was not needed as there was no order at first and [laughs] the more 'orderly' it is, the greater the problems that arise as where there is such regulatory order the ones at the bottom of that ladder suffer more.*

It is therefore unsurprising that Sajić often references *synergies between producers and customers, the meeting of needs of parties at each end of the exchange process,* as pivotal to the effective delivery of improved practice and effective policy that benefits the most.

6   GMOs, Industry, Ethics and Ideology

The sixth subgroup of new foods[33] consists of the foodstuffs built from cell cultures, or tissues derived from animals, plants or microorganisms, fungi and algae, or was isolated out of them. That cell culture is very problematic—unless I am mistaken, it was the French who first did experiments on producing a meat by methods of cell-cultivation and following a three-year research project they found they did not progress much they got a gelatinous mass that was squidgy, tasteless and so on. . . I would not know how much more was done around the world in that field. Particularly problematic is the seventh group of these new foods; *that is the food consisting of synthesised nanomaterials.* . . therefore, what is being achieved by means of nanomaterials? Firstly, we need to define nanomaterials. Nowhere in our legislation have I found *what is being understood as 'nanomaterial'.* Especially then, how would food be created by means of synthesis of nanomaterials? . . .

(Prof. Miladin Ševarlić, interview 1, 2021 with reference to the 2019 amendments to national food legislation)

Prof. Ševarlić is a vocal opponent of GMO crops and any form of genetic *modification* that is in current or planned use. In the early stages of his 5+ hour interview, he cites Prof. Séralini's pioneering research[34] into GMO crops (Arjó et al., 2013; Butler, 2012; De Vendômois et al., 2010; European Food Safety Authority,

---

[33] E.g. Guiné et al. (2016).

[34] As a groundbreaking venture into probable harm from GMO crops it continues to influence the hotly debated discussions on genetic modification of crops and foods. This study, which is seen as either groundbreaking or controversial, depending on one's vantage point. The 'Séralini Affair' (2012–2014, referenced above) involved the high-profile publication of an article connecting GMOs and a herbicide and malignant tumours in rats, attracting criticism, retraction and republication [see main body above for sources cited].

2012a, b; Fagan, Traavik & Bøhn, 2015; Loening, 2015; Resnik, 2015; Séralini, Cellier & de Vendomois, 2007; Séralini et al., 2011, 2012, 2013, 2014a, b, c; Vidal, 2012). Ševarlić's approach *to what food is,* is distinctly traditional. The means utilising technology are not frowned upon as such by him. However, Ševarlić's sentiment towards genetic modification experiments in human health and nutrition appear as alien to him as does the proposed cannibalistic solution to the concerns over climate change.

As he states in the long interview, the drive towards the explicit banning of GMO foods and crops in Serbia was achieved in 2009 in a feat of activism, as the local/ municipal authorities have one by one adopted a strong stance against it. He is a strong proponent of direct local representation. However, he also raised concerns over the insertion of amendments that seem to make the notion of natural origin of foodstuffs more blurred than before. This was hotly debated in several instances within the national media since 2012 and a BBC featured report from 2019 seems to imply that a space was created for interpretations that would eventually allow genetically modified foodstuffs and crops into legal use in the land (Антељ, 2014; Galetin, 2020; Gulan, 2020a, b; Janković, 2019). In a reference concerning one of his televised debates, Ševarlić elaborates on the statistical sampling of the information available to him at the time, concerning the malpractice where farmers procure GMO soya imported and sown illegally. In order to avoid prosecution when the inspection arrives, they scold those parts of the soya plants after the herbicide test treatment performed by the sanitary inspection. The damaged leaves leave the impression of being shrivelled through the use of herbicide, not hot water. In his estimation based on what was reported by the authorities from the tested sample, he extrapolated that as much as 5000 ha would have been used for illegal GMO soya plants. This remains debated and open to interpretation though the statistical sampling cited by him seems plausible.

Another hotly contested topic at the time of fieldwork as well as writing of this piece is the contract concerning extensive exploitation of the lithium ore in the Serbian countryside (Ambrose, 2022; Anđelković, 2022; Bojić, 2019; Cvetković, 2022; De Launey, 2022; Roze, 2021; Paunović, 2021; Tatalović, 2022; Trpeski, Šmelcerović & Jarevski, 2021). While Prof. Ševarlić is a strong opponent of this, and there seems to be a popular stance against it, the authorities in place at the time of writing report that the agreement in principle was granted as early as 2004 when the current opposition was in power. As stated in Simić's 2001 text, interpreting current events is a thankless task whereby the author of this piece will refrain from it as objectivity is impossible within the eye of the storm or the position at the interior of a social equivalent of the Heisenbergian uncertainty model (Katz & Csordas, 2003), metaphorically adopted and not unheard of in social sciences.[35] The fact remains that this prolific respondent is consistent in his professed belief systems of direct activism, hands-on engagement and a remarkable blending of modern and traditional

---

[35] Not least as elaborated at length in Todorovic (2016).

approaches to environmental subjects and the notion of health foods to which he seems committed as a practicing farmer as well.

Another somewhat different take on ideology and ethics, belief systems and their practical meaning comes in the guise of a traditional spirituality gaining ground in Serbia over the past two decades. A decisively pacifist and environmentalist reading of traditional spirituality referred to earlier in the text, albeit briefly, stands at the heart of it.[36]

In the words of Dragan Sajić, when asked about the relative benefits of vegetarian and vegan diets where the produce is highly dependent on novel technologies:

it is not as widespread here to the same extent [as in the West], as we are lower on the scale of such economic development, and it has not yet reached that level but we, for example in the ministry when we came for open debates, public discussions, people on the highest level totally do not have a clear awareness of what health foods really are. We had that, we tried to make them aware of what it actually means to have health foods. What it really represents. Sadly, here we don't yet have, in fact it will take time for that awareness, on the other hand—that tradition, we have been 'breastfed' our tradition of [Christian] Orthodoxy so we know the difference between the fast and the feast,[37] and we were explaining that [abroad] that this vegetarian way we have much deeper, stronger and more embedded, that notion of fast and feast. We have a natural [veganism], much healthier, especially when one takes into account the spiritual and bodily aspects brought together, we were constantly making people [abroad?] aware of folks, we have that perfect cross-section of health, the balance of animal and vegan produce, grease and fast that we do not really need that vegetarian—especially the unhealthy aspects of it, as you say, do you know to what extent that is unhealthy, [vegetarians] consider industrially produced champignon a health food. Which is treated so much that nothing can ever grow there [where champignons are industrially grown]. When I explain to them how champignons are produced, and how many times it is intoxicated, that not a single blade of grass can grow there. And I tell them, you eat those manufactured champignons, that is so full of toxin that nothing natural can grow there. You'll see, try to take that sack spill it out—take it outdoors and no grass ever grows out of it. So that sack, the grow bag is so poisoned that . . .and then you take that mushroom and eat it, thinking you're eating something healthy.

This may appear like an antiquated definition of healthy living, and some actually uphold it. Indeed, over the past decade or so as part of restitution of land and resources the monastic communities have started producing on a larger scale their own crops and health foods, remedies, wines, brandies, or beer (e.g. е-продавница, Патријаршија (2014a, b)), much like in the West's uninterrupted tradition that was not affected by external conquest or extreme discourses of the totalitarian left. That is not to say these represent views of any degree of extremity as a personal choice and individual responsibility are all that matters in such discourses. A practising faithful

---

[36] Cvetković (2015) presents a dependable overview of the discursive complexities and perceptions within the Serbian Orthodox Church concerning Europe, the West, modernity and (post)communist social forces. A subject oft-argued, passionately if ill-informedly in at times an unhelpfully charged manner, so his treatise sheds much needed light. Intricacies of discursive contests of tradition and progress deserve further research.

[37] The term 'grease' would use in literal translation from Serbian.

would abstain from meat for up to 200 days per year; when combined with inner reflexion and withdrawal from societal strife, the practise resembles that of mindfulness.

From a more secular and socially engaged angle, the respondent elaborates:

> And then when all those are done, the analysis that was recently conducted say with dairy products which are the only ones regulated and the peasant farmers can actually sell the produce in the green market, as that is regulated, we told them [the ministry] to use that regulatory framework to model on them all the other produce that farmers deliver in agriculture. Including vegetables and fruits and all the rest so that the farmers can work them, process- and sell. And we have a situation that now when you do the analysis, young people when they are asked would they have cheese or kajmak[38] *from the green market* they respond, 'no way, that is not safe at all'. They will buy a mozzarella at the mall—not knowing what they eat—but cheese or kajmak they won't dare buy from peasants because they have a prejudice in their minds, their consciousness that this is 'goodness knows what'. That is therefore a big problem which I do not know how to, when a quality product may be offered, in the right sort of way, that would easily find its way to the customer—but how does that pay, well that is a matter of... see, how long we had to work to... and now there are some individual cases, the interconnections between the farmers and city folk, directly. And that is largely connected to the people from the Church, those who possess the awareness of this and who care to feed their families with better quality foods, and who have an option to establish better communication with those from rural areas, who know where, say, is Ljig or some small village and who do not live solely in Belgrade or the bigger towns and cities.
>
> . . .
>
> . . . And that is the quality that is sought after so people, when they do consume it, they grasp this is *the quality* and they grasp the fact that *health has no alternative*. All of that search and pursuit and chasing of something—what is the point, when we then utilise those means that we chased around so that we may then regain the health thus lost... as I work, I know, I apply the very same principles, I do the same thing but in my case I do it in horticulture, I've been doing it for 30 years, that is my fundamental... see, we work on the projects of designing, dressing and decorating those green areas, and the past 2–3 years has seen a major expansion of young, highly educated people who are financially sound and who buy... there is a 76% increase of real estate in rural areas, last year alone. Those are the people who buy estates, we do the arranging, and the same model applies to agriculture and that functions well.
>
> . . .
>
> There are people who do those projects [with the EU, when referring to a friend who does] but do not live the life. I do it and live it. I live in a village. I have chosen so, not for being unable to live in a city but because I do not consider that good life.

7   Political Economy and the Weight of History

For all the difficulties faced by the people and especially small landowners as these were the only ones during the 45 years of communist rule, Sajić observes:

---

[38] *Pronounced as 'kaymak', Serbian clotted cream, similar in texture to that of mascarpone.*

We have examples in Serbia where I had the opportunity to watch and see the difference between, say, I was in Russia and could see their farms—Kolkhoz[39] in those days—what a massive difference between the Russian householder and a Serbian *householder*[40] [farm owner, farm landlord, settled farmer, manager etc.]—in fact, there is no such thing, there, as a term of "Russian householder farmer". That is the first thing. As it totally does not exist. Because in their case, with the confiscation of all ownership [of land] during the transition to (real)socialism they only had 5 are, 500 sqm, around the house, and they were allowed no machinery. No one there knew how to drive a tractor; they had large Kolkhoz and big *cooperatives*, just as we had, and in those *cooperatives* [and Kolkhoz's] people knew how to run a tractor, but regular people [small farmers] neither had one nor did they know how to run it. That's when I saw what advantage we had at the time. Here every house [in Serbia] has a tractor, it has all the machines, and we could all . . . we could in the 80s when that was in bloom, we would, that was maybe—it all went strange, we analysed it, in the post-war period (2nd WW) the countryside had bad economic position because of the war, . . . there was a linear growth of the financial potential of [Serbian] countryside, it may have peaked sometime in the 80s, but at that time already at least one member of a village household worked in the state sector, had a salary plus they worked the land. That gave them massive potential as they could produce and offer goods in the marketplace.

Regardless of the causal sequences and forces at play since the 1980s, the transformation of Serbian cities, towns and villages was unprecedented: A new way of living, an ageing society and its shortening lifespan, a damaged economy and a hurt world status desire much rebuilding and rethinking, rebranding and reviving. In hindsight, what looked like an ideological discourse that limited people's choices now feels like a simple life to reflect on. Whatever enterprising spirit existed at the time is now redefined beyond recognition.

Market forces and strategic shifts in the economy have spelled the transformation of society and its fortunes. The search for an old/new identity affects societies far more stable in recent experience. Tensions between a postmodern cosmopolitan approach to life and a return to values rooted in a reimagined past are not new in modern experiences. As balanced against the shocks of history which are being construed and redefined as an eternal now, infinitesimally sliced up from the miniscule components of a perpetually changing reality of lockdowns, travel restrictions and health policies that seem to have abolished and suspended any sense of continuity, yet this may well change almost overnight but not without a lasting impact on humanity's self-perception to which no community is immune. Structural forces of political economy, the balance of power within the increasing gap between the *'glocal' haves and have nots* are spreading across borders just as germs seem to do, and the very notion of global/local, cosmopolitan-local dichotomies that occupied so much attention of theorists (e.g. Hannerz, 2002) less than a generation ago, prior to the 2008 subprime implosion are seeming as trivial as

---

[39] The devastating impact of forced collectivisation on food production in the USSR is identified in a number of treatises, including Arendt, Stojanović & Bajazetov-Vučen, A. (1998, xxxiv, xxxvii, 328, 402).

[40] Serbian word 'домаћин': more than householder, host; it also means farmer, family man and landlord.

they appear virtual. *The new glocal* may well 'stay put' and this is likely to affect the very notion of enterprise as a physical pursuit. Whether we will be forced to endorse food printing machines to stave off starvation is less of a concern than is the return to something that does not resemble the mediaeval cycles of plague and blame. No single place remains immune to such discursive tremors so the conditional nature of reality is revealed in its fullness beyond any artificial boundaries of identity and status.

8 Is Future Cyclical: Fact and Metaphor

Grand societal structures surpassing the size of any individual state have caught every nation-state and trade bloc in their whirlwind since early 2020. Prof. Ševarlić wittily sums up thus his view of the quality of rural life and a possible future that does not diverge greatly from some of the visions of green living in the West, aided by high technology tools:

> The pandemic has shown" that "a village is the healthiest environment for human living" and so did the property prices that trebled or increased fivefold in Serbian villages since 2020" ... "and the village, as a settlement will grow in significance ever more, especially as the result of IT developments, and you can now run a business from a meadow.

Whether one speaks of the concept of renewables and cyclical economies; whether or not the notion of sharing is endorsed or imposed. Whether surviving replaces healthy living, is an unknown. As was the case throughout history though, tending a plot of land remains a privilege.

References

Aditya, A., & Kim, N. P. (2022). 3D printing of meat following supercritical fluid extraction. *Food, 11*(4), 554.

Agro TV Srbija. (2019, June 6). *Autohtone vrste Srbije: Balkanski magarac, ep. 16.* [online]. *Agro TV Srbija.* Accessed Mar 28, 2022, from https://www.youtube.com/watch?v=IhUH-tuVooo 2015.

Албуновић, И. (2015, November 28). Ко су власници српске земље. [online]. Београд: Политика. Accessed Mar 27, 2022, from https://www.politika.rs/scc/clanak/344264/Ekonomija/Ko-su-vlasnici-sccpske-zemlje

Албуновић, И. (2017, January 19). *Српске оранице јефтине за странце.* [online]. Београд: Политика. Accessed Mar 27, 2022, from https://www.politika.rs/scc/clanak/372330/Ekonomija/Srpske-oranice-jeftine-za-strance

Alexander, E. (2017). Yugosphere insiders or Croatian outsiders: The reception of serbian films in Croatia since the breakup of Yugoslavia. *Image & Narrative, 18*(1).

Alispahić, B. (2020). PROPAGANDA KAO OBLIK SPECIJALOG RATA. *Društvena i tehnička istraživanja, 2,* 71–90.

Alvesson, M., & Sköldberg, K. (2017). *Reflexive methodology: New vistas for qualitative research.* Sage.

Ambrose, J. (2022, January 21). *Serbia scraps plans for Rio Tinto lithium mine after protests.* [online]. *The Guardian.* Accessed Jan 22, 2022, from https://www.theguardian. com/world/2022/jan/20/serbia-scraps-plans-for-rio-tinto-lithium-mine-after-protests. Society (Institut za filozofiju i drustvenu teoriju, editor: Božidar.

Anđelković, N. (2022, January 21). *Jadarit, Rio Tinto i životna sredina: Šta treba da znate o nalazištu litijuma u Srbiji-BBC News na srpskom.* [online]. BBC News in Serbian. Accessed Mar 27, 2022, from https://www.bbc.com/serbian/lat/srbija-55646178

Anđelković, P. (2018). *REGIONALIZACIJA I (ILI) BEOGRADIZACIJA SRBIJE. 1. Regionalni razvoj i prekogranična saradnja.* PK Pirot.

Angelkova, T. (2010). *Specifični oblici povezivanja turizma i poljoprivrede na Staroj Planini.* (Doctoral dissertation). University Singidunum.

Антељ, J. (2014, September 19). *Нисмо довољно заштићени од ГМ хране.* [online]. Politika. Accessed Mar 27, 2022, from https://www.politika.rs/scc/clanak/296815/Nismo-dovoljno-zasticeni-od-GM-hrane

Arendt, H., Stojanović, S., & Bajazetov-Vučen, A. (1998). *Izvori totalitarizma.* Feministička izdavačka kuća 94.

Arjó, G., Portero, M., Piñol, C., Viñas, J., Matias-Guiu, X., Capell, T., . . . Christou, P. (2013). Plurality of opinion, scientific discourse and pseudoscience: An in depth analysis of the Séralini et al. study claiming that roundup™ ready corn or the herbicide roundup™ cause cancer in rats. *Transgenic Research, 22*(2), 255–267.

Arsić, M. (2015). Privatizacija javnih preduzeća u Srbiji. u. *Restrukturiranje javnih preduzeća u uslovima institucionalnih ograničenja,* 105-122.

Askew, A., Askew, M. A. J. D. C. L., & Askew, C. (1916). The stricken land: Serbia as we saw it. E. Nash.

Aung, M. M., & Chang, Y. S. (2014). Traceability in a food supply chain: Safety and quality perspectives. *Food Control, 39,* 172–184.

Avramović, Z. (2012). 'Brain drain' from Serbia: One face of globalization of education? *Sociološki pregled, 46*(2), 189–202.

B92 & Tanjug. (2013, January 22). 'Zemlja nije predmet restitucije–IZVOR: Tanjug'. *B92.net.* Accessed Mar 27, 2022, from https://www.b92.net/biz/vesti/srbija/zemlja-nije-predmet-restitucije-679591

Babić, M., & Primorac, E. (1986). Some causes of the growth of the Yugoslav external debt. *Soviet Studies, 38*(1), 69–88.

Bajt, A. (1986). Economic growth and factor substitution: What happened to the Yugoslav miracle?: Some comments. *The Economic Journal, 96*(384), 1084–1088.

Bakir, A., & Todorovic, M. (2010). *A Hermeneutic Reading into. Qualitative Report, 15*(5), 1037–1057.

balkaniym.tv & b92.net. (2008, June 25). Društvo i ekonomija: Pet tajkuna ima 100.000 hektara. [online]. *balkaniym.tv. Izvor: B92.* Accessed Mar 27, 2022, from https://www.balkaniyum.tv/srpski/vesti/21242.shtml; https://www.b92.net/biz/vesti/srbija/pet-tajkuna-ima-100-000-hektara-305452?version=amp

Baričević, H. (2018). Tatjana Šarić, U vrtlogu komunizma: Mladi Hrvatske 1945.-1954. (Zagreb-Hrvatski državni arhiv, 2017), 541 str. *Arhivski vjesnik, 61. no. 1.*

Bataković, D. T. (2013). Storm over Serbia: Rivalry between Civilian and military authorities (1911–1914). *Balcanica,* (XLIV), 307-356.

Bauer, E. (1970). Die jugoslawischen Gastarbeiter in Westeuropa. *Der Donauraum, 15*(3-4), 140–151.

Bazdulj, M. (2015, February 2). Beleška o (auto)šovinizmu. [online]. *Beograd: Politika.* Accessed Dec 13, 2021, from https://www.politika.rs/scc/clanak/317868/Beleska-o-auto-sovinizmu

Beall, A. (n.d.) *The green sludge that could transform our diets.* BBC: Follow the food. Accessed Mar 27, 2022, from https://www.bbc.com/future/bespoke/follow-the-food/the-green-sludge-that-could-transform-our-diets.html

Begović, B., Živković, B., & Mijatović, B. (2000). Novi model privatizacije u Srbiji. *Centar za liberalno-demokratske studije.*

Bendix, A. (2019). A Swedish scientist suggested the climate crisis could lead people to consider eating human flesh. It's not the first time a scientist has suggested the idea. *Business insider.* Accessed Sep 5, 2021, from https://www.businessinsider.com/cannibalism-eating-human-flesh-climate-change-2019-9?r=US&IR=T

Beograd.rs. (2022). *Stanovništvo.* Accessed Mar 08, 2022, from https://www.beograd.rs/lat/upoznajte-beograd/1199-stanovnistvo/

BETA. (2022, March 28). *U Beogradu 1.198 biračkih mesta i više od milion i po glasača.* [online]. *Danas.* Accessed Mar 28, 2022, from https://www.danas.rs/vesti/politika/izbori22/u-beogradu-1-198-birackih-mesta-i-vise-od-milion-i-po-glasaca/

BIZLife. (2015). Delhaize kupuje preostale akcije C marketa. [online] *BIZLife.* Accessed Mar 28, 2022, from https://bizlife.rs/82610-delhaize-kupuje-preostale-akcije-c-marketa/

Blic. (2017, October 17). *Beograd pravi 40 ODSTO srpskog BDP, a izjave ljudi iz drugih gradova u Srbiji pokazuju koliko je to PORAŽAVAJUĆE. Blic.* Accessed Mar 27, 2022, from https://www.blic.rs/biznis/beograd-pravi-40-odsto-srpskog-bdp-a-izjave-ljudi-iz-drugih-gradova-u-srbiji-pokazuju/5kk3bmy

Boarov, D. (2005). Junkeri ponovo jašu. *Vreme, BROJ 756.* Accessed Mar 27, 2022, from https://www.vreme.com/vreme/junkeri-ponovo-jasu/

Bobić, M. (2009). Dijaspora kao ekonomski i socijalni kapital Srbije. *Sociološki pregled, 43*(3), 361–378.

Boeckh, K. (2016). The rebirth of pan-Slavism in the Russian empire, 1912–13. In *The Balkan wars from contemporary perception to historic memory (pp. 105-137).* Palgrave Macmillan.

Bogdanov, N. L., Đorđević-Milošević, S., & Clark, L. (2007). *Mala ruralna domaćinstva u Srbiji i ruralna nepoljoprivredna ekonomija.* UNDP.

Bojić, S. (2019, September 3). Srbija raspolaže najvećim zalihama litijuma. [online]. DW. Accessed Mar 27, 2022, from https://www.dw.com/sr/srbija-raspola%C5%BEe-najve%C4%87im-zalihama-litijuma/a-50269312

Bojičić, V. (2018). Slowed serbian development in the early 19th century until world war II. *Knowledge-International Journal, 26*(6), 1879–1884.

Brdar, I., & Spasić, V. (2010). AVIO-SAOBRAĆAJ U SRBIJI SA POSEBNIM OSVRTOM NA NISKOBUDŽETNE AVIO-KOMPANIJE. *Singidunum Scientific Review/Singidunum Revija, 7*(2).

Brown, M. B. (1993). The war in Yugoslavia and the debt burden: A comment. *Capital & Class, 17*(2), 147–160.

Bukowski, C. (1987). Politics and the prospects for economic reform in Yugoslavia. *East European Politics and Societies, 2*(1), 94–114.

Bulić, R. (2017). Na marginama jednog skandala ili šta je kome bosanski jezik. *Bosanski jezik, 14,* 137–143.

Business Insider. (2021, March 27). *Why the world's rarest cheese (Pule Donkey Cheese) is so expensive.* Accessed Dec 21, 2022, from https://www.youtube.com/watch?v=KZAp4r0n0B0

Butler, D. (2012). Hyped GM maize study faces growing scrutiny. *Nature, 490*(7419), 158.

Ć. G/Novosti (2018, September 15). *U selu Ratina kod Kraljeva 11. Projada: Proja–narodna đakonija.* [online]. Novosti. Accessed Mar 27, 2022, from https://www.novosti.rs/vesti/srbija.73.html:749700-U-selu-Ratina-kod-Kraljeva-11-Projada-Proja%2D%2D-narodna-djakonija

Calic, M. J. (2011). The beginning of the end: the 1970s as a historical turning point in Yugoslavia. *The crisis of socialist modernity. The Soviet Union and Yugoslavia in the* 1970s, 66-86.

Caporgno, M. P., & Mathys, A. (2018). Trends in microalgae incorporation into innovative food products with potential health benefits. *Frontiers in Nutrition, 58*(5). Accessed Dec 17, 2022, from https://www.frontiersin.org/articles/10.3389/fnut.2018.00058/full

Caranović, B (2021, May 31). *Beograd "Daje" Skoro Pola Bdp: Dve trećine radnika platu zaradi u glavnom gradu ili u Novom Sadu.* [online]. Novosti. Accessed Mar 27, 2022, from https://www.

novosti.rs/vesti/ekonomija/1002140/beograd-daje-skoro-pola-bdp-dve-trecine-radnika-platu-zaradi-glavnom-gradu-ili-novom-sadu.

Chittle, C. R. (1975). *The industrialization of Yugoslavia under the workers' self-management system: Institutional change and rapid growth* (No. 26). Kiel Working Paper.

Chossudovsky, M. (2015). *How the IMF dismantled Yugoslavia*. The Globalization of Poverty.

Christitch, A. (1922). A new Dawn in Serbia. *The Irish Monthly, 50*(592), 424–427.

Ćirjaković, Z. (2021). *Otkrivanje autošovinizma*. Catena Mundi.

Ćoćkalo, D., Đorđević, D., Kavalić, M., & Bešić, C. (2019). Implementation of certification schemes in the Balkan agro-food sector. *Економика пољопривреде, 66*(1), 77–88.

Ćosin, J (2021, September 15). *Cela krda umiru danima od žeđi: Na Suvoj planini presušio jedini izvor, 800 krava i 200 konja pred skapavanjem*. [online] Novosti. Accessed Dec 17, 2021, from https://www.novosti.rs/srbija/vesti/1036305/cela-krda-umiru-danima-zedji-suvoj-planini-presusio-jedini-izvor-800-krava-200-konja-pred-skapavanjem

Cvetković, L. J. (2022, January 21). *Od dobrodošlice do raskida Srbije sa Rio Tintom*. [online]. Slobodna Evropa. Accessed Mar 27, 2022, from https://www.slobodnaevropa.org/amp/srbija-rio-tinto-jadarit-litijum/31665241.html

Cvetković, V. (2015). *From "merciful angel" to "fortress Europe": The perception of Europe and the west in contemporary Serbian orthodoxy*. Universität Erfurt, Religionswissenschaft (Orthodoxes Christentum).

Cvikl, M. M., & Mrak, M. (1996). *Former Yugoslavia's debt apportionment*. World Bank.

Davidović, M. (1990). Migracije u Srbiji: Rezultati jednog istraživanja. *Migracijske i etničke teme, 6*(2), 157–172.

De Launey, G (2022, January 21) *Serbia revokes Rio Tinto lithium mine permits following protests*. [online]. BBC.

De Vendômois, J. S., Cellier, D., Vélot, C., Clair, E., Mesnage, R., & Séralini, G. E. (2010). Debate on GMOs health risks after statistical findings in regulatory tests. *International Journal of Biological Sciences, 6*(6), 590.

Dimitrijević, D. (2011). Zahtevi pripadnika nekadašnje nemačke manjine za restituciju imovine u Srbiji. *Međunarodni problemi, 63*, 126–159.

Dinkić, M., & Antić, S. (1999). *Završni račun: Ekonomske posledice NATO bombardovanja*. Stubovi kulture.

Dondur, N., Radojević, S., & Veljković, Z. (2007). Efekti privatizacije i restrukturiranja u industrijskim preduzećima u Srbiji. *Industrija, 35*(3), 13–25.

Đorđević, M. (2009). *Proces privatizacije u Srbiji u periodu od 2000. Do 2008. Godine*. ŠKOLA BIZNISA Naučnostručni časopis.

Drašković, B. (2010). *Kraj privatizacije i posledice po razvoj*. Univerzitet u Beogradu.

Друштво српских домаћина. (2022). Accessed Apr 03, 2022, from http://www.drustvosrpskihdomacina.org.rs/

Đurić, M. V. (2007). Obnova manastira Banjska 1938-1940. godine. *Baština, 22*, 173–182.

Dyker, D. A. (2013). *Yugoslavia: Socialism, development and debt*. Routledge.

Dyker, D. A. (2014). The degeneration of the yugoslav communist party as a managing elite–a familiar east european story? In *Yugoslavia and after: A study in fragmentation, despair and rebirth* (p. 48).

Džubur, A (2017, March 19) 'Europejci i šeici na srbijanskim oranicama'. *Al Jazeera Balkans*. Accessed Mar 27, 2022, from https://balkans.aljazeera.net/teme/2017/3/19/europejci-i-seici-na-srbijanskim-oranicama

Ejdus, F. (2017). Memories of empire and Serbia's entry into international society. In *Memories of empire and entry into international society* (pp. 121–137). Routledge.

ekspres.net. (2016, November 8). 'SSP I Poljoprivredno Zemljište: Može li Srbija da odloži prodaju oranica strancima?!'. ekspres.net. Accessed Mar 27, 2022, from https://www.ekspres.net/ drustvo/ssp-i-poljoprivredno-zemljiste-moze-li-srbija-da-odlozi-prodaju-oranica-strancima

е-продавница, Патријаршија. (2014a). *Манастирски мед из Дуљева, Фенека. . . Ливадски мед, мед од жалфије, сунцокретов мед, мед од уљане репице. . . Патријаршијски управни одбор*. Accessed Mar 28, 2022, from http://www.patrijarsija-puo.rs/kategorije-proizvoda/manastirski-prozivodi/manastirski-med/

е-продавница, Патријаршија. (2014b). *Ракија дуња 0.7 литара–Манастир Ковиљ. Патријаршијски управни одбор*. Accessed Mar 28, 2022, from https://www.patrijarsija-puo.rs/e-prodavnica/manastirski-prozivodi/manastirska-rakija/rakija-od-dunja-manastir-kovilj/

Ersado, L. (2006). Rural vulnerability in Serbia. *World Bank Policy Research Working Paper*, (4010).

ETNOHIP.RS. (2021, July 1). *Највећа смотра прозвођача традиционалних производа намењених тржишту у Србији и региону Југоисточне Европе: 15. Сајам етно хране и пића, 25–28. новембар 2021. 10:00–19:00 часова, Хала 3, Београдски сајам*. [online]. ETNOHIP. Accessed Mar 27, 2022, from https://etnohip.rs/

European Food Safety Authority. (2012a, September 19). Final review of the Séralini et al. (2012a) publication on a 2-year rodent feeding study with glyphosate formulations and GM maize NK603 as published [online]in Food and chemical toxicology. *EFSA Journal, 10(11)*, 2986.

European Food Safety Authority. (2012b). Review of the Séralini et al. (2012) publication on a 2-year rodent feeding study with glyphosate formulations and GM maize NK603 as published online on 19 September 2012 in food and chemical toxicology. *EFSA Journal, 10*(10), 2910.

Eyer, S. H. (2020). *The Eastern Europe Britain wanted: Serbian Independence as recorded in British newspapers, 1867*. University of Dayton.

Fagan, J., Traavik, T., & Bøhn, T. (2015). The Seralini affair: Degeneration of science to re-science? *Environmental Sciences Europe, 27*(1), 1–9.

Farkas, E. (2014). *Yes for the Yugosphere, no for Yugoslavia. A case study of domestic music and identity in Serbia and Croatia*. Central European University.

Farnam, R. S. (1918). *A nation at bay: What an American woman saw and did in suffering Serbia*. Bobbs-Merrill.

Galetin, Ž. (2020, April 2). *Da li je borba protiv GMO u Srbiji–borba sa vetrenjačama?* [online]. Agroklub.rs. Accessed Sep 5, 2021, from https://www.agroklub.rs/poljoprivredne-vesti/da-li-je-borba-protiv-gmo-u-srbiji-borba-sa-vetrenjacama/58638/

Gaynor, G. K. (2019, September 9). *Scientist suggests eating human flesh to fight climate change*. [online]. New York Post. Accessed Sep 5, 2021, from https://nypost.com/2019/09/09/scientist-suggests-eating-human-flesh-to-fight-climate-change/

Gligorov, V. (2018). Demokratija, pluralizam i ekstremizam. *37 helsinške sveske*, 15.

Glintic, M. (2012). Challenges for the republic of Serbia in the process of law harmonization in the field of food safety and veterinary polices. *Strani Pravni Zivot*, 215.

Glišin, V. (2018). *Kontinuitet dekonstrukcije srpskog nacionalnog identiteta od 1918*. Do 2018. Godine–Slučaj AP Vojvodine. *Viteška kultura, god. VII (2018)*, 305.

Gnjatović, D. (2012). Restitucija oduzete imovine kao jedno od otvorenih pitanja tranzicije u Srbiji. *Evropska Unija i srbija*, 121.

Gordy, E. (2013). *Guilt, responsibility, and denial*. University of Pennsylvania Press.

Gotev, G. (2017, March 2). *Visegrád readies attack against 'double standards' for food*. [online]. EURACTIV.com with Reuters. Accessed Sep 5, 2021, from https://www.euractiv.com/section/agriculture-food/news/visegrad-readies-attack-against-double-standards-for-food/

Gubić, J. M., Šarić, L. C., Šarić, B. M., Mandic, A. I., Jovanov, P. T., Plavšić, D. V., & Okanović, D. G. (2014). Microbiological, chemical and sensory properties of domestic donkey's milk from autochthones serbian breed. *Journal of Food and Nutrition Research, 2*(9), 633–637.

Guiné, P. F., Ramalhosa, E., & Paula Valente, L. (2016). New foods, new consumers: Innovation in food product development. *Current Nutrition & Food Science, 12*(3), 175–189.

Gulan, B. (2011, May 16). *Obradiva zemlja: šta će biti sa poljoprivrednim zemljištem u procesu pridruživanja eu*. Dragas.biz. Accessed Mar 27, 2022, fromhttps://www.dragas.biz/obradiva-zemlja/

Gulan, B. (2016, November 17). *Čija je zemlja, zemlje Srbije!?*. Makroekonomija.org. Accessed Mar 27, 2022, from https://www.makroekonomija.org/0-branislav-gulan/cija-je-zemlja-zemlje-srbije/

Gulan, B. (2020a, August 10). *Restitucija pre licitacije*. Makroekonomija.org. Accessed Mar 27, 2022, from https://www.makroekonomija.org/0-branislav-gulan/restitucija-pre-licitacije/

Gulan, B (2020b, May 20). *GMO pred vratima Srbije. . .* [online]. Makroekonomija.org. Accessed Sep 5, 2021, from https://www.makroekonomija.org/0-branislav-gulan/gmo-pred-vratima-srbije/

Hadžić, S. (2008). Refleksije sukoba Informbiroa i KPJ-a na zaostajanje poljoprivrede u Bosni i Hercegovini (1949–1952). *Istraživanja, 03*, 75–90.

Hannerz, U. (2002). The cultural role of world cities. In *Transnational connections* (pp. 137–149). Routledge.

Hatzivassiliou, E. (2005). Cyprus at the Crossroads, 1959–63. *European History Quarterly, 35*(4), 523–540.

Hirshfield, C. (1986). In search of Mrs. Ryder: British women in Serbia during the great war. *East European Quarterly, 20*(4), 387.

Holjevac Tuković, A. (2003). Društveno-gospodarske reforme 1950-1952. i njihov odraz na upravu narodne republike hrvatske. *Arhivski vjesnik, 46*(1), 131–146.

Houser, K. & Tangerman, V. (2019, September 10). *Scientist suggests new response to climate change: Cannibalism.* He says he'd be open to tasting it. Would you? [online]. Futurism.com. Accessed Sep 5, 2021, from https://futurism.com/the-byte/scientist-cannibalism-climate-change

Howlett, M. (2021). Looking at the 'field' through a zoom lens: Methodological reflections on conducting online research during a global pandemic. *Qualitative Research, 1468794120985691.*

Hutton, I. G. E. (1928). *With a woman's unit in Serbia*. Salonika and Sebastopol.

IFOAM. (2022a). *Explore our affiliates from Serbia-membership directory.*

IFOAM. (2022b). *IFOAM–Organics international* Accessed Sep 5, 2021, from https://ifoam.bio/

IMDB. (n.d.). *Soylent green. Internet movie database: IMDB.* Accessed Mar 27, 2022, from https://www.imdb.com/title/tt0070723/

Isić, M. (2008). *Seljanka u Srbiji u prvoj polovini 20. veka.* Helsinški odbor za ljudska prava u Srbiji.

Istinomer. (2020, May 11). Akteri: Miladin Ševarlić. [online]. *Istinomer; Center for research, transparency and accountability (CRTA).* https://www.istinomer.rs/akter/miladin-sevarlic/

IT Serbia. (2022). *Category: IT Serbia facts.* Accessed Mar 27, 2022, from https://itserbia.info/category/serbia-facts/

izletjada.rs. (2022). *Kupusijada u Mrčajevcima. izletjada.rs.* Accessed Mar 27, 2022, from https://www.izletjada.rs/posts/kupusijada-u-mrcajevcima/

Jagiełło-Szostak, A. (2013). The Phenomenon of Yugo-nostalgia in post-Yugoslav countries. In A. Jagiełło-Szostak (Ed.), *Idea narodu politycznego kontra etnonacjonalizmy* (Vol. 2013, pp. 275–296). Od jugoslawizmu do jugonostalgii.

Jakovina, T. (2011). Tito's Yugoslavia as the pivotal state of the non-aligned. In *Tito-Viđenja i tumačenja* (pp. 389–404). Beograd.

Jakovina, T. (2017) Part III: Yugoslavia from a historical perspective (1918–1991)-Yugoslavia on the international scene: The active coexistence of non-aligned yugoslavia. In *Yugoslavia from a historical perspective* (pp. 461-514). Helsinški odbor za ljudska prava u Srbiji.

Jakovljević, M. (2017). Population ageing alongside health care spending growth. *Srpski arhiv za celokupno lekarstvo.*

Jandyal, M., Malav, O. P., & Chatli, M. K. (2021). 3D printing of meat: A new frontier of food from download to delicious: A review. *International Journal of Current Microbiology and Applied Sciences, 10*, 2095–2111.

Janković, M. (2019, March 14). Novi Zakon o bezbednosti hrane: Uvodi li Srbija GM hranu?. [online]. *BBC News na srpskom.* BBC News in Serbian. Accessed Mar 27, 2022, fromhttps://www.bbc.com/serbian/lat/srbija-47496566

Jelavich, B. (2004). *Russia's Balkan entanglements, 1806-1914.* Cambridge University Press.

Jenkins, E. (1883). Young Serbia-1883. *The Contemporary review, 1866-1900*(44), 437–459.

Judah, T. (2009). *Yugoslavia is dead: Long live the Yugosphere good news from the Western Balkans*. Papers on South Eastern Europe, 1.

Jurić, T. (2021). *Gastarbeiter millennials*. Verlag Dr. Kovač.

Juriša, S. (1983). Agrarna politika i problemi kolektivizacije u Jugoslaviji u vrijeme sukoba KPJ s Informbiroom. *Časopis za suvremenu povijest, 15*(1), 55–73.

Justman, M., & Teubal, M. (1991). A structuralist perspective on the role of technology in economic growth and development. *World Development, 19*(9), 1167–1183.

Handral, K., Hua Tay, S., Wan Chan, W., & Choudhury, D. (2022). 3D printing of cultured meat products. *Critical Reviews in Food Science and Nutrition, 62*(1), 272–281.

Kalenjuk, B., Timotić, D., Tešanović, D., Gagić, S., & Banjac, M. (2016). Analysis of the state and offer of the hospitality facilities in special nature reserves in Vojvodina. *Monitoring and Management of Visitors in Recreational and Protected Areas*, 498–500.

Katz, J., & Csordas, T. J. (2003). Phenomenological ethnography in sociology and anthropology. *Ethnography, 4*(3), 275–288.

Kežić, D. (2017). Политичке последице економских реформи 60-их година у СФРЈ. Од дезинтеграције економског система до конфедерализације Југославије (1961–1971). *Tokovi istorije, (2),* 11-36.

Kišgeci, J. (2015, December 29). *Državno zemljište ne prodavati pre restitucije*. [online]. Politika. Accessed Mar 27, 2022, from https://www.politika.rs/sr/clanak/346179/Pogledi/Drzavno-zemljiste-ne-prodavati-pre-restitucije

Kišjuhas, A. (2020, December 27). Autošovinizam je etiketa koju su nacionalisti izmislili za kritičare velikosrpskog naciona. [online] *Danas*. Accessed Dec 13, 2021, from https://www.danas.rs/vesti/drustvo/kisjuhas-autosovinizam-je-etiketa-koju-su-nacionalisti-izmislili-za-kriticare-velikosrpskog-naciona/

Kljajić, S. (2015, June 18). *Restitucija: Oštra rasprava o vojvođanskim oranicama*. [online] RTS. Accessed Mar 27, 2022, from https://www.rtv.rs/sr_lat/vojvodina/novi-sad/restitucija-ostra-rasprava-o-vojvodjanskim-oranicama_611653.html

Kneafsey, M., Venn, L., Schmutz, U., Balázs, B., Trenchard, L., Eyden-Wood, T., . . . Blackett, M. (2013). Short food supply chains and local food systems in the EU. A state of play of their socio-economic characteristics. *JRC scientific and policy reports, 123*, 129.

Knežević, V. (2021, November 28). *Smisao metroa: Zaustavite beogradski metro!*. [online]. RTS. Accessed Feb 12, 2022, from https://www.rts.rs/page/oko/sr/story/3205/politika/4607782/zaustavite-beogradski-metro.html

Knipfel, J. (2018, May 9). *Why soylent green is more relevant now than ever: Background becomes foreground in Richard Fleischer's eco-dystopian gem, Soylent Green*. Denofgeek.com. Accessed Mar 27, 2022, fromhttps://www.denofgeek.com/movies/why-soylent-green-is-more-relevant-now-than-ever/

Ko, H. J., Wen, Y., Choi, J. H., Park, B. R., Kim, H. W., & Park, H. J. (2021). Meat analog production through artificial muscle fiber insertion using coaxial nozzle-assisted three-dimensional food printing. *Food Hydrocolloids, 120, 106898*.

КОБАСИЦИЈАДА ТУРИЈА. (2022). *КОБАСИЦИЈАДА ТУРИЈА*. Accessed Mar 22, 2022, from https://kobasicijada.rs/

Kolin, M. (2010). Novo zadrugarstvo i mogućnosti zapošljavanja žena. In: *Tematski zbornik Agrarna i ruralna politika u Srbiji–3–Održivost agroprivrede, zadrugarstva i ruralnog razvoja*, Društvo agrarnih ekonomista Srbije i Poljoprivredni fakultet Univerziteta u Beogradu, 155-167.

Končar, M. S. (2020). Isabel Emslie Hutton's account of world war I in Serbia. The faculty of philology, University of Belgrade. BELLS90 Proceedings. 2, 11 (p. 143–151).

Kostić, C. (1970). Oblici porodice i zemljišnog vlasništva u našem selu. *Sociologija i prostor: časopis za istraživanje prostornoga i sociokulturnog razvoja, 29-30*, 141–149.

Kostić, D. (2010). Konflikti u procesu privatizacije i njihov uticaj na industrijalizaciju Srbije. *Ekonomski vidici, 15*(2), 257–264.

Kostić, M. (2007, April 22). *(Evropa) Miodrag Kostić–Dobio sam ponudu od milijardu dolara.* [online] Ekapija. Accessed Mar 27, 2022, from https://www.ekapija.com/people/101691/miodrag-kostic-dobio-sam-ponudu-od-milijardu-dolara

Kovačević, D., & Lazić, B. (2012). Modern trends in the development of agriculture and demands on plant breeding and soil management. *Genetika, 44*(1), 201–216.

Kozinets, R. V. (2006). Netnography. In R. W. Belk (Ed.), *2007* (pp. 129–142). Edward Elgar Publishing.

Kozinets, R. V. (2015). *Netnography: Redefined.* Sage.

Krivokapić, B. (2018, April 12). 'Čiji su naši hektari?'. *Slobodne novine.* Accessed Mar 27, 2022, from https://slobodna.rs/ekonomija/2696-2/

Kršev, B. (2012). *PRIKAZ. Fakultet za pravne i poslovne studije Dr Lazar Vrkatić, Novi Sad Ranko Končar*, Dimitrije Boarov, Stevan Doronjski–odbrana autonomije Vojvodine Novi Sad- Muzej Vojvodine, 2011, str. 879.

Krstić, M. (2011). The SFR Yugoslavia during the cold war and current Serbian foreign policy. *Antropologija, 11*(1), 21–44.

Kržišnik-Bukić, V. (1989). Agrarno I Seljačko Pitanje U Jugoslaviji S Osvrtom Na Međuratni I Ratni Period 1945-1953. *Istorija 20. veka,* (1+ 2), 89-108.

Lampe, J. R., Prickett, R. O., Adamovic, L. S., Adamović, L. S., & Adamović, L. S. (1990). *Yugoslav-American economic relations since world war II.* Duke University Press.

Lazić, M., & Cvejić, S. (2007). Class and values in postsocialist transformation in Serbia. *International Journal of Sociology, 37*(3), 54–74.

Leipold, S., Feindt, P. H., Winkel, G., & Keller, R. (2019). Discourse analysis of environmental policy revisited: Traditions, trends, perspectives. *Journal of Environmental Policy & Planning, 21*(5), 445–463.

Loening, U. E. (2015). A challenge to scientific integrity: A critique of the critics of the GMO rat study conducted by Gilles-Eric Séralini et al. (2012). *Environmental Sciences Europe, 27*(1), 1–9.

Lucas, R. E., Jr. (1993). Making a miracle. *Econometrica: Journal of the Econometric Society,* 251–272.

Lyon, J. M. (1997). " a peasant mob": The Serbian army on the eve of the great war. *The Journal of Military History, 61*(3), 481.

MacGuill, D. (2019, September 18). *Did a swedish scientist propose cannibalism as a solution to climate change?.* Snopes.com. Accessed Sep 5, 2021, from https://www.snopes.com/fact-check/swedish-scientist-cannibalism/

MacKenzie, D. (1994). Serbia as Piedmont and the Yugoslav idea, 1804-1914. *East European Quarterly, 28*(2), 153–183.

Majdin, Z. (2006, June 1). *Restitucija imovine: Crkva na državnom tasu.* [online] Vreme. BROJ 804. Accessed Mar 27, 2022, from https://www.vreme.com/vreme/crkva-na-drzavnom-tasu/

Manastiri u Srbiji. (2022). *Južna srbija: Manastir banjska.* Accessed Mar 04, 2022, fromhttps://manastiriusrbiji.com/manastir-banjska/

Mandić, I. (2014, October 29). *Igor Mandić: Narodi eks-Ju u magarećoj klupi istorije.* [online]. Blic. Accessed Mar 27, 2022, from https://www.blic.rs/vesti/drustvo/igor-mandic-narodi-eks-ju-u-magarecoj-klupi-istorije/peqlbbe

Marković, P. (2005). Gastarbeiters as the factor of modernization in Serbia. *Istorija, 20. veka,* (2), 145–163.

Mazzucchelli, F. (2012). What remains of Yugoslavia? From the geopolitical space of Yugoslavia to the virtual space of the web Yugosphere. *Social Science Information, 51*(4), 631–648.

Mićić, V., & Zeremski, A. V. (2011). Deindustrijalizacija i reindustrijalizacija privrede Srbije. *Industrija, 39*(2), 51–68.

Mihajlović, T. (2013). Demographic ageing of the old population of Serbia. *Zbornik radova-Geografski fakultet Univerziteta u Beogradu, 61*, 73–102.

Mijailović, A. (2021). *Beogradizacija Srbije, i gde su tu ostali gradovi?.* [online] Zapad Info, redakcija Šabac. Accessed Mar 27, 2022, from https://zapad-info.rs/1547/

Miladinović, Z. (2019, November 6). *Avio-kompanije obustavljaju i komercijalne linije iz Niša* [online]. Danas. Accessed Dec 12, 2021, from https://www.danas.rs/vesti/ekonomija/avio-kompanije-obustavljaju-i-komercijalne-linije-iz-nisa/

Milićević, S., & Petrović, J. (2017, June). Tourist products in the function of improving competitiveness of Serbia as a tourist destination. In *Tourism International Scientific Conference Vrnjačka Banja-TISC* (Vol. 2, No. 1, pp. 167-183).

Miljković, A. A. (2004). Osuda komunističkog režima, uslov rehabilitacije žrtava komunističkog terora. *Hereticus-Časopis za preispitivanje prošlosti, 02*, 37–46.

Mills, R. (2011). Book review: Vesna drapac, constructing Yugoslavia: A transnational history., Palgrave Macmillan: Basingstoke, 2010; 335 pp., 5 maps; 9780333925546. *European History Quarterly, 41*(3), 519–521.

Minehan, P. B. (2006). *Civil war and world war in Europe: Spain, Yugoslavia and Greece, 1936-1949*. Palgrave Macmillan.

Mondo/BETA. (2015, December28). *Beograd i Vojvodina–dve trećine Srbije.* [online]. Mondo/BETA. Accessed Feb 12, 2022, from https://mondo.rs/Info/Ekonomija/a861086/Beograd-i-Vojvodina-ostvare-dve-trecine-BDP-a-Srbije.html

Monro, A. (2017). *Non-aligned movement (NAM)*. Britannica Academic.

MPC Properties. (2019, October 1). *Usce tower two*. MPC Properties. Accessed Mar 27, 2022, from https://mpcproperties.rs/portfolio-items/usce-tower-two/

Nedeljković, M., Vukonjanski, J., Nikolić, M., Hadžić, O., & Šljukić, M. (2018). A comparative analysis of Serbian national culture and national cultures of some European countries by GLOBE project approach. *Journal of the Geographical Institute" Jovan Cvijic", SASA, 68*(3), 363–382.

Nikolić, I. D. (2014). Efekti privatizacije na performanse industrijskih preduzeća u Srbiji. *Универзитет у Београду.*

Nikolić, L. (2019). Politika zaštite konkurencije u Srbiji-dometi i ograničenja. *Zbornik radova Pravnog fakulteta u Nišu, 58*(84), 151–170.

Njegovan, Z. (Ed.). (2017). *Teorija i praksa agrara u istorijskoj perspektivi.* C. Z. A. ISTORIJU-CAI.

Nova Video. (2021). Obradović: Ubija nas "beogradizacija" Srbije. Accessed Dec 21, 2021, from https://www.youtube.com/watch?v=ZKEp4JqTkQA

Novakovic, N. (1994). Disintegration of the Yugoslav working class. In B. Jakšić (Ed.), *Philosophy and society* (Vol. VI, p. 309). Institut za filozofiju i drustvenu teoriju.

Novaković, N. G. (1993). Uloga menadžerske i tehnokratske elite u privrednom razvoju Jugoslavije. u: Bolčić Silvano. *Milošević Božo i Stanković Fuada [ur.] Preduzetništvo i sociologija, Novi Sad: Sociološko društvo Srbije-Ogranak za Novi Sad, str*, 135, 136.

Novaković, S. (1892). *Manastir Banjska: zadužbina Kralja Milutina u srpskoj istoriji.* Državna Štamparija Kraljevine Srbije.

O'Rourke, K. H., & Williamson, J. G. (2017). *The spread of modern industry to the periphery since 1871* (p. 410). Oxford University Press.

Palić, S. (2011, July 5). *Ko poseduje srbiju? tajkuni vlasnici najplodnije zemlje u srbiji.* [online] Blic. Accessed Mar 27, 2022, from http://web.arhiv.rs/develop/vesti.nsf/feaee540dc011162c12 56e7d0032cb98/058c75f8f04635eec12578c3006c1967?OpenDocument

Paull, J. (2020). Translations of Rudolf Steiner's agriculture course (Koberwitz, 1924): The seminal text of biodynamic farming and organic agriculture. *International Journal of Environmental Planning and Management, 6*, 94–97.

Paunović, P. (2021, April 14). *Srbija ima litijum–novu naftu ekonomije 21. veka: Zašto umesto prilike možemo da zamislimo katastrofu?*. [online]. startit.rs portal. Accessed Mar 27, 2022, from https://startit.rs/srbija-ima-litijum-novu-naftu-ekonomije-21-veka-zasto-umesto-prilike-gradani-vide-ekolosku-katastrofu/

Pavkovic, A. (2000). *The fragmentation of Yugoslavia: Nationalism and war in the Balkans.* Springer.

Pavlowitch, S. (2008). *Hitler's new disorder: The second world war in Yugoslavia.* Hurst & Company Limited.

PC Press. (2020, November 30). Microsoft Razvojni centar u Beogradu (MDCS) dobija najveće pojačanje do sada. *PC Press.* Accessed Mar 27, 2022, from https://pcpress.rs/microsoft-razvojni-centar-u-beogradu-mdcs-dobija-najvece-pojacanje-do-sada/

Pejanović, R. (2011). O Agrarnim krizama., *Letopis naučnih radova. Godina 35, broj I, strana 5-16.* Novi Sad: Poljoprivredni fakultet.

Пејовић, Б. (2019, February 10). *На Пршутијади у Мачкату месо из Гренобла и Парме.* [online] Politika. Accessed Feb 28, 2022, from https://www.politika.rs/sr/clanak/422421/ Srbija/Na-Prsutijadi-u-Mackatu-meso-iz-Grenobla-i-Parme

Penev, G. (1998). Trends of demographic ageing and feminization of aged population in Serbia. *Stanovnistvo, 36*(3-4), 43–60.

Penev, G. (2014). Population ageing trends in Serbia from the beginning of the 21st century and prospects until 2061: Regional aspect. *Zbornik Matice srpske za drustvene nauke, 148,* 687–700.

Percival, A (2019,March 15). *Your meat will soon come from algae (and it will be delicious).* [online] *livekindly.* Accessed Dec 17, 2021, from https://www.livekindly.co/food-of-future-made-from-algae/

Pijanović, P. (2012). Everyday culture and private life in Serbia during late 19th and early 20th century. *Novi Sad: Serbian studies research: 71.*

Pivac T D., Blešić I, Anđelović A, Stamenković I, & Besermenji V S (2011). Attitudes of local population about maintaining the economic and touristic event "slaninijada" (bacon fest) in kacarevo. *Conference on contemporary trends in tourism and hospitality, Novi Sad, 2011.*

Plavšin, A. (2013). Privatizacija u "privatnoj" Srbiji. In *Danubius.* Peace and Development (ECPD).

Polajžer, T. (2013). *Jugosfera: fikcija ali resničnost? (Doctoral dissertation).* University of Ljubljana.

Popović-Filipović, S. (2018). Elsie Inglis (1864-1917) and the Scottish women's hospitals in Serbia in the great war. Part 2. *Srpski Arhiv za Celokupno Lekarstvo, 146*(5-6), 345–350.

Posavec, D., Tomiša, M., & Šarkanj, B. (2022). Importance of labelling biodynamic product packaging in Croatia. *Tehnički glasnik, 16*(1), 60–66.

Poslovni.hr & HINA. (2013, February 27). Zaštićeni slavonski kulen mora biti proizveden od svinje iz Slavonije. [online] *Zagreb: Poslovni.hr & HINA.* Accessed Mar 27, 2022, from https://www. poslovni.hr/hrvatska/zasticeni-slavonski-kulen-mora-biti-proizveden-od-svinje-iz-slavonije-232089

Првуловић, В. (2019, April 2). *Аутошовинизам у Србији.* [online] Београд: Политика. Accessed Dec 13, 2021, from https://www.politika.rs/scc/clanak/426053/Pogledi/ Autosovinizam-u-Srbiji

PSD. (2012, January 10). *Deseci tisuća hektara oranica promijenili vlasnike što je izazvalo ogorčenje seljaka buna u vojvodiNI: "Hrvatski tajkuni pokupovali srpsku zemlju".* [online]. Split: Slobodna Dalmacija. Accessed Mar 27, 2022, from https://slobodnadalmacija.hr/vijesti/ svijet/buna-u-vojvodini-hrvatski-tajkuni-pokupovali-srpsku-zemlju-153935

Rakić-Vodinelić, V. (2007). Uvod u tranzicionu pravdu: osnovni pojmovi. *Genero: časopis za feminističku teoriju i studije kulture, (10-11), 11-31.*

Rakić, Z. V. (2018). *Odnos tradicije i modernizacije u oblastima priključenim Srbiji posle Berlinskog kongresa; Leskovac (1878-1941).* Универзитет у Београду.

Ramet, S. P. (1996). Nationalism and the 'idiocy' of the countryside: The case of Serbia. *Ethnic and Racial Studies, 19*(1), 70–87.

Rastović, A. P. (2004). Мери Дарам о Србима. *Београд: Историјски часопис, 51,* 124–150.

Ratner, P. (2019, September 9). Swedish scientist advocates eating humans to combat climate change. *Big Think.* Accessed Sep 5, 2021, from https://bigthink.com/the-present/swedish-scientist-eating-humans-climate-change/

Републички завод за статистику/РЗС. (2022a). *Годишњи национални рачуни*. Београд: Републички завод за статистику, Република Србија. [online] Accessed Mar 27, 2022, from https://www.stat.gov.rs/sr-Cyrl/oblasti/nacionalni-racuni/godisnji-nacionalni-racuni

Републички завод за статистику/РЗС. (2022b). *Званична интернет презентација*. Београд: Републички завод за статистику, Република Србија. [online] Accessed Mar 27, 2022, from https://www.stat.gov.rs/

Републички завод за статистику/РЗС. (2022c). *Регионални рачуни*. Београд: Републички завод за статистику, Република Србија. [online] Accessed Mar 27, 2022, from https://www.stat.gov.rs/sr-Cyrl/oblasti/nacionalni-racuni/regionalni-podaci

Републички завод за статистику/РЗС. (2022d). *Регистар индустријских зона*. Београд: Републички завод за статистику, Република Србија. [online] Accessed Mar 27, 2022, from https://www.stat.gov.rs/vesti/20201123-registar-industrijskih-zona/

Resnik, D. B. (2015). Retracting inconclusive research: Lessons from the Séralini GM maize feeding study. *Journal of Agricultural and Environmental Ethics, 28*(4), 621–633.

Reuters. (2017a, February 23). *Central European leaders to call for EU action against food "double standards"*. [online]. Reuters.com. Accessed Sep 5, 2021, from https://www.reuters.com/article/slovakia-food-quality-idUSL8N1G87BX

Reuters. (2017b, March 1). *East Europeans decry "double standards" for food, seek change to EU law*. [online]. Reuters.com. Accessed Sep 5, 2021, from https://www.reuters.com/article/centraleurope-food-idUSL5N1GD4N4

Reuters. (2017c, February 27). *Hungary, Slovakia to ask EU to take action against food "double standards"*. [online]. Reuters.com. Accessed Sep 5, 2021, from https://www.reuters.com/article/hungary-slovakia-food-quality-idUSL5N1GC253

Rosmino, C. (2021, May 25). Is microalgae the sustainable food of the future? *Euronews*. Accessed Mar 28, 2022, from https://www.euronews.com/green/2021/05/25/is-microalgae-the-sustainable-food-of-the-future

Roze, J (2021, November 17). *Litijum u Srbiji: belo zlato ili prokletstvo?*. [online]. DW. Accessed Mar 27, 2022, from https://www.dw.com/sr/litijum-u-srbiji-belo-zlato-ili-prokletstvo/a-59845710

RTV. (2021, November 25). *U Beogradu otvoren 15. Sajam etno hrane i pića*. Novi Sad: Radio-televizija Vojvodine. Accessed Mar 27, 2022, from https://www.youtube.com/watch?v=qSl9cf6Nu3U

Sanborn, J. A. (2014). *Imperial apocalypse: The great war and the destruction of the russian empire*. Greater War.

Sapir, A. (1980). Economic growth and factor substitution: What happened to the Yugoslav miracle? *The Economic Journal, 90*(358), 294–313.

Šarić, L. Ć., Šarić, B. M., Mandić, A. I., Hadnađev, M. S., Gubić, J. M., Milovanović, I. L., & Tomić, J. M. (2016). Characterization of extra-hard cheese produced from donkeys' and caprine milk mixture. *Dairy Science & Technology, 96*(2), 227–241.

Savić, M. (2001). 'Event and narrative (on judging the character of contemporary events)', *revolution and order* (Institute for Philosophy and Social Theory, Belgrade, editor: Ivana Spasić, Milan Subotić), pp. 11–20.

Séralini, G. E., Cellier, D., & de Vendomois, J. S. (2007). New analysis of a rat feeding study with a genetically modified maize reveals signs of hepatorenal toxicity. *Archives of Environmental Contamination and Toxicology, 52*(4), 596–602.

Séralini, G. E., Clair, E., Mesnage, R., Gress, S., Defarge, N., Malatesta, M., . . . de Vendômois, J. S. (2014a). Republished study: Long-term toxicity of a roundup herbicide and a roundup-tolerant genetically modified maize. *Environmental Sciences Europe, 26*(1), 1–17.

Séralini, G. E., Clair, E., Mesnage, R., Gress, S., Defarge, N., Malatesta, M., ... & De Vendômois, J. S. (2012). *Retracted: Long term toxicity of a Roundup herbicide and a Roundup-tolerant genetically modified maize*.

Séralini, G. E., Mesnage, R., Clair, E., Gress, S., De Vendômois, J. S., & Cellier, D. (2011). Genetically modified crops safety assessments: Present limits and possible improvements. *Environmental Sciences Europe, 23*(1), 1–10.

Séralini, G. É., Mesnage, R., Defarge, N., & de Vendômois, J. S. (2014b). Conclusiveness of toxicity data and double standards. *Food and Chemical Toxicology: An International Journal Published for the British Industrial Biological Research Association, 69*, 357–359.

Séralini, G. E., Mesnage, R., Defarge, N., & Spiroux de Vendômois, J. (2014c). Conflicts of interests, confidentiality and censorship in health risk assessment: The example of an herbicide and a GMO. *Environmental Sciences Europe, 26*(1), 1–6.

Séralini, G. E., Mesnage, R., Defarge, N., Gress, S., Hennequin, D., Clair, E., . . . De Vendômois, J. S. (2013). Answers to critics: Why there is a long term toxicity due to a roundup-tolerant genetically modified maize and to a roundup herbicide. *Food and Chemical Toxicology, 53*, 476–483.

Ševarlić, M., & Nikolić, M. (2010). Stavovi direktora zadruga o zadružnim savezima u Srbiji. In *Tematski zbornik Agrarna i ruralna politika u Srbiji–3–Održivost agroprivrede, zadrugarstva i ruralnog razvoja* (pp. 75–92). Društvo agrarnih ekonomista Srbije i Poljoprivredni fakultet Univerziteta u Beogradu.

Sič, M. (2013). Заштита животне средине: заштита здравља и безбедна храна–ЕУ и Србија. *Zbornik radova Pravnog fakulteta u Novom Sadu, 47*(1).

Sinkević, I. (2022). 2 Serbian Royal Mausolea. *Sense, matter, and medium*, 57. In eclecticism in late medieval visual culture at the crossroads of the Latin, Greek, and Slavic Traditions, de Gruyter.

SLANINIJADA. (2022). *Slaninijada*. Accessed Mar 27, 2022, from http://slaninijada.org/

Šljukić, S. (2006). Agriculture and the changes of the social structure: The case of Serbia. *Sociologija, 48*(2), 137–148.

Šljukić, S. L. (2004). The social structure in the former socialist countries: Transition or transformation? *Sociološki pregled, 38*(1-2), 267–279.

Slobodna Bosna. (2017, March 6). *Eliminacija "duplih standarda": Poziv zemalja "Višegradske grupe" EU na borbu protiv razlika u kvaliteti hrane*. [online]. Slobodna Bosna: Nezavisni internet portal. Accessed Sep 5, 2021, from https://www.slobodna-bosna.ba/vijest/47605/eliminacija_duplih_standarda_poziv_zemalja_visegradske_grupe_eu_na_borbu_protiv_razlika_u_kvaliteti_hrane.html

Stančić, I., & Stančić, B. (2013). Animal genetic resources in Serbia. *Slovak Journal of Animal Science, 46*(4), 137-140.

Štetić, S., & Pavlović, S. (2014). Tourist events important economic and cultural generators of sustainable development in Serbia. *Sustainable agriculture and rural development in terms of the republic of Serbia strategic goals realization within the Danube region. Rural development and (un) limited resources*, 870.

Štiblar, F. (2019). External indebtedness of Yugoslavia and its federal units. *The transition from command to market economies in East-central Europe: The vienna institute for comparative economic studies Yearbook Iv*, 159.

Stojanović, D. (2020). Imagining the zadruga. Zadruga as a political inspiration to the left and to the right in Serbia, 1870-1945. *Revue des études slaves, 91*(XCI-3), 333–353.

Stojanović, S. Đ., & Milošević, S. (2017). Management of animal genetic resources in Serbia-current status and perspective. *Slovak Journal of Animal Science, 50*(4), 154–158.

Stojanović, V., Milić, D., Obradović, S., Vanovac, J., & Radišić, D. (2021). The role of ecotourism in community development: The case of the zasavica special nature reserve, Serbia. *Acta Geographica Slovenica, 61*(2), 171–186.

Stojilković, J., & Devedžić, M. (2010). Retirees and employees ratio in the context of demographic ageing in Serbia. *Zbornik Matice srpske za drustvene nauke, 131*, 177–186.

Stošić, I. (2014). Uticaj postojećeg modela privatizacije na stanje industrijskog sektora Srbije. *U: "Izazovi i perspektive strukturnih promena u Srbiji: Strateški pravci ekonomskog razvoja i usklađivanje sa zahtevima EU" (OI 179015), finansirano od strane Ministarstva za prosvetu, nauku i tehnološki razvoj Republike Srbije.*

Subotica.com. (2022) *Zvanična Internet prezentacija.* Accessed Dec 17, 2021, from https://www.subotica.com/tag/terras

Suddaby, R. (2006). From the editors: What grounded theory is not. *Academy of Management Journal, 49*(4), 633–642.

Sudžum, M. (2021, July 19) 'Umesto farmera stvorili nadničare: Sela su prazna i zbog tranzicijskih grešaka sa početka veka'. *Nedeljne novine.* Accessed Mar 27, 2022, from https://backapalankavesti.com/drustvo/privreda/umesto-farmera-stvorili-nadnicare/

Svetličić, J. (2022). *U hodu-Krug dvojke.* RTS.RS. Accessed Mar 27, 2022, from https://www.rts.rs/page/tv/sr/story/2688/rts-zivot/4751025/u-hodu-krug-dvojke.html

Tadić, L. J. (1993). 'Broz Ruled By Series of Coup D' Etat', *Philosophy and Society (Institut za filozofiju i drustvenu teoriju,* editor: Božidar Jakšić), *4(IV),* pp. 200–201.

Tatalović, Ž (2022, January 9). *Iskopavanje litijuma u Srbiji i Nemačkoj se potpuno razlikuje.* [online]. N1 Srbija. Accessed Mar 27, 2022, from https://rs.n1info.com/vesti/iskopavanje-litijuma-u-srbiji-i-nemackoj-se-potpuno-razlikuje/

Tayaben, J. L. (2018). Reflecting on the book of Alvesson and Skoldberg 'Reflexive methodology: New insights and its importance in qualitative studies. *Qualitative Report, 23*(10).

Teng, X., Zhang, M., & Mujumdar, A. S. (2021). 4D printing: Recent advances and proposals in the food sector. *Trends in Food Science & Technology, 110,* 349–363.

Terras Subotica. (2022). *Zvanična Internet prezentacija.*

Timpf, K. (2019, September 13). Scientist suggests climate change could pave the way for cannibalism. [online]. *National Review.* Accessed Sep 5, 2021, from https://www.nationalreview.com/2019/09/scientist-suggests-climate-change-could-pave-the-way-for-cannibalism/

Todorović, M. (2004). *The underground music scene in Belgrade, Serbia: A multidisciplinary study (doctoral dissertation).* Brunel University.

Todorovic, M. (2016). *Rethinking strategy for creative industries: Innovation and interaction.* Routledge.

Тодоровић, В. (2021). *Model primene identifikacionih tehnologija u procesu praćenja prehrambenih proizvoda.* Докторска дисертација. Београд.

Tomanović, S., & Ignjatović, S. (2006). The transition of young people in a transitional society: The case of Serbia. *Journal of Youth Studies, 9*(3), 269–285.

Tomić, I. D. (2020). Prilog proučavanju životnog standarda i društvenog položaja radnika uoči privredne reforme 1965. *Istorija 20. veka,* (1), 249-272.

Trpeski, P., Šmelcerović, M., & Jarevski, T. (2021). The impact of lithium mines on the environment. *Knowledge-International Journal, 46*(3), 455–458.

Turinek, M., Grobelnik-Mlakar, S., Bavec, M., & Bavec, F. (2009). Biodynamic agriculture research progress and priorities. *Renewable Agriculture and Food Systems, 24*(2), 146–154.

Turudija Živanović, S. L. (2015). *Organizacija proizvodnje i prerade lekovitog i aromatičnog bilja u Srbiji* (Doctoral dissertation, Univerzitet u Beogradu-Poljoprivredni fakultet).

Untappd.com. (2022, February 26). *Monastery Beer–Manastirsko Banjsko PivoЧ Manastir Banjska, Pale Ale–International.* Accessed Mar 04, 2022, from https://untappd.com/b/manastir-banjska-monastery-beer-manastirsko-banjsko-pivo/2453653

Varvara, R. A., Szabo, K., & Vodnar, D. C. (2021). 3D food printing: Principles of obtaining digitally-designed nourishment. *Nutrients, 13*(10), 3617.

Vidal, J. (2012, September 28). *Study linking GM maize to cancer must be taken seriously by regulators.* [online]. The Guardian. Accessed Dec 17, 2021, from https://www.theguardian.com/environment/2012/sep/28/study-gm-maize-cancer

Visit Subotica (2021) *17. Međunarodni festival organskih proizvoda BIOFEST.* Accessed Dec 17, 2021, from https://visitsubotica.rs/en/event/17-medjunarodni-festival-organskih-proizvoda-biofest-2021/

Vujičić, M., & Ristić, L. (2016). Branding strategy and positioning of Gruža microregion in Serbian tourism industry. *Megatrend revija, 13*(1), 241–264.

Vujović, S. (2014). Socioprostorni identitet Beograda u kontekstu urbanog i regionalnog razvoja Srbije. *Sociologija, 56*(2), 145–166.

Vukić, M., Kuzmanović, M., & Gligorijević, M. (2014). Attitudes of tourists towards gastronomic tourism in Serbia: Empirical research. In *IV International Symposium Engineering Management and Competitiveness* (pp. 315–320).

Vukov, T. (2013). Seven thesis on neobalkanism and ngoization in transitional Serbia. In A. Choudry & D. Kapoor (Eds.), *NGOization: Complicity, contradiction and prospects* (pp. 163–184). Zed Books.

Vuković, S. (2011). Економски узроци разбијања/распада Југославије. *Београд: Социолошки преглед, 45*(4), 477–504.

Wedgwood, R. (2002). Gallant delusions. *Foreign Policy, 132*, 44–46.

Wilson, A., Anukiruthika, T., Moses, J. A., & Anandharamakrishnan, C. (2021). Preparation of fiber-enriched chicken meat constructs using 3D printing. *Journal of Culinary Science & Technology*, 1-12.

Woodward, S. (2003). The political economy of ethno-nationalism in Yugoslavia. *Socialist register, 39*.

Yarashevich, V., & Karneyeva, Y. (2013). Economic reasons for the break-up of Yugoslavia. *Communist and Post-Communist Studies, 2*(46), 263–273.

Завод за интелектуалну својину. (2022). *Statistics, studies and overview of registered geographical indications*. The intellectual property office of the Republic of Serbia https://www.zis.gov.rs/en/rights/indications-of-geographical-origin/statistics/

Задужбине Немањића. (2022). *Манастир Бањска*. Accessed Mar 04, 2022, from https://www.zaduzbine-nemanjica.rs/Banjska/index.htm. Accessed: 04/03/2022.

Zdravković, A., Domazet, I., & Nikitović, V. (2012). Impact of demographic ageing on sustainability of public finance in Serbia. *Stanovnistvo, 50*(1), 19–44.

Zec, M., & Radonjić, O. (2012). Ekonomski model socijalističke Jugoslavije-saga o autodestrukciji. *Sociologija, 54*(4), 695–720.

Zubović, J., & Domazet, I. (2008). Organic food market development: The case of Serbia. *Part of Research project No.159004, Ministry of Science and Technological Development of Republic of Serbia: "The Integration of Serbian Economy into the EU–Planning and Financing of Regional and Rural Development and Enterprise Development Policy"*. Institute of Economic Sciences.

Dr Milan Todorovic contributes a singular combination of skill, experience, expertise, and interest. Early involvement in science including a qualification in Laser Physics continues to inform his research and practice as can be evidenced from his 2016 monograph on Creative Industry Strategies (Routledge). Classically trained in Fine Arts, Milan also practices across visual and audio-visual media and nowadays primarily focuses on figurative painting and drawing. This includes the 2014 group show "A Europe for all, by all" in London, organised by the *European Greens London* in which he explored the notions of genetic modification in visual art works and *fictional science* narratives. He continues to paint, with ethics a major theme in his art. Worthy of mention are also his *Sonic Painting* and video film projects.

As a young entrepreneur he made a lasting impact on the music, media, and art scenes in his native Belgrade (Serbia). Part of Milan's early creative business legacy is explored in his multidisciplinary doctoral study of the city's creative scenes, combining cultural strategy, ethnomethodology, semiotics, and political economy (2004). Milan also wrote the UK's very first MA research addressing the challenges facing the then emerging matter of internet music and strategic difficulties the new medium imposed on practitioners (1998/99).

He has been a university academic since 1999, teaching predominantly in the areas of Creative and Cultural Industries, Music and Media Management, Marketing and Creative Enterprise. Milan nurtured an interest in Sustainability and Business Ethics since the early days of his business career, engaging with the subject as a researcher in recent years (Springer, Routledge). Looking at *Higher*

Education as a Creative Industry in its own right, his research includes critical ethical debates around the implementation of change in curricular design. As a course leader in his field, Milan engages in practical educational innovation.

His research interests in Sustainability are aligned with his creative practice and experience: post-industrial societies, genetically modified organisms, extreme utilitarian and neoliberal outlooks, media ethics, health strategies, food and agriculture for the new millennium, the internet/digital impact on everyday life, entrepreneurial drives for innovation both disruptive and incremental, corporatisation of leisure and the Self, personal choice and the position of individuals in strategic social change, political economy of hunger, resources as rights and utilities, and consumerism in marketing/PR.

His multidisciplinary projects and methodologies have long advocated the bridging of artificial boundaries of *quantitative and qualitative, science and art, belief and reason.*

Sustainability Challenges and the Way Forward in the Tea Industry: The Case of Sri Lanka

Y. Alahakoon, Mahendra Peiris, and Nuwan Gunarathne

Abstract

In the global food and beverage market, tea as a natural drink represents the non-alcoholic beverages segment. The tea industry in the tea-growing developing countries such as Sri Lanka plays a crucial role in earning foreign exchange, providing employment opportunities for women, and acting as a custodian of the sensitive ecosystems. Despite the economic, environmental, and social significance, the tea industry in Sri Lanka (that produces world-famous Ceylon tea) has faced many challenges from various fronts. However, these challenges and the responses of the tea industry have not been systematically explored. Therefore, this chapter explores the sustainability challenges under economic, environmental, and social dimensions while discussing the strategies adopted by the Ceylon tea industry. It then presents how the tea plantation companies in Sri Lanka have adopted various measures to address them, including cost-saving initiatives, climate-smart agricultural practices, mechanisation of operations, new product development, community engagement and development, and conservation of biodiversity. While some of these measures have shown positive impacts, many of the Ceylon tea industry's sustainability challenges, such as the effects of climate change, declining land productivity, increase in production cost, low-value addition of tea, social issues, and unequal profit distribution along the global value chain still remain partially solved. The Ceylon tea industry needs to focus on resolving these issues through multi-stakeholder engagement, innovations, and better marketing and value-added product development.

Y. Alahakoon · N. Gunarathne (✉)
University of Sri Jayewardenepura, Nugegoda, Sri Lanka
e-mail: nuwan@sjp.ac.lk

M. Peiris
Maskeliya Plantations PLC, Maharagama, Sri Lanka

271

Keywords

Central highlands · Climate-smart agriculture · Food and beverage · Plantation companies · Sri Lanka · Tea industry

1   Introduction

Food and drinks have been an essential requirement since the origin of human existence. Since then, they have come a long way towards the contemporary discussions on food and beverage as an industry contributing to the world economy. Major events in world history, from the industrial revolution through the World Wars to globalisation, have all contributed to the dynamics of this industry (Risch, 2009). For instance, begun initially as production by people for self-consumption and later, for local consumption, today, the food and beverage industry has become a global business with the ever-growing human needs and advancing technology. Furthermore, this industry 'has major implications for sustainability, such as the fulfilment of basic human needs, generation of employment opportunities and livelihoods, impacts on the natural environment through its use of various resources, and dealing with poverty' (Gunarathne et al., 2018, p. 291).

The food and beverage industry is currently a growing global economic segment with an anticipated 7% growth rate in the post-COVID-19 context (GlobeNewsWire, 2020). With this growth rate, it is expected that the global food and beverage market would reach $7527.5 billion in 2023. In the global market, Asia-Pacific is considered the major region in terms of consumer spending with a 42% share, while Africa has the smallest percentage (Cushman & Wakefield, 2017). The Coronavirus outbreak has negatively impacted the overall industry in 2020 from the disturbed supply chains due to border closures worldwide. This impact is expected to continue in 2021.

The global food and beverage market is fragmented into several sections as alcoholic beverages, non-alcoholic beverages, cereals, bakery products, confectionery, frozen fruits and vegetables, meat, syrup, spice, oil and general food, food pet, and tobacco products (Chandrasekaran et al., 2015; Maxime et al., 2006). In the non-alcoholic beverages segment, natural drinks, such as tea (*tea plant is botanically named as Camellia sinensis and the beverage prepared by brewing dried tea leaves is technically called as "Camellia tea"*), have witnessed an increasing demand, particularly with the emerging trends in herbal tea and organic tea consumption. For instance, the Organic Tea Global Market Report (2020) predicts that the organic tea market will grow by 7.27% annually due to the increasing health concerns of consumers worldwide after the COVID-19 pandemic experience. Apart from the health concerns, this growth could also be driven by the positive attitude of the socially responsible behaviours of the tea manufacturers (Mzembe & Lindgreen, 2018). Educated consumers are concerned about the environment and labour well-being of the manufacturing processes of the products they consume. Hence, the industry's sustainability practices and the industry's labour well-being have recently become positive marketing aspects within the tea industry.

Tea, commercially grown in over 35 countries, is the most widely consumed and manufactured drink globally (Chang, 2015). Sri Lanka is amongst the world's top four tea producing and exporting countries, competing in the list behind China, India, and Kenya (Forum for the Future, 2014). The tea industry represents one of the oldest industries in Sri Lanka, with its origin traced back to the mid-1800s when the British rulers first introduced the tea plant to the island during the colonial period. During this period, tea grown in Sri Lanka (then Ceylon) was taken in bulk to London auctions for trading (Kasturiratne, 2008). Even after the country's independence in 1948, for a considerably long period since then, exports took the form of bulk tea, without any value additions done within the country. In more recent times, the industry has taken steps to export value-added tea; however, the value-added segment's contribution still lies below that of the bulk tea exports, which amounts to 60% of the total tea exports of Sri Lanka; a signifier of the lack of innovative capabilities of the industry (Pilapitiya et al., 2020).

The tea industry being a higher contributor to Sri Lanka's economy in terms of foreign exchange earnings and employment has been facing numerous sustainable challenges in myriad facets. Tea plantations in Sri Lanka, which are historically situated in the country's highlands, heavily contribute to soil erosions. Further, the energy consumption of tea processing facilities is high, which leads to environmental issues such as global warming. Besides, social problems faced by the labourers, such as occupational health hazards, malnourishment, and poor living conditions, are frequent (Kamalakkannan et al., 2020). Apart from these environmental and social issues, the industry is currently facing numerous economic challenges as well, as the output of the sector has experienced a downward trend since 2017 (Central Bank of Sri Lanka, 2019) (see Sect. 3 for more details of these challenges).

In response to these growing sustainability challenges, the Sri Lankan tea industry has pursued many strategies to address its economic, environmental, or social problems. They include innovative agriculture and land management practices, herbicide-free weed management techniques, soil conservation practices, biodiversity conservation approaches, new product and market development, and actions aimed at uplifting the social development of the plantation workers. Although the tea industry's sustainability challenges have been previously examined to some extent, the strategies adopted at the field level have not been systematically explored so far from a sustainability dimension. This chapter thus aims to discuss the sustainability challenges, strategies adopted, and their effects on the tea industry in Sri Lanka.

The rest of the chapter is organised as follows: Section 2 provides an overview of the tea industry by explicitly highlighting the economic, social, and environmental importance. Section 3 discusses the sustainability challenges the Ceylon tea industry faces, followed by a section on the strategies adopted to overcome these challenges. The last section provides the conclusions and the way forward.

2   The Tea Industry in Sri Lanka

2.1   Tea Production and Sales in Sri Lanka

Tea is not a seasonal product, as it is grown throughout the year under tropical climatic conditions in Sri Lanka. However, its yield can fluctuate periodically due to variations in other climatic conditions. Depending on the climate and the elevation in which the tea plantations are situated, Ceylon tea is categorised into three tea-growing regions in the country: low-grown, medium-grown, and high-grown teas (see Table 1). This variation gives unique features and tastes to tea produced in each respective area. The high-grown teas, especially Dimbula and Nuwara Eliya areas, are the most sought-after tea by the blenders in the export market (Sri Lanka Export Development Board, 2019).

Sri Lanka is the world's largest producer and exporter of orthodox black tea. Besides, other black tea types such as Orthodox—Rotorvane tea and CTC (Cut Tear and Curl) tea are produced in considerable quantities in the country. The country's most-produced tea categories are several grades of whole leaf, wiry leaf, broken leaf, fannings, and dust grades. In producing the orthodox tea, withered fresh tea shoots are subjected to rupture and twisting by a rolling process achieved through a rotary movement. Once the tea leaves are damaged/crushed, chemicals naturally present in tea leaves termed "polyphenols" undergo an inborn enzymatic oxidation process, with the influence of polyphenol-oxidase enzyme. This process is usually called the fermentation of tea leaves, where the colour strength and flavour of tea are developed. Then the fermented tea leaves are put into a drying process (Sri Lanka Export Development Board, 2019). In green tea manufacturing, the above-mentioned natural fermentation process is ceased by destroying the polyphenol-oxidase enzyme by exposing tea leaves to a heat treatment prior to rolling. Thus, green tea retains raw polyphenols, while black tea contains mellow and more palatable forms of oxidised polyphenols.

Followed by the said black tea types, Sri Lanka also produces a variety of different teas such as semi-fermented tea, bio tea, and flavoured tea, but in comparatively lower quantities. The production output of tea in these three major categories from January to November 2020 are: orthodox black tea = 226,691 metric tonnes,

Table 1 Geographical locations of tea growing in Sri Lanka

Tea type	Elevation (feet)	Geographical area of plantations
Low-grown	Below 2000	• Sabaragamuwa Province (Rathnapura)
		• Southern Province (Galle)
Mid-grown	2000–4000	• Central Province (Kandy)
High-grown	4000–6000	• Central Province (Dimbula)
		• Uva Province (Badulla)
		• Central Province (Nuwara Eliya)

Source: Adapted from Jayasinghe and Kumar (2019)

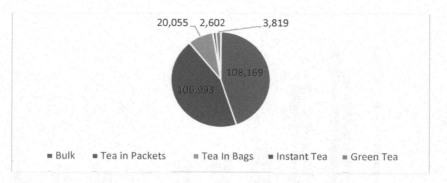

Fig. 1 Tea export categories in Sri Lanka in 2020. Source: Tea Exporters Association Sri Lanka (2020) (quantities are in metric tonnes)

CTC = 21,375 metric tonnes, and green tea = 1838 metric tonnes, respectively (Tea Exporters Association Sri Lanka, 2020).

The majority of tea production in Sri Lanka is exported to the global market. From January to November 2020, Sri Lanka has exported 241.6 million kilograms of tea in all its exporting categories. This export volume indicates a decrease from the total quantity shipped the previous year, which amounted to nearly 268.9 million kilos (Tea Exporter's Association Sri Lanka, 2020). The composition of the exports is tea packets, tea bags, tea in bulk, instant tea, green tea, and flavoured tea (see Fig. 1). Bulk tea is the largest export category, followed by tea in packets. The value-added products such as green tea, instant tea, flavoured tea, organic tea, iced tea, and ready-to-drink tea, as well as recent innovative products (e.g. tea-based soap, bath gel, shampoo, and cosmetic products), do not represent a major portion of the tea exports in Sri Lanka.

Most of the Ceylon tea is exported to the Middle East region and channelled to many other countries from there. Sri Lanka's major export markets are presented in Fig. 2.

In Sri Lanka, tea plantations are owned by two major types of commercial cultivators (i.e. the large-scale regional plantation companies and state-owned enterprises) and individual farmers who operate on a small scale. The output of both these sectors is supplied to collecting centres operated by the tea processors. Then the manufactured tea is auctioned in the Colombo Tea Auction, the largest tea auction globally. At the end of the local supply chain, some exporters and traders sell tea to overseas buyers and distributors before reaching the end consumers. A simplified global tea industry supply chain can be presented as follows (see Fig. 3).

Usually, a significant part of the blending, packing, and marketing activities are carried out by overseas buyers and distributors. These are usually the most lucrative stages of the global tea industry value chain (Bloomfield, 2020).

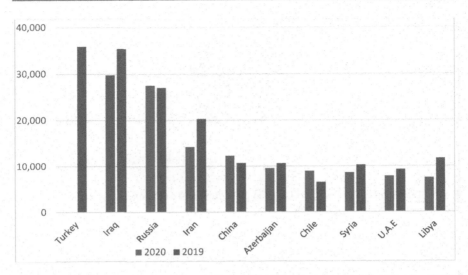

Fig. 2 Export markets of Sri Lanka's tea. Source: Tea Exporters Association Sri Lanka (2020) (quantities are in metric tonnes)

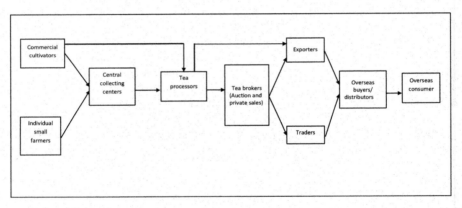

Fig. 3 A simplified global tea industry supply chain. Source: Adapted from Sri Lanka Export Development Board (2019)

2.2   Significance of the Tea Industry in Sri Lanka

Over an extended period, 'Ceylon tea' (tea produced in Sri Lanka) has earned a reputation for its high quality and demanded premium prices in the world markets. Like many other developing tea-growing countries, the tea industry forms an important economic sector in Sri Lanka's economy by providing a valuable source of employment and foreign exchange earnings (Gunarathne & Peiris, 2017). For instance, the tea industry contributes to over 65% of the country's agricultural revenue and 13.5% of the export revenue, while accounting for about 2% of GDP (Bloomfield, 2020). Further, this industry offers over one million direct jobs and 2.5 million indirect jobs (Central Bank of Sri Lanka, 2019). The sector also promises future growth prospects to Sri Lanka, as it is estimated that global tea consumption is gradually increasing at a rate of approximately 5% per year (Chang, 2015). However, as described in the next section, the Sri Lankan industry has a long way to develop in order to harness this potential.

The economic significance of the tea industry also extends, especially towards smallholder farmers. As depicted in the supply chain above, smallholder farmers are a vital player in the tea industry. For them, the tea industry provides opportunities for economic advancement and growth potential at regional levels. In Sri Lanka, almost over 78% of the green leaf processed in tea factories are produced and supplied by the tea smallholder sector (International Labour Organization, 2018). Therefore, the tea industry provides more employment opportunities locally and enhances the sustainability of their livelihoods. As the economic reports indicate, the average yield is much higher in the smallholder sector than in the large plantation sector (Central Bank of Sri Lanka, 2019).

Labour and employment opportunities are essential considerations within the tea industry. The tea industry value chain's initial stages are highly labour intensive, as advanced technologies are not widely used in the activities in this area; hence, it opens up many job opportunities for the local communities (Munasinghe et al., 2017). For instance, in both commercial plantations and well in the smallholder sector, plucking of the tea crop is done manually. The statistics state that 75–85% of the tea industry workforce comprises female workers (Kotikula & Solotarof, 2006). Hence, the industry enables women's participation and their contribution to economic activities.

Tea is grown over a land area exceeding 203,020 hectares within the island (Jayasinghe & Kumar, 2019). Amongst these, most of the large tea plantations are located in the Central Highlands of the island, which is also home to an extraordinarily diverse range of flora and fauna. Therefore, the tea industry is also crucial from an environmental perspective as well (Peiris & Gunarathne, 2022a). The Central Highlands of Sri Lanka is a critically important land area on the planet. Therefore, said landscape is declared a UNESCO World Heritage Site. For instance, ICUN has identified the country's Central Highlands as a 'biodiversity super-hotspot' in the world (International Union for Conservation of Nature, IUCN, United Nations Environment and World Conservation Monitoring Centre, UNEWCMC, 2017). With these natural rain forests receiving high rainfall, the Central Highlands is home to the rivers' water catchments that provide water for cultivations and other human activities all over the island (Wickramagamage, 1998). Therefore, the tea industry's agricultural practices, such as the use of fertiliser and chemicals and soil management, have a significant impact on the millions of people in this island nation.

The Central Highland has a tremendous social significance since many culturally, and religiously important sites are located here. For example, "Dalada Maligawa"

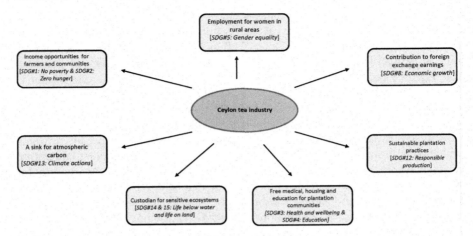

Fig. 4 Ceylon tea industry's contribution to sustainable development. Note: Although the tea industry contributes to many other SDGs, this figure explicates the main contribution only

(Temple of the Sacred Tooth Relic) can be considered as the religiously most significant site for the majority of Buddhist devotees of the country. Further, Sri Pada (Adam's Peak) is a sacred mountain peak venerated by people of all faiths in the country; therefore, it is important from religious and tourism perspectives. The region is also home to many other sites, which is important from a cultural and recreational perspective.

The tea industry in Sri Lanka contributes to the sustainable development of the country in numerous ways. Figure 4 presents the main contributions of the Ceylon tea industry towards sustainable development goals.

3 Sustainability Challenges in the Tea Industry in Sri Lanka

Over the last decade, the tea industry in Sri Lanka has faced many challenges both locally at the production levels and internationally at the buyer markets. From a sustainability dimension, these challenges can be analysed from economic, environmental, and social perspectives. In this chapter, although these challenges are discussed separately from the three different sustainability perspectives, practically, there are mostly intertwined.

One of the significant challenges the tea industry faces is the rising cost of production, mainly due to the continuous series of biannual wage hikes imposed connected to a collective agreement between the regional plantation companies and the worker trade unions. As a result, the worker benefit portion, including worker wages, statutory benefits, worker housing, health care, free water supply, childcare, and education, has gone up to 88% of Ceylon tea's production cost at present (Poholiyadde, 2018). This implies that the cost of all other inputs such as fertiliser, pesticides, energy, and machinery upkeep only count for less than12% of the total cost of production, at the plantation level. Despite the increase in wages, a majority of the workforce works only for 20–25 hours per week, with an effective plucking time of only 40% of the working time (Thibbotuwawa et al., 2019). Therefore, they reach a low productivity level of only 16-18 kg plucked tea per day whereas the actual ability is about 25–35 kg per day on average during a 6–7 hour working spell if the adequate crop is available in the field. As a result, it is identified that their wages are still below the average daily earning levels at the national level despite the recent wage hikes (Thibbotuwawa et al., 2019). Therefore, the tea plantation companies have attempted to address the living wage and current pay by providing various in-kind benefits such as free housing, medical services, and education. However, due to the low productivity levels and dwindling international tea prices, plantation companies alone cannot afford a further wage hike. The international tea brands, tea consumers and retailers in developed countries, international and national level tea buyers should support the local tea producers to provide a wage increase while ensuring the sustainability of the whole industry.

In addition to the increase in cost, another major challenge faced by the tea industry in Sri Lanka is the falling tea prices in the world market. Tea prices have decreased considerably since 2014 due to the broader global trend of declining commodity prices (Oxford Business Group, 2016a). These declining prices are also attributable to the low-quality tea and insufficient demand from Sri Lanka's major tea export markets. Although the government has subsidised the tea plantation companies as a temporary solution, some industry experts fear that falling tea prices will threaten the sustainability of the whole tea industry in Sri Lanka. The increased cost of tea production, together with the declining international prices, has led to severe economic, social, and environmental repercussions, such as lower profit margins, for producers and tea suppliers, inadequate investments in efficient technologies, low incomes for workers, and issues in soil quality management (Munasinghe et al., 2017). Not only do these affect the quality of the tea in the long run, but they also affect a range of actors in the supply chain and the economic activities reliant on the tea industry adversely.

Furthermore, the tea industry in the country has also started to witness the adverse impacts of climate change. Studies have shown that tea yield in tea-growing countries in Asia and Africa is highly dependent on climatic factors such as temperature, precipitation, rainfall patterns, water availability, and solar radiation (Chang & Brattlof, 2015; Jayasinghe & Kumar, 2020). Especially adverse weather conditions such as droughts can significantly lower tea production as many tea estates depend on rain-fed agriculture (Munasinghe et al., 2017). On the other hand, heavy rain causes heavy surface runoff and flash floods, eroding the nutrient-rich topsoil cover and washing off the fertiliser and other chemical inputs away. This leads to water source pollution that originates from the Central Highlands in Sri Lanka (Dharmasena & Bhat, 2011; Peiris & Gunarathne, 2022b).

Further, due to widespread allegations that Glyphosate causes kidney disease in some parts of the country, the government imposed a ban on weedicide (Jayawardana, 2018; Rajapakse et al., 2016), leaving the tea industry with limited options to control weeds in their tea estates (Gunarathne & Lee, 2020). However, instead of adopting other integrated weed management techniques to control weed in the tea estates, some plantation companies have resorted to the option of applying chemicals that are not permitted to use in the tea industry. Moreover, the health concerns developed at the consumer end due to the exceeding levels of permitted maximum residue levels (MRLs) of agrochemicals in the tea have also created tremendous pressure on the industry. This issue has led to a situation where Ceylon tea is rejected in highly lucrative markets such as Japan and Germany. For instance, in 2018, tea shipments worth one billion Sri Lankan Rupees were refused by Japanese authorities due to the presence of MCPA, a weedicide used as a substitute for Glyphosate (Fernando, 2018).

Another issue associated with the tea industry is the large amount of household waste generated by the plantation sector workers and their families. Owing to various reasons, such as lack of proper infrastructure for waste collection, rugged mountain landscapes, and poor community support and awareness, waste management has become a severe issue in the Central Highlands of Sri Lanka, which is highly ecologically sensitive (Gunarathne & Peiris, 2020). As a result, many environmental and social problems, including pollution of water sources, loss of biodiversity, the spread of diseases, and other health issues, are inevitable (Atupola & Gunarathne, 2022; Peiris & Gunarathne, 2022b).

Thus, the future of the commercial tea industry in Sri Lanka has become uncertain and volatile, which urgently calls for innovative strategies aimed at improving productivity, conserving the environment, and ensuring social development (Oxford Business Group, 2016a). Table 2 provides a summary of these challenges under the economic, environmental, and social dimensions.

4 Sustainability Strategies Adopted by the Sri Lankan Tea Industry

In response to these growing sustainability challenges, the Sri Lankan tea industry has pursued many strategies to address its economic, environmental, or social problems. These include innovative agriculture and land management practices, herbicide-free weed management techniques, promotion of soil conservation practices, productivity improvement initiatives in harvesting, biodiversity conservation approaches, new product and market development, crop diversification, and actions to uplift the social development of the plantation workers. These solutions are described in detail in this section (Table 2 presents a summary of the challenges and strategies adopted).

4.1 Strategies to Address the Increasing Costs of Production

As explained in the previous section, the main contributor to the rocketing-up costs of production in the Ceylon tea industry over the past two decades is the series of

Table 2 Summary of tea industry challenges and strategies adopted

Sustainability dimension	Tea industry challenges in Sri Lanka	Strategies adopted to defeat challenges
Economic	• Increasing costs of production	• Controlling the cost items in every possible way
	• Declining land productivity of tea lands	• Development of 'climate-smart agricultural' techniques to improve land productivity
	• Labour scarcity	• Mechanisation of labour-intensive field operations
	• Fluctuations in world tea market prices	• Promotion of niche market-oriented high-value tea and product diversification
	• Unfair sharing of benefits along the value/supply chain	• Canvassing and influencing for a fair share of benefits gained in the value chain to the producer while showcasing the unique plus points of the Ceylon tea industry
	• Ban of certain weedicides (e.g. glyphosate)	• Development of more effective sustainable integrated weed management techniques
Environmental	• Occurrence of frequent extreme weather conditions	• Adoption of climate-smart agriculture methods
	• Short but intense rainy periods	• Intensification of conventional periodic agricultural practice (e.g. 'deep envelop forking')
	• Soil erosion and degradation and water scarcity	• Adoption of field agricultural operations such as 'SSTB,' 'deep envelope forking', and 'Herbicide-free integrated weed management'.
	• Loss of biodiversity due to extreme weather, human activity, and habitat loss	• Conduct of community awareness programmes targeting youth and schooling children as the change agents
Social	• Underperformance of the workforce	• Negotiations on including productivity improvements based merits to wage structure
	• Unattained earning opportunities due to ignorance	• Conducting worker awareness campaigns aiming at win-win situations
	• Health concerns developed at the consumer end due to the violations of pesticide MRLs	• Imposition of regulatory measures to maintain pesticide residues in made tea below particular MRLs and enhancement of monitoring facilities
	• Low standard of living of industry workers	• Continuous collaborative efforts by the state government and the regional plantations companies to further improve their living standards
	• Waste management in the plantation sector and associated health concerns	• Introduction of community-driven waste management systems

periodic wage hikes without an accompanying productivity gain (Poholiyadde, 2018; Thibbotuwawa et al., 2019). In order to minimise the negative impacts of this cost increase, the tea plantation companies were forced to explore every possible cost reduction opportunity for their survival. Some of these methods led to the curtailment of certain inputs that are essential for long-term survival. For instance, the regional plantation companies were compelled to compromise on vital inputs such as the material, energy, up keeping of cultivation, machinery, and supervisory aspects.

These measures initially enlightened and guided the industry to minimise wastage and losses in their operations and improve productivity. However, the labour productivity did not witness a corresponding improvement, primarily due to the resistance of the trade unions motivated by political reasons. Despite the initial signs of success of the above-mentioned cost reduction mechanisms, later, they have led to a vicious cycle of further cost increases. The reduction of essential inputs over the long term had a detrimental impact on both the product quality and quantity while decreasing the operational efficiencies of the entire plantation structure. For example, reducing the number of fertiliser applications per annum resulted in a lower yield.

Further, the low-quality harder leaf negatively affects the quality of the tea produced. Besides, low-quality tea also results in lower auction prices. Additionally, the insufficient nutrient levels debilitated the tea bushes, and inadequate canopy cover speeds up the weed growth, increasing weeding costs. Processing harder leaf incurs higher machinery wear and tear, which increases machinery maintenance and energy costs. Eventually, the actions pursued to reduce the cost itself have resulted in a degradation of the entire system, leading to severe wastage of resources. Finally, the unit cost of production has further increased, aggravating the industry problems.

Moreover, in the long term, the curtailment of investments in cultivation upkeeping has led to the deterioration of plantations in many ways. Inadequate attention paid to soil conservation resulted in soil degradation, which in turn resulted in reduced water holding capacity and nutrient retention ability of the soil. As a result of these unhealthy management practices, tea bushes have debilitated, and a drop in the crop performance levels have been experienced. The impact of this condition has been worsened by the frequent extreme weather conditions, such as cyclones, thunderstorms, and prolonged droughts caused due to climate change.

4.2   Strategies to Address the Declining Land and Worker Productivity of Tea Estates

As mentioned above, another leading cause of the issues faced by the Ceylon tea industry is the declining productivity of land and labour. Over the years, the tea industry has developed certain novel initiatives to address these twin issues.

4.2.1 Initiatives to Improve Land Productivity

To improve land productivity, mainly due to the vacancies of tea bushes in tea fields, the plantation companies have traditionally resorted to replanting and infilling (*filling up the vacancies due to the casualties in cultivation with new plants of the same crop or another crop species*) vacancies of the existing cultivation with new plants. Although these standard productivity enhancement solutions are available and adopted by many plantation companies, they attract considerable investment and recovery period. However, such costly investments have become highly unaffordable in the present context due to the industry's loss-making status. Therefore, certain plantation companies have developed some innovative leapfrogging techniques in addition to these standard solutions. One such method is the stripe-spreading of tea bushes (SSTB) (Gunarathne, 2020).

Since tender tea foliage is the economically vital part of the tea bush, the canopy coverage and healthiness are essential to optimise crop production (Snapir et al., 2018). The SSTB is an alternative to standard replanting and infilling. Following SSTB, canopy gaps could be filled within months. This method allows the tea bushes to grow, after pruning, up to 120 days. Then, a radial spread of shoots is done using tight parallel stripes arranged along the tea rows, instead of the traditional practice of tipping the tea shoots after pruning (See Gunarathne & Peiris, 2017 for more details). Thereby, a quick re-establishment of the tea bush canopy is attained with substantially higher yields, while suppressing the weed growth (see Fig. 5).

Recent multisite studies have provided evidence that the SSTB technique can improve tea yields considerably with minimal investment (Peiris & Nissanka, 2021). Although the SSTB is a low-cost and high-impact technique that solves many issues inherited in the Ceylon tea industry with early signs of potential, adoption of this technique has not been, so far, widespread in the industry.

Fig. 5 SSTB Method. Note: Application of the SSTB method (left) involves the radial spread of shoots using parallel stripes. This method is capable of re-establishing bush canopy also filling canopy gaps within months boosting up tea yield with the existing tea bushes with minimum investment (right)

4.2.2   Initiatives to Improve Labour Productivity

Due to worker migration from the plantation sector, the plantation companies face a severe shortage of workers, especially during the cropping months of the year. This situation has pushed the plantation companies to mechanise their field operations such as harvesting, pruning, and mechanical weeding. This move significantly reduced the labour requirements for the labour-intensive processes improving operational productivity. It has also tremendously enhanced the resource use efficiency of the costliest input, the labour, while markedly reducing production cost (see Fig. 6).

As stated before, the Ceylon tea industry is on the verge of collapse due to the frequent wage increments without any productivity-related improvements from part of the employee. However, there are intense negotiations between the plantation companies and worker trade unions since 2020 on wage increments based on productivity improvements. Many of these trade unions represent political parties, and hence, apart from the trade union orientation, they have their political agendas and rivalries, which has led to many difficulties in the negotiation processes. Therefore, the government has to frequently intervene in these processes to push both parties towards an early amicable agreement.

4.3   Strategies to Address the Fluctuations in World Tea Market Prices

Periodic changes in the global tea market prices are the other significant factors that directly affect the sustainability of the Ceylon tea industry. As explained before, there is an evident decline in the world market tea prices since 2015 (Oxford Business Group, 2016b). Therefore, some initiatives were adopted by the tea industry to minimise the impact of decline. One crucial strategy was developing value-added products with a diverse range of teas with unique characteristics. Teas that promote their region-specific natural aroma, flowery flavours and cup (liquor) colour, health-orientation, and organic production process are a few examples of

Fig. 6 Conventional Vs mechanised methods of tea plucking. Note: Traditional tea harvesting method involves plucking only the tender young shoots at the top of the canopy by hand (Left). Mechanised harvesting methods automate this process (Right) while improving productivity and reducing operational cost. It leads to enhanced worker benefits in many folds

these value additions. Most of these high-priced speciality teas cater to niche markets in Japan, China, Eastern Europe, the Middle East, and the European Union. Due to the diversity of natural tea flavours found in Sri Lanka, a significant market opportunity exists for these speciality products.

Since these recent initiatives are few and operated on a small scale in the industry, they need to be expanded with government support. However, such a move needs a great effort, dedication, creativity, and technical know-how on novel manufacturing techniques to cater to these unique niche markets' demands. This requires the whole Ceylon tea industry to be vigilant on the changing global market trends, and adjust accordingly to meet the consumer demands with novel product developments and marketing initiatives.

4.4 Strategies to Address Unfair Sharing of Benefits along the Value/Supply Chain

The Ceylon tea industry has exhibited its commitment to improving its sustainability performance by obtaining various certifications such as Rainforest Alliance, Fairtrade Labelling Organisation, and Universal Trade Zone certification programmes. These sustainability certifications need substantial cost and administrative inputs to be borne by the plantation companies. Unfortunately, the tea auction prices have been on a declining trend for the last few years, suggesting that the plantation companies have not enjoyed the benefits of extra sustainability measures. Therefore, it is also the responsibility of such certification bodies to include all the levels of the value chains and develop mechanisms to share the gains achieved according to pain and the responsibilities held by each party. However, these mechanisms are still at a negotiation level and have not yet come into practice.

4.5 Strategies to Address the Ban of Glyphosate

Sri Lanka was the first country to ban the use of systemic weedicide glyphosate on tea and paddy crops in the world (Jayawardana, 2018). However, the effectiveness of Glyphosate has been questioned, as recent studies and field-level evidence have shown that over 90% of the harmful tea weeds found in Sri Lanka have developed resistance against the chemical by now (Gunarathne & Peiris, 2017). Therefore, applying increased chemical dosages to control tea weeds has resulted in the weakening and early debilitation of tea bushes, reducing their performance.

With the ban of Glyphosate, some integrated weed management techniques had been developed within the industry, which promotes natural weed control methods (Gunarathne, 2020; Peiris & Nissanka, 2021). For instance, a 'herbicide-free integrated weed management' (HFIWM) technique was developed within the Ceylon tea industry parallel to the Glyphosate ban imposed by the Sri Lankan government. Based on natural weed control methods, HFIWM promotes the establishment of friendly weed cover over the ground by selective manual removal of harmful weeds from the tea fields.

These integrated weed management methods offer many merits, such as enhanced soil fertility, moisture and nutrient retention, higher yields with improved quality, and climate resilience for the tea bushes. The methods such as SSTB also rapidly enhance the tea yield, which improves the tea industry's land and labour productivity. Both HFIWN and SSTB are capable of minimising the wastage of resources in tea fields. The minimising wastage of costly synthetic inputs such as fertiliser and weed killer chemicals offers massive financial gains to the tea industry. Most importantly, these techniques reduce the product residue levels, which protects the consumer's health as well.

Thus, foreseeing the possible reduction in the fertiliser and weed killer chemical usage in plantations, the synthetic input dealers have already launched silent propaganda campaigns against both SSTB and HFIWM techniques to stop their adoption by the tea industry. As a result, despite the many benefits of these techniques to consumers, plantation workers, and the tea industry, the Ceylon tea industry's acceptance of these sustainable solutions has been hindered (Gunarathne, 2020).

4.6   Strategies to Address Climate Change Impacts

Since tea is grown in open ground under rain-fed conditions, potential losses due to crop damage are comparatively high. Additionally, the extreme weather conditions cause soil erosion and degradation by the violent surface runoff during the wet/rainy season while minimising groundwater recharge due to low infiltration (Han et al., 2018; Jayasinghe & Kumar, 2020; Munasinghe et al., 2017). To overcome the negative impacts of climate change, plantation companies have adopted many strategies, including crop diversification, maintenance of forest covers in the tea estates, and soil conservation practices. Around the world, soil conservation practices have become the most common form of measures adopted by the tea industry to mitigate climate change effects (Nowogrodzki, 2019). In order to preserve the groundwater infiltration and recharging ability, some plantation companies have also followed an agricultural practice known as 'deep envelope forking'; i.e. breaking-up of the compacted upper soil layer by inserting an agricultural fork to a depth of about 40 cm–50 cm in the tea field (Peiris & Gunarathne, 2022a). This method also improves nutrient retention, aerates the soil, and enhances subsoil biological activity while minimising soil erosion due to surface water runoff.

Although some plantation companies have given much attention to the climate change impacts, the Ceylon tea industry has yet to adopt comprehensive climate-smart agriculture practices; a concept to enhance the production capacity of agricultural systems through improved resource use efficiency and to build resilience to climate change to support food security and farmer development while reducing the greenhouse gas emissions of agriculture practices (Campbell et al., 2014; Food and Agriculture Organization of the United Nations, 2013).

4.7   Strategies to Address the Loss of Biodiversity

One of the main threats to biodiversity is the human activities of the plantation communities. Over 95% of the plantation community in Sri Lanka are descendants of imported workers from arid zones of South India (Galgamuwa et al., 2017). Thus, they are unaware of the biodiversity and the importance of its conservation for the Central Highlands of Sri Lanka. Hence, spreading conservation awareness among this community is of paramount importance. Therefore, under the influence of Rainforest Alliance, Fairtrade Labelling Organisation, and Universal Trade Zone certification programmes, the plantation companies on the island have commenced conservation awareness programmes for the plantations community as a biodiversity conservation move (Atupola & Gunarathne, 2022). Initially, most of these conservation awareness programmes had targeted only adult worker groups. However, those early efforts were not proved successful due to the low education levels, lack of interest, attitude issues, and language barriers. Therefore, further initiatives have been taken to channel the conservation awareness of the plantation communities via programmes that target school children and community opinion leaders as the change agents. To overcome the language barriers, these programmes are conducted in their mother tongue, Tamil.

4.8   Strategies to Address Worker's Income Levels

All the initiatives taken to improve crop availability and land productivity in the Ceylon tea industry also directly impact the worker's income, as they are paid a piece rate based on the number of additional kilograms of tea leaves plucked over and above the standard. Therefore, the above initiatives oriented towards increasing the production levels create extra earning opportunities (Gunarathne & Peiris, 2017). Additionally, on-the-job training sessions on productivity improvement techniques are also conducted regularly for the workforce.

These training programmes had motivated the workers to tap more earning opportunities that they were unaware of or neglected before, which resulted in increased monthly earnings of the workforce. Further, even with little initial resistance to change, mechanisation of field operations has paid back. Both employers and employees have gained from the win–win situations created. Such initiatives have improved both land productivity and worker productivity levels.

4.9   Strategies to Address Residue Levels in Tea

Restrictions and bans had been imposed on Ceylon tea in certain buyer countries due to health concerns, particularly the violations of pesticide MRLs (Fernando, 2018; Gunarathne & Lee, 2020). The Sri Lankan government has also taken action to arrest the breach of international MRLs by imposing stern legal measures to control the situation. This includes flying squads appointed to conduct frequent inspections in tea plantations, enhancing the analytical facilities to detect such chemicals and their concentrations, and cancelling the operators' industrial licence if MRLs violations are detected.

4.10 Strategies to Improve the Low Standard of Living of Industry Workers

Although the living standards of the plantation communities are comparatively low to the general living standards in Sri Lanka, the facilities of the Sri Lankan plantation worker community are much superior compared to the plantation sectors in many other countries. This is due to the provision of many other benefits to the plantation communities by the tea companies, which include free housing and potable water supply, free healthcare, childcare, maternal care, and free feeding ration for children. Further, the plantation sector has the highest density of schools in Sri Lanka. As Galgamuwa et al. (2017) mentioned, some companies conduct health education programmes to improve the nutritional status among the plantation communities.

Some plantation companies have also made serious efforts to solve the waste management problem in the plantation sector, which is relatively ineffective, compared to other parts of the country. This is due to many reasons unique to the region and the attitudes and value systems associated with plantation communities (Peiris & Gunarathne, 2022b). Therefore, community-driven waste management techniques have been developed within the plantation sector, in which the use of household waste is segregated in situ at source and directed for systematic recycling via an organised collection network. The recovery of nutrients present in food waste and fresh greenery is promoted by feeding them to domestic livestock and using them for home gardening purposes. Since the traditional municipal waste management practices are not effective in the plantation sector, these alternative waste management methods that emphasise the socio-cultural context of the communities are gaining popularity in the Ceylon tea industry.

5 Conclusions and Way Forward

This chapter discusses the sustainability challenges of the Ceylon tea industry, under three categories: economic, environmental, and social, and the strategies adopted by the plantation companies with a view to addressing them. The chapter also presents how the tea plantation companies in Sri Lanka have adopted various measures to overcome or manage these challenges. While some of these measures have shown positive impacts, many of the Ceylon tea industry's sustainability challenges still remain unresolved or partially solved. These challenges include the effects of climate change, declining land productivity, increase in production cost, low-value addition of tea, social issues, and unequal profit distribution along the global value chain.

The Ceylon tea industry needs to focus on resolving these issues through multi-stakeholder engagement, innovations, and better marketing and product development. Particularly, improved branding and value-added product development offer one of the best paths for the Ceylon tea industry to address much of its interconnected economic and other problems. Besides, international consumer associations, sustainability certification organisations, public media, health authorities, and the governments should come forward to voice against some of the unethical and unfair business practices that affect the sustainability of the Ceylon tea industry.

References

Atupola, U., & Gunarathne, N. (2022). Institutional pressures for corporate biodiversity management practices in the plantation sector: Evidence from the tea industry in Sri Lanka. *Business Strategy and the Environment*. https://doi.org/10.1002/bse.3143

Bloomfield, M. J. (2020). South-south trade and sustainable development: The case of Ceylon tea. *Ecological Economics, 167*, 106393. https://doi.org/10.1016/j.ecolecon.2019.106393

Campbell, B. M., Thornton, P., Zougmoré, R., Van Asten, P., & Lipper, L. (2014). Sustainable intensification: What is its role in climate smart agriculture? *Current Opinions on Environmental Sustainability, 8*, 39–43. https://doi.org/10.1016/j.cosust.2014.07.002

Central Bank of Sri Lanka. (2019). *Annual report*. https://www.cbsl.gov.lk/en/publications/economic-and-financial-reports/annual-reports/annual-report-2019

Chandrasekaran, M., Basheer, S. M., Chellappan, S., Krishna, J. G., & Beena, P. S. (2015). Enzymes in food and beverage production: An overview. In M. Chandrasekaran (Ed.), *Enzymes in food and beverage processing* (pp. 117–137). Taylor & Francis.

Chang, K. (2015). World tea production and trade: Current and future development. *Food and Agriculture Organization of the United Nations*. http://www.fao.org/3/a-i4480e.pdf

Chang, K., & Brattlof, M. (2015). Socio-economic implications of climate change for tea producing countries. *Food and Agriculture Organization of the United Nations*. http://www.fao.org/3/a-i4482e.pdf

Cushman & Wakefield. (2017). *The global food & beverage market: What's on the menu?* https://www.cushmanwakefield.com/-/media/cw/insights/asia-pacific/2017/global_food___beverage_market_summer_2017.pdf

Dharmasena, P., & Bhat, M. S. (2011). Assessment of replacement cost of soil erosion in Uva high lands tea plantations of Sri Lanka. *Current World Environment, 6*(2), 241–246.

Food and Agriculture Organization of the United Nations. (2013). *Climate-smart agriculture sourcebook*. http://www.fao.org/3/i3325e/i3325e00.htm

Fernando, N. (2018, May 2). Ceylon tea risks losing Japanese market. *Daily mirror online*. http://www.dailymirror.lk/article/Ceylon-Tea-risks-losing-Japanese-market-149375.html

Forum for the Future. (2014). *The future of tea: Aa hero crop for 2030*. https://issuu.com/forum4thefuture/docs/future-tea-report_b083545dfd3fca

Galgamuwa, L. S., Iddawela, D., Dharmaratne, S. D., & Galgamuwa, G. L. S. (2017). Nutritional status and correlated socio-economic factors among preschool and school children in plantation communities. *Sri Lanka. BMC Public Health, 17*(1), 1–11. https://doi.org/10.1186/s12889-017-4311-y

Globe Newswire. (2020). *Food and beverages global market report 2020-30: COVID-19 impact and recovery*. https://www.globenewswire.com/news-release/2020/05/19/2035458/0/en/global-food-and-beverage-stores-market-report-2020-to-2030-covid-19-impact-and-recovery.html

Gunarathne, A. D. N. (2020). Making sustainability work in plantation agriculture: The story of a sustainability champion in the tea industry in Sri Lanka. In P. Flynn, M. Gudić, & T. K. Tan (Eds.), *Global champions of sustainable development* (pp. 49–63). Routledge.

Gunarathne, A. D. N., & Peiris, H. M. P. (2017). Assessing the impact of eco-innovations through sustainability indicators: The case of the commercial tea plantation industry in Sri Lanka. *Asian Journal of Sustainability and Social Responsibility, 2*(1), 41–58. https://doi.org/10.1186/s41180-017-0015-6

Gunarathne, A. D. N., & Lee, K. H. (2020). Eco-control for corporate sustainable management: A sustainability development stage perspective. *Corporate Social Responsibility and Environmental Management, 27*(6), 2515–2529. https://doi.org/10.1002/csr.1973

Gunarathne, A. D. N., & Peiris, H. M. P. (2020). Sustainable waste Management for the Plantation Sector in Sri Lanka. In T. K. Tan, M. Gudic, & P. M. Flynn (Eds.), *Struggles and successes in the pursuit of sustainable development* (pp. 70–83). Routledge.

Gunarathne, A. D. N., Navaratne, D. G., Pakianathan, A. E., & Perera, N. Y. T. (2018). Sustainable food supply chain management: An integrated framework and practical perspectives. In H. Qudrat-Ullah (Ed.), *Innovative solutions for sustainable supply chains* (pp. 289–315). Springer.

Han, W. Y., Li, X., & Ahammed, G. J. (2018). *Stress physiology of tea in the face of climate change.* Springer.

International Labour Organization. (2018). *Future of work for tea smallholders in Sri Lanka.* https://labordoc.ilo.org/discovery/delivery/41ILO_INST:41ILO_V2/1256499620002676?lang=en

International Union for Conservation of Nature & UN Environment World Conservation Monitoring Centre. (2017). *Central Highlands of Sri Lanka.* https://yichuans.github.io/datasheet/output/site/central-highlands-of-sri-lanka

Jayasinghe, S. L., & Kumar, L. (2019). Modeling the climate suitability of tea [Camellia sinensis (L.) O. Kuntze] in Sri Lanka in response to current and future climate change scenarios. *Agricultural and Forest Meteorology, 272*, 102–117. https://doi.org/10.1016/j.agrformet.2019.03.025

Jayasinghe, S. L., & Kumar, L. (2020). Climate change may imperil tea production in the four major tea producers according to climate prediction models. *Agronomy, 10*(10), 1536. https://doi.org/10.3390/agronomy10101536

Jayawardana, S. (2018). *Glyphosate ban lifted.* Sunday times. http://www.sundaytimes.lk/180715/news/glyphosate-ban-lifted-302600.html

Kamalakkannan, S., Kulatunga, A. K., & Kassel, N. C. (2020). Environmental and social sustainability of the tea industry in the wake of global market challenges: A case study in Sri Lanka. *International Journal of Sustainable Manufacturing, 4*(2–4), 379–395. https://doi.org/10.1504/IJSM.2020.107129

Kasturiratne, D. (2008). An overview of the Sri Lankan tea industry: An exploratory case study. *The Marketing Review, 8*(4), 367–381. https://doi.org/10.1362/146934708X378659

Kotikula, A., & Solotarof, J. (2006). *Gender analysis of labor in Sri Lanka'S estate sector.* World Bank. https://paa2007.princeton.edu/papers/71572

Maxime, D., Marcotte, M., & Arcand, Y. (2006). Development of eco-efficiency indicators for the Canadian food and beverage industry. *Journal of Cleaner Production, 14*(6–7), 636–648. https://doi.org/10.1016/j.jclepro.2005.07.015

Munasinghe, M., Deraniyagala, Y., Dassanayake, N., & Karunarathna, H. (2017). Economic, social and environmental impacts and overall sustainability of the tea sector in Sri Lanka. *Sustainable Production and Consumption, 12*, 155–169. https://doi.org/10.1016/j.spc.2017.07.003

Mzembe, A. N., & Lindgreen, A. (2018). On the infusion of ethics in entrepreneurial and managerial action: Reconciling actors' CSR-related perceptions in the Malawian tea industry. In A. Lindgreen, F. Maon, C. Vallaster, S. Yousafzai, & B. Palacios Florencio (Eds.), *Sustainable entrepreneurship: Discovering, creating and seizing opportunities for blended value generation* (pp. 301–329). Routledge Taylor & Francis.

Nowogrodzki, A. (2019). How climate change might affect tea. *Nature, 566*(7742), S10–S11. https://doi.org/10.1038/d41586-019-00399-0

Oxford Business Group. (2016a). *Growth prospects for Sri Lanka's tea industry.* http://www.oxfordbusinessgroup.com/news/growth-prospects-sri-lanka%E2%80%99s-tea-industry

Oxford Business Group. (2016b). *Orthodox thinking: Tea remains vital economic contributor.* https://www.oxfordbusinessgroup.com/analysis/orthodox-thinking-tea-remains-vital-economiccontributor

Peiris, M., & Gunarathne, N. (2022a). The changing landscape of the plantation sector in the central highlands of Sri Lanka. In U. Schickhoff, R. B. Singh, & S. Mal (Eds.), *Mountain landscapes in transition: Effects of land use and climate change* (pp. 539–554). Springer.

Peiris, M., & Gunarathne, N. (2022b). A community-driven household waste management system in the tea plantation sector: Experiences from Sri Lanka towards a circular economy. In C. Baskar, S. Ramakrishna, S. Baskar, R. Sharma, A. Chinnappan, & R. Sehrawat (Eds.), *Handbook of solid waste management: Sustainability through circular economy* (pp. 847–876). Springer.

Peiris, H. M. P., & Nissanka, S. P. (2021, February 10). *An innovative climate-smart approach to boost p yield in the commercial tea industry: Stripe-spreading of tea bushes* [Paper presentation]. 02nd National Symposium on Sustainable Plantation Management.

Pilapitiya, H. M. C. G., De Silva, S., & Miyazaki, H. (2020). *Innovative value addition in tea industry: Sri Lanka vs.* Japan. https://doi.org/10.21203/rs.3.rs-27492/v1

Poholiyadde, S. (2018). *The evolution of Sri Lanka's plantation sector into diversified, vertically integrated, globally aligned, agri-businesses.* https://medium.com/the-planters-association-of-ceylon/the-evolution-of-sri-lankas-plantation-sector-into-diversified-vertically-integrated-glob ally-b5983cb01cca

Rajapakse, S., Shivanthan, M. C., & Selvarajah, M. (2016). Chronic kidney disease of unknown etiology in Sri Lanka. *International Journal of Occupational and Environmental Health, 22*(3), 259–264. https://doi.org/10.1080/10773525.2016.1203097

Risch, S. J. (2009). Food packaging history and innovations. *Journal of Agricultural and Food Chemistry, 57*(18), 8089–8092. https://doi.org/10.1021/jf900040r

Snapir, B., Waine, T. W., Corstanje, R., Redfern, S., De Silva, J., & Kirui, C. (2018). Harvest monitoring of Kenyan tea plantations with X-band SAR. *IEEE Journal of Selected Topics in Applied Earth Observations and Remote Sensing, 11*(3), 930–938. https://doi.org/10.1109/JSTARS.2018.2799234

Sri Lanka Export Development Board. (2019). *Industry capability report: Tea.* https://www.srilankabusiness.com/ebooks/tea%2D%2D-industry-capability-report%2D%2D%2D%2Ddecember-2019.pdf

Tea Exporters Association Sri Lanka. (2020). *Sri Lanka tea production-January to December 2020.* http://teasrilanka.org/market-reports

Thibbotuwawa, M., Jayawardena, P., Arunatilake, N., & Gunasekera, N. (2019). Living wage report Sri Lanka: Estate sector-context provided in the tea sector. *Global living wage coalition.* https://www.globallivingwage.org/wp-content/uploads/2019/07/Sri-Lanka-Living-Wage-report.pdf

Wickramagamage, P. (1998). Large-scale deforestation for plantation agriculture in the hill country of Sri Lanka and its impact. *Hydrological Processes, 12*(13–14), 2015–2028. https://doi.org/10.1002/(SICI)1099-1085(19981030)12:13/14<2015::AID-HYP716>3.0.CO;2-3

Yasanthi Alahakoon is a Senior Lecturer in the Department of Business Administration, University of Sri Jayewardenepura, Sri Lanka. She completed her PhD at the Queensland University of Technology, Australia, in knowledge management discipline. She has an MBA from the Postgraduate Institute of Management (PIM), Sri Lanka. Her Bachelor's degree is in Business Administration from the University of Sri Jayewardenepura, Sri Lanka. Yasanthi has obtained professional qualifications from the Chartered Institute of Management Accounting (UK) and Chartered Institute of Marketing (UK). She has teaching experience in undergraduate and postgraduate courses on Knowledge Management, Organisational Behaviour, Contemporary Issues in Management and Research Methodology. She pioneered in introducing the Knowledge Management course unit to the BSc Business Administration degree programme, University of Sri Jayewardenepura which has been continuing successfully for nine consecutive years. She has authored, co-authored, and edited several publications on knowledge management, organisational studies and sustainability management in international journals, conference proceedings and professional journals. She has administrative experience in curriculum development and quality assurance programmes at the University of Sri Jayewardenepura. Her research interests are in epistemological stances of knowledge management and sustainable management practices in developing countries.

Mahendra Peiris is a professional planter, an inventor, and a trainer in the field of tea. He holds an honours degree in plantation crop management—specialised in tea, and a master's degree in biodiversity, ecotourism, and environment management both from the University of Peradeniya, Sri Lanka. Mahendra is the winner of the Presidential Green Award 2016 for the best Sustainable Farming Model in Sri Lanka and the winner of twin awards at Merrill J. Fernando Eco-innovation Awards 2016 for 'Climate-smart breakthrough technologies' developed by him for the commercial tea industry. Further, his work towards a climate-smart tea industry has also been published in international science/management journals. His innovative prototype models gathered worldwide attention when nominated for the Global Tea Sustainability Award 2016 by the Rainforest Alliance. Further, Rainforest Alliance's "General Inspirational Video-2017" is based on his work. Currently, Mahendra is the Manager Compliance and Project Management at Maskeliya Plantations PLC, Sri Lanka.

Dr. Nuwan Gunarathne is a Senior Lecturer in the Department of Accounting, University of Sri Jayewardenepura, Sri Lanka. He is a fellow member of the Institute of Certified Management Accountants of Sri Lanka and a member of the Chartered Institute of Management Accounting (UK). He obtained his doctorate in corporate sustainability and sustainability accounting from Griffith University, Australia. He has an MBA from the Postgraduate Institute of Management (PIM), Sri Lanka, and a Business Administration degree from the University of Sri Jayewardenepura. He has authored and co-authored many national and international publications in different spheres of management accounting, sustainability accounting, waste management and accounting education. His articles have been published in several international journals such as *Journal of Cleaner Production, Accounting, Auditing & Accountability Journal, Meditari Accountancy Research, Business Strategy and the Environment, Corporate Social Responsibility and Environmental Management, Sustainability, Resources, Conservation and Recycling, Tourism Economics, Journal of Accounting and Organizational Change, Accounting Research Journal,* and *Managerial Auditing Journal.* He has been a sought-after speaker on sustainability, sustainability accounting and reporting, and integrated reporting. Nuwan has presented papers in Finland, Australia, Sri Lanka, Indonesia, Germany, and South Korea. He also spearheaded the launch of 'Environmental Management Accounting (EMA) Guidelines for Sri Lankan Enterprises' in 2014 and is also a committee member of the Environmental and Sustainability Management Accounting Network (EMAN) Asia Pacific (AP) and country representative of the Sri Lanka Chapter of EMAN-AP.

Index

A

Affordability, 152, 187
Agrarian reform, 234, 236
Agriculture, 2–5, 7–11, 13–17, 19, 21, 25–29,
    35, 37, 38, 50–77, 102, 104, 111, 112,
    121, 122, 127, 159, 160, 162–165, 187,
    196, 204, 208, 242, 243, 248, 250, 254,
    273, 279, 280, 286
Agriculture and health, 230
Agriculture sector, 62–64, 66, 68–74
Agriculture and tourism, 21, 29
Agrifood, 102, 103, 106–109, 113, 114, 123,
    197
Animals, 22, 23, 26, 29, 35, 102, 104,
    108–111, 114, 160, 177–179, 199,
    207, 209, 214, 220–222, 225, 242, 243,
    251, 253
Antibiotics, 16, 102, 109
Apartheid, 137, 139, 141, 143, 145, 154
Austria, 2–39, 205, 250

B

Bees, 4, 11–14, 18, 29, 30, 37
Belgradisation, 237, 241
Biodiversity, 2–7, 10–12, 14, 17, 18, 24, 32, 34,
    37, 38, 55, 56, 106, 159, 164, 165, 169,
    172, 177, 179, 187, 192, 273, 277, 280,
    281, 287
Biodynamic farming, organic (certification,
    farming, foods), 26, 249
Bio-waste, 204
Black Africans, 137, 140, 143
Blockchain technology, 197, 198
Bulk tea, 273, 275

C

Camellia sinensis, 272
Cantabria, 121–124, 129–131
Carbon footprint, 54, 204, 208, 222
Case study, 103, 105, 106, 110, 113, 114, 197
Central highlands, 277, 279, 280, 287
Certifications, 20, 21, 27, 87, 104–106, 109,
    110, 113, 122, 167, 168, 173, 174, 180,
    189, 192–194, 196, 197, 199, 242, 244,
    248, 250, 285, 287
Ceylon tea, 273–276, 278, 280–289
Changing diets, 2
Changing lifestyles, 188
Chickens, 103, 106, 108, 111, 249
Circular, 28, 32, 35, 108, 109, 165
Circular economy, 38, 102, 103, 111, 113, 120,
    131, 192, 205
Climate change, 2, 3, 5–7, 10, 16, 17, 55, 56,
    102, 113, 159, 165, 173, 187, 188, 220,
    244, 252, 279, 282, 286, 288
Climate smart agriculture, 281, 286
Colombo Tea Auction, 275
Commodities, 138, 165, 167, 192, 199, 279
Companies Act, 147–149, 151, 153
Competitive advantage, 50, 104, 113, 161
Composting, 208, 211, 212
Conservation, 85, 160, 161, 165, 169, 170, 172,
    176, 178, 179, 273, 277, 280, 282, 286,
    287
Constitution, 83, 144, 145
Consumer, 11, 16, 20, 21, 30, 35, 38, 51, 52, 57,
    64, 84, 85, 87, 89, 92, 102–104, 106,
    108, 111, 113, 114, 120, 123–129, 148,
    160–162, 165, 166, 169, 170, 186, 188,
    189, 191, 193, 195–199, 204–206, 208,

209, 211, 215, 216, 219, 223–225, 242,
    272, 275, 279, 280, 285, 286, 289
Consumer behaviour, 191, 208, 249
Consumerism, 210, 213, 224
Cooperatives, 109, 114, 124, 166, 239, 243,
    255
Corporate social responsibility (CSR), 65, 110,
    125, 126, 129, 136–155, 195, 214, 215,
    217, 218, 220
Corruption, 54, 57, 82, 84, 145, 152, 154, 196,
    235, 236
COVID-19, 16, 83, 136–155, 189, 199, 272
Crop diversification, 280, 286
Cut Tear and Curl (CTC) tea, 274, 275

D
Date labelling, 215
Date marking, 205, 206, 213, 224
Deep envelope forking, 281, 286
Designation of origin, 105, 106
Developing countries, 17, 82, 84, 85, 87, 95, 96,
    120, 189
Development, 5, 15–21, 27–29, 33, 34, 39,
    52–54, 56–59, 61, 73, 74, 82, 84, 85,
    104, 113, 121, 124, 128, 129, 145, 146,
    148–152, 160–162, 165–180, 186, 188,
    191, 193, 194, 196–199, 208, 216, 218,
    253, 256, 273, 274, 276, 278, 280, 281,
    285, 286, 289
Discourse, 138, 139, 233, 234, 237, 239–241,
    243, 253, 255
Distributor, 170, 197, 275
Double standards, 240

E
Eco-innovation, 103, 104, 110
Economic and financial barriers, 66, 69, 75
Economic pillar, 55
Economy, 12, 35, 51–53, 57, 82–86, 95, 96,
    102, 104, 106, 127, 136, 146–149, 152,
    153, 163, 165, 166, 189, 192, 231,
    234–236, 238, 240, 242–244, 249, 255,
    256, 272, 273, 276
Ecosystem, xii, 5, 6, 10–13, 15, 17, 18, 29, 106,
    113, 162–165, 170–172, 174, 177, 188
Edible food, 189, 191, 204
Efficiency, 10, 50, 84, 96, 121, 122, 179, 192,
    194, 197, 282, 284, 286
Enlightened shareholders, 148

Entrepreneurship, 51–53, 61, 62, 74, 75, 84, 85,
    152
Environment, 2, 4, 14, 15, 19, 28, 31, 53, 54,
    56–58, 68, 70, 72, 73, 77, 82, 85, 102,
    103, 106, 110–113, 124, 129, 148, 150,
    151, 170, 175, 187, 192, 198, 199, 204,
    205, 208, 218, 221, 243, 249, 256, 272,
    277, 280
Environmental Pillar, 17, 54
European Food Banks, 216
European Green Deal, 11, 17, 18, 162
Eurosceptic, 240, 245
Exports, 74, 163, 188, 247, 273–277, 279

F
Farms, 9, 10, 13, 17, 19, 20, 24, 26–28, 32, 35,
    36, 52, 103, 111, 160, 162, 163, 166,
    170–175, 177–180, 188, 189, 194, 197,
    207, 220, 242, 243, 247, 248, 250, 255
Fast, 52, 53, 62, 253
Fermentation, 247, 274
Fight against food waste, 214
Fileni, 102–114
Food, 2–5, 10–13, 15–18, 20, 23, 25, 30–39,
    50–77, 82, 84–86, 95, 96, 102–107,
    110–114, 120–129, 136–147, 149,
    152–154, 159, 162, 164–166, 170,
    185–195, 197–199, 204–225, 232, 234,
    238–254, 256, 272, 286
Food and Agriculture Organization of the
    United Nations (FAO), 2, 13, 204, 286
Food and beverage industry, 272
Food bank, 206, 209, 210, 214, 216, 218,
    220–223, 225
Food chain, 51, 105, 113, 141, 194, 195, 197
Food donations, 123, 124, 205, 211, 217,
    219–224
Food donors, 209, 211, 222
Food industry, 50–53, 62–64, 66–68, 72, 74,
    161, 186–188, 190, 191, 196, 198, 213,
    214
Food insecurity, 15, 16, 102, 136–141, 143,
    145, 153, 189
Food recycling, 4, 18, 38
Food retailers, 166, 214–225
Food safety, 12, 15, 102–107, 109, 110, 113,
    114, 193, 195, 197, 198, 205, 251
Food safety management, 104, 105
Food security, 2, 5, 9, 10, 12, 15, 39, 55,
    102–105, 109, 110, 113, 137–141, 143,

145–147, 149–155, 186, 187, 190, 199, 216, 286

Food sovereignty, 138, 139, 141, 146, 149, 152

Food supply chain, 52, 104, 125, 186, 189, 190, 194, 204, 215, 220, 223

Food system, 2–4, 6, 11, 15, 17, 32, 38, 138, 152, 153, 160, 163, 170, 186–188, 190, 194, 199

Food traceability, 104, 109

Food wastage, 11, 18, 187, 190, 191, 204

Food waste, 11, 16, 19, 25, 31, 32, 64, 120–131, 186, 189–191, 194, 195, 204–225, 288

Food waste combat, 208, 212–225

Food waste in romania, 206–209, 211, 213–215, 217

Forest cover, 286

G

Genetically modified organism (GMOs), 4, 6, 14–15, 18, 20, 30, 231, 241, 244, 248, 251–254

Geographical indications, 105, 106

Global sustainability, 64, 65, 73

Glyphosate, 12, 20, 27, 175, 279, 280, 285–286

GMO - genetically modified (crops, foodstuffs, organisms), 4, 25, 252

Greenhouse gas emissions, 31, 102, 188, 286

H

Halal, 186, 190, 193–194, 196–197, 199

Health, 4, 5, 14–16, 39, 52, 55, 57, 84, 102–104, 110, 113, 114, 121, 136, 137, 145, 147, 151, 187, 189, 199, 208, 214, 218, 232, 238–241, 244, 248, 249, 252–255, 272, 273, 278, 280, 281, 286–289

Herbicide free integrated weed management, 281, 285

Households, 20, 120, 136–145, 152, 198, 204–207, 209, 213, 214, 217, 218, 225, 234, 250, 255, 280, 288

Household survey, 139, 141–144

Human population, 187

Human rights, 56, 102, 136, 161, 187

Hunger, 9, 15, 17, 55, 120, 136, 137, 139–141, 187, 190, 205, 221

I

Improprieties, 186

Industrialization, 53, 56, 233, 235, 236

Industrial reform, 235

Inedible food, 205

Inequality, 136, 137, 139–141, 143, 146, 149, 152–154

Inequality, 56, 82, 83, 189

Innovation, 35, 50, 51, 53, 55–59, 66–75, 77, 82, 84, 109, 110, 112, 113, 122, 199, 204, 220, 241, 289

Intensive agriculture, 2–4, 7, 9, 11–14, 28, 29

International sustainability frameworks, 75

Italy, 12, 37, 102, 106–108, 113, 125, 128, 206, 214, 250

J

JSE listing requirements, 150, 151, 153

K

King Codes, 150–151, 153, 154

L

Landowners, 234, 236, 242, 254

Land ownership, 139, 234

Land reform, 139, 141, 146, 149

Law, 25, 38, 104, 105, 113, 137, 144, 147–149, 151, 153, 154, 161, 162, 179, 206, 209, 211, 213, 243

Leftovers, 120, 127, 207–209, 211, 218, 224

Legal regulatory and institutional barriers, 68, 71, 75

Less waste, 121, 122

Livestock, 2, 13, 52, 53, 105, 124, 129, 197, 242, 288

Local communities, 54, 57, 58, 103, 108, 110–112, 146–150, 152, 153, 170, 198

Lockdown, 136, 137, 139–141, 145, 189, 210, 255

Low meat diet, 6, 11

M

Malaysia, 186, 189–194, 196–199

Maximum residue levels, 280

Meals, 32, 52, 110, 113, 127, 145, 187, 207, 208, 213, 224

Meat alternatives, 9, 34, 35

Multi-stakeholder, 130, 154, 160, 168, 169, 196, 289

Multi-stakeholder processes, 120

N

Natural resources, 16, 29, 54, 65, 84, 102, 186–188, 190, 192

Networking strategies, 49

New foods, 251
Nigeria, 50, 53, 60, 61, 64, 65, 71–75
Nonaligned Movement (NAM), 235

O
Oil palm, 186, 190, 192, 194, 196
Organic agriculture, 2–39, 109, 162, 163
Organic feed, 109, 111
Orthodox (Rotorvane tea Pesticide), 274

P
Pandemic, 76, 77, 83, 136–141, 143, 145, 146,
    149, 152–154, 194, 210, 217–221, 223,
    256, 272
People, 2, 4, 15, 17, 29, 32, 35, 52–55, 65,
    71–75, 102, 103, 107, 110–112, 114,
    120, 121, 129, 131, 136–139, 141–145,
    154, 170, 187–189, 191, 192, 196, 199,
    204–211, 214, 217–225, 242, 244,
    247–251, 253–255, 272, 277, 278
Permaculture, 2, 7, 11, 18, 28–29, 39
Pesticides, 3, 9–14, 16–18, 20, 25, 27, 31, 102,
    167, 174, 175, 177–180, 193, 244, 248,
    278, 281, 287
Planet, 5, 14, 28, 37, 54, 55, 277
Planetary boundaries, 3–8, 10, 11, 14, 39
Plantation companies, 275, 278–280, 282–288
Pollution, 164, 174, 188, 279, 280
Polyphenols, 274
Printed meat, 241
Privatisation, 245
Processing, 23, 38, 52, 53, 105, 108, 111, 112,
    120, 123, 149, 167, 187–190, 194, 273,
    282
Producer, 13, 14, 23, 26, 30, 84, 103, 107, 109,
    123–126, 129, 138, 163, 164, 166, 173,
    174, 177, 191, 193, 240, 242–245, 248,
    249, 251, 274, 279, 281
Product certifications, 104, 105, 189
Production, 2, 4, 6, 9, 11–13, 15–20, 22–24, 27,
    28, 30, 33–36, 38, 51–53, 56, 58, 65, 74,
    83, 86, 92–96, 102–112, 120, 123, 126,
    138, 141, 144, 152, 153, 159, 160,
    162–165, 167, 168, 170, 172, 175, 177,
    180, 185–190, 192, 194, 196, 211, 225,
    232, 234, 238–245, 248, 250, 255, 272,
    274, 275, 278–284, 286, 287
Production cost, 168, 278, 284, 288
Productivity, 31, 53, 58, 84, 85, 113, 141, 187,
    279–284, 286–288

Profit, 54, 55, 72, 74, 108, 215, 216, 220–224,
    245, 279, 288
Protein rich crops, 9, 11
Public policy, 5, 15, 33

Q
Quality, 19, 22, 23, 27, 34, 52, 54, 55, 57, 71,
    85, 86, 94, 95, 102–106, 110, 113, 121,
    124, 137–139, 166, 179, 187, 191, 193,
    195, 198, 204–206, 240, 245, 247, 254,
    256, 276, 279, 282, 286

R
Rainforest Alliance, 192, 285, 287
Regulations, 13, 15, 19, 22, 23, 26, 38, 52, 68,
    70, 71, 76, 77, 82–84, 87, 105, 111, 122,
    126, 136, 138, 148, 149, 152, 161, 189,
    231, 239–244, 248
Regulatory and environmental strategies, 72,
    73, 75
Resource efficient, 188
Responsible consumption, 104, 120
Responsible production, 278
Restitution, 240, 253
Retail, 21, 23, 27, 31, 52, 53, 62, 64, 120, 123,
    124, 160, 166, 207, 243, 245, 247

S
Serbia, 232–256
Shared value, 50, 58, 72
Small and medium enterprises (SMEs), 20, 52,
    53, 58–60, 62–64, 71, 75
Small farms, 243, 249
Social and ethics committee, 148, 151
Social development goals (SDGs), 2, 5, 9, 55,
    56, 64, 73, 104, 111, 120, 160, 161, 205
Social entrepreneurship, 58
Social partners, 146, 153–155
Social pillar, 54
Soil erosion, 281, 286
Solidarity Fund, 141, 145, 146
South Africa, 136–152, 154, 155
Spoilage, 189
Sri Lanka, 272–289
Stakeholders, 21, 28, 50, 54, 57, 58, 84, 89,
    113, 147, 148, 150, 151, 162, 168, 170,
    191, 195, 198, 215, 219, 222
Standards, 19, 23, 25–27, 30, 33, 37, 72, 77, 82,
    91, 103–105, 112, 113, 121, 126, 127,

150, 166–168, 174, 189, 192, 196, 197, 245, 281, 283, 287, 288
Strategic planning, 92, 121
Strategy, 2, 6, 11, 17, 28, 30, 33, 51, 58, 60–62, 71–75, 102–105, 109, 112, 114, 121, 122, 124, 126, 129–131, 150, 151, 160, 162, 171, 172, 180, 197, 198, 204, 205, 210, 213–215, 218, 249, 273, 280–288
Stripe spreading of tea bushes, 283
Superfoods, 4, 18, 36
Supply chains, 50–52, 55–61, 64–71, 73, 75–77, 82–96, 102, 108–111, 113, 114, 120, 126, 127, 160–162, 170, 171, 180, 187, 188, 191–197, 199, 204, 209, 219, 222, 245, 272, 275–277, 279, 281
Supply chain sustainability innovations, 50–77
Sustainability, 3, 4, 21, 24, 27, 29, 32, 33, 37, 39, 50, 51, 54–58, 60, 61, 64, 65, 70–73, 75–77, 82–85, 96, 103, 105–114, 146, 150, 160, 165–167, 186–190, 192, 194, 195, 199, 217–219, 221–224, 239–241, 272, 273, 277–289
Sustainability and Sustainable Development, 54–57
Sustainability barriers, 49–77
Sustainability certifications, 168, 196, 285, 289
Sustainability challenges, 166, 199, 273, 280, 288
Sustainability initiatives, 55, 60, 62, 64, 65, 72, 73, 75, 186
Sustainability strategies, 51, 58, 61, 62, 71–76, 110
Sustainable development goals, 2, 55, 56, 64, 104, 120, 121, 150, 160, 278
Sustainable food systems, 4, 5, 7, 8, 11, 15, 17, 18, 30, 31, 33, 34, 39, 187
Sustainable issues, 185, 186, 190
Sustainable production and consumption patterns, 2, 4, 10, 33
Sustainable technology development, 72, 73, 75
Systemic change, 199

T
Tea foliage, 283

Tea industry, 273, 275–289
Technological barriers, 66, 68, 75
Technology advancement, 59, 71, 103
Ten principles of the global impact, 56, 64
Tick box CSR, 149, 153
Traceability, 85, 104–106, 110, 186, 187, 191, 193, 195, 197, 199, 242, 243
Tradition/traditional/traditionalist, 15, 19, 28, 71, 110, 147, 148, 232, 233, 237, 239–242, 248, 249, 252, 253, 283, 288
Transversal strategies, x, 120, 124–129
2030 agenda, 204

U
United Nations Sustainable Development Goals (UN SDGs), 2, 5, 55, 56, 64, 73
Urban agriculture, 4, 32, 38

V
Vaccine, 136
Value addition, 188, 273, 285

W
Wage hikes, 278, 279, 282
Waste, 2, 31, 35, 38, 54, 64, 66, 67, 69, 74, 103, 108, 120, 122, 123, 125, 127–129, 180, 186, 189–192, 194, 198, 204–225, 280, 281, 288
Waste Framework Directive (WFD), 204, 207
Weed management, 273, 280, 281, 285, 286
Well-being, 10, 55, 104, 110, 113, 272, 273
Wholesalers, 123, 124, 197
Workers' Self-Management, 235

Y
Yugoslavia, 232, 235

Z
Zadruga, 243
Zero hunger, 104, 205

Printed by Printforce, United Kingdom